ELEMENTS OF INORGANIC PHOTOCHEMISTRY

ELEMENTS OF INORGANIC PHOTOCHEMISTRY

G. J. Ferraudi
University of Notre Dame
Notre Dame, Indiana

A WILEY-INTERSCIENCE PUBLICATION

JOHN WILEY & SONS
New York • Chichester • Brisbane • Toronto • Singapore

***Library of Congress Cataloging in Publication Data*:**

Ferraudi, G. J. (Guillermo J.), 1942–
Elements of inorganic photochemistry.

"A Wiley-Interscience publication."
Bibliography: p.
Includes index.
1. Photochemistry. 2. Chemistry, Inorganic.
I. Title.
QD708.2.F47 1987 541.3′5 87-23144
ISBN 0-471-81325-7

Printed in the United States of America

10 9 8 7 6 5 4 3 2 1

PREFACE

Although the photochemical reactivity of coordination complexes was discovered long ago, it was not investigated under well-defined experimental conditions or viewed from the standpoint of physical chemistry until the 1960s. A considerable body of experimental and theoretical information has accumulated since then. Most of this material has been covered in review articles and in the books of V. Balzani and V. Carasiti (*Inorganic Photochemistry*) and A. Adamson and P. D. Fleischauer (*Concepts of Inorganic Photochemistry*). Today research in inorganic photochemistry has grown in size and diversity and shows continuous progress in fields within basic and applied science. In this regard, it was necessary to compromise between depth and breadth of coverage while focusing on those subjects that are considered essential for understanding inorganic photochemistry. The level to which subjects have been developed in this book will better suit beginner scholars than specialists. For example, the quantum mechanical treatments related to principles of spectroscopy and chemical dynamics should be readily accessible to graduate students with a solid background in physical chemistry. Yet it is the author's hope that tabulated data, equations, and general information will make the book a helpful complement to the literature required in the daily planning of photochemical work.

G. J. Ferraudi

Notre Dame, Indiana
December 1987

CONTENTS

LIST OF ABBREVIATIONS

Symbols

ϵ	Molar extinction coefficient
α	Specific cross-sectional area for a photon–molecule collision
k	Boltzmann constant
$h, \hbar$	Planck's constant
h_0	Symmetry group's order
$\boldsymbol{J}$	Photonic density
μ	Reduced mass
η	Viscosity
β	Probability that a pair of primary products will recombine
D	Bond dissociation energy
D_i	Diffusion coefficient of spherical particle with radius r_i
u	Ionic mobility
$U(q_i)$	Potential energy
U	Electrostatic potential
v	Velocity of a particle
$\mathcal{V}_i$	Volume of the photolysis cell
$\mathcal{V}_d$	Volume of the photolyzed solution
$\mathcal{K}, \mathcal{K}_s$	Dielectric constant
$\mathcal{K}_{op}$	Optical dielectric constant
R_D	Photomultiplier dynode resistors

R_L	Load resistor
θ	Photomultiplier quantum efficiency
$\hat{\mathcal{H}}$	Hamiltonian operator
ρ_i	Mass weighted coordinates
$E_n(Q)$	Potential energy of the nuclei for a given electronic configuration
P_n	Normal coordinate
I	Reducible representation
I_A, I_a	Absorbed light intensity
I(M–L)	Bond energy
I^*(M–L)	Excited state bond energy
α_i, $\alpha_{i,n}$	Expansion coefficients of the diagonalized (harmonic) potential energy expression in normal coordinates
β_e	Electron Bohr magneton
$\hat{l}_j$	Inertia momentum operator = $\mathbf{r} \hat{} \hat{p}$
$\hat{S}_j$	Spin angular momentum operator for electron j
λ	Spin–orbit coupling constant
τ_i	Lifetime of excited state, fluorescence phosphorescence
ξ^2_{CR}	Proportionality factor (exchange mechanism)
$\Delta\epsilon$	Energy gap between electronic states
S	Overlap integral
χ_L, χ_M	Optical electronegativities of the ligand L and metal M
D_{SP}	Spin pairing parameter
k_p	Rate constant of phosphorescence
k_f	Rate constant of fluorescence
k_{rad}	Radiative rate constant
k_{isc}	Rate constant for intersystem crossing
k_n, $k_{n'}$	Internal conversion rate constant
E_{max}	Energy for a maximum absorption
E_{th}	Threshold energy for photochemical reactivity

Ligand Abbreviations

bipy	2,2′-Bipyridine
COT	Cyclooctatetraene
NBD	Norbornadiene
phen	1,10-Phenanthroline
ppy	Polypyridine ligands
PR_3	Phosphines

ELEMENTS OF INORGANIC PHOTOCHEMISTRY

1

BASIC PRINCIPLES

1-1 THE NATURE OF LIGHT AND THE UNCERTAINTY PRINCIPLE

Many experiments show that radiation can be treated either as an electromagnetic wave or as corpuscles called photons.[1-3] The diffraction of light suggests that radiation is a wave, but photoelectric phenomena and the Compton effect point to a corpuscular nature. In these contexts, the use of an electron for probing a beam of light, that is, a stream of the corpuscles called photons, allows us to determine the position of the photons with almost complete exactness. However, the impact between the photon and the electron changes the momentum of the photon and the wavelength energy is then undefined. In diffraction experiments, the grating or narrow slit used for creating diffraction patterns causes the direction of the photons to be undefined and their position becomes uncertain, while the energy or wavelength can be determined with almost complete exactitude.

These basic principles apply to other particles.[4,5] For example, consider an electron, whose position and momentum we wish to determine. We will illuminate the particle with light of wavelength λ and collect the light with a microscope (Fig. 1). It is possible to demonstrate according to optical theory that the position of such a particle can be determined with an accuracy

$$\Delta x = \frac{\lambda}{\sin \theta} \tag{1}$$

where θ is the optical aperture of the microscope. To maximize the accuracy of the determination, λ must be small, a condition that makes the Compton

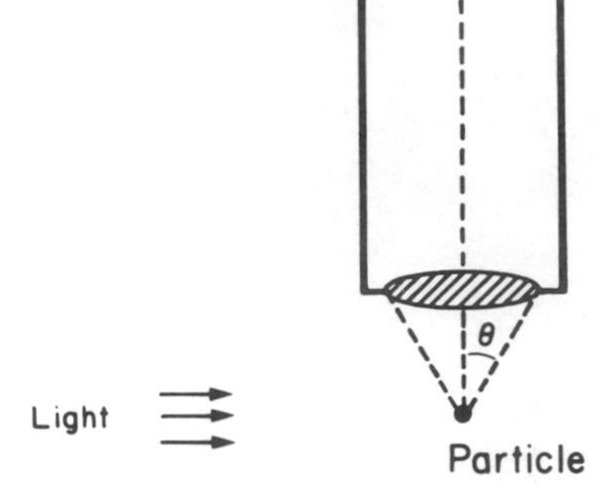

Figure 1. The Compton effect. A beam of light striking a particle (electron) is scattered and a fraction of it is collected by the microscope objective.

effect appreciable.[6,7] Indeed the particle undergoes a recoil as a consequence of the impact with the photon, which acts as a particle of mass m and energy E (Appendix I). Since this is the relativistic energy $E = mc^2$, the relationship between the momentum of the photon and its energy $\hbar\nu$ is

$$p = mc = \frac{E}{c} = \frac{\hbar}{\lambda} \tag{2}$$

After the impact, the momentum of the photon is changed, and if it is going to be collected by the microscope it must be between the values $(1 \pm \sin\theta)\hbar/\lambda$. Therefore the transfer of momentum between photon and particle must be between the values $\pm(\hbar/\lambda)\sin\theta$, which determine a momentum uncertainty

$$\Delta p = \frac{2\hbar \sin\theta}{\lambda} \tag{3}$$

It is possible, then, to combine the uncertainties in the momentum and position of the particle in a relationship

$$\Delta p \cdot \Delta x = 2\hbar$$

This is an expression of the Heisenberg uncertainty principle, which establishes that the product of these uncertainties must be of the order of magnitude of Planck's constant.[8] Since we are usually interested in the orders of magnitude of Δp and Δx, the expression is reduced to

$$\Delta p\, \Delta x \simeq \hbar \tag{4}$$

1-2 ABSORPTION OF LIGHT: THE LAMBERT–BEER LAW

In both undulatory or corpuscular descriptions of light, light is described as flowing energy, so there is some parallel between the movement of a fluid and a beam of light, that is, a stream of photons moving along trajectories determined by optics. In this context, the light intensity I can be defined as the number of photons crossing a surface A (Fig. 2a) in unit time,

$$I = \frac{dn}{dt} \tag{5}$$

This intensity is (according to a vectorial model) the integral of the vector photonic density $\mathbf{J}$ over the surface A crossed by the photons,

$$I = \int_A \mathbf{J} \cdot \breve{n} \cdot da \tag{6}$$

where da is an infinitesimal element of area, $\breve{n}$ is the unitary vector perpendicular to the element of area, and $\mathbf{J}$ is the vector photonic density defined as the number of photons per unit time and unit area that cross the surface at a given point. In the undulatory description of light this vector will be perpendicular to the wavefront.[3] Let us consider a beam of light with origin in a point source (Fig. 2b) that intersects two planes a and b separated by a distance l. Let I_0 be the light intensity at surface a where the cross section of the beam is a circle of radius r. Because of the symmetry around the optical axis 0, the flux across the plane a is

$$\mathbf{J}_a = \frac{I_0}{\pi r^2}\,\breve{n} \tag{7}$$

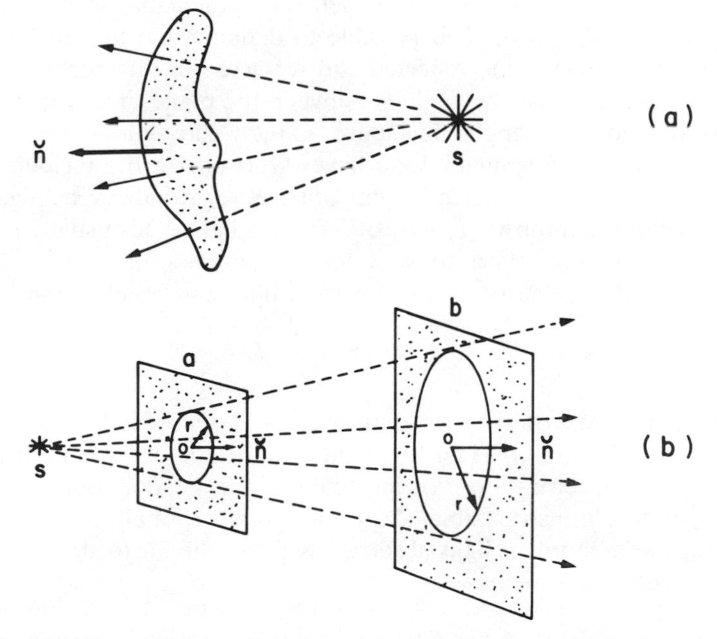

Figure 2. Intensity and flux in a light beam intersecting surfaces defined by unity vectors $\breve{n}$ perpendicular to each of them.

The density across plane b can be estimated by using the angle θ for the divergency of the beam, with respect to the optical axis 0, and the fact that the intensity of light at plane b is the same as the intensity I_0 at plane a. Therefore the flux is given by

$$\mathbf{J}_b = \frac{I_0}{\pi(r^2 + 2rl \tan\theta + l^2 \tan^2\theta)} \check{n} \tag{8}$$

and for small values of angle θ we can use the approximation

$$\mathbf{J}_b = \mathbf{J}_a \frac{1}{(1 + 2(l/r) \tan\theta)} \tag{9}$$

Let us consider now that instead of two planes a and b (Fig. 2b) we have an optical cell containing a compound that absorbs part of the light. Therefore, the intensity I_0 of the light incident on the front surface of the cell must be equal to the summation of the intensities for the light absorbed, I_A; transmitted, I_T; and reflected, I_R:

$$I_0 = I_R + I_A + I_T \tag{10}$$

Indeed, when a beam of light arrives at the surface of separation between two different media it splits into reflected and refracted beams according to the ordinary laws of reflection and refraction. Figure 3 describes an optical wave by a vector, the optical vector, which is perpendicular to the direction of propagation of the wave. It is possible to demonstrate that the distribution of intensity between the reflected and refracted beams depends on the angle of incidence, θ, and the angle θ_v between the plane of vibration of the optical vector and the plane of incidence, namely the plane containing the three rays.[9] It is therefore possible to have directions of the incident beam where the intensity of reflected light can be regarded as negligible in comparison with the intensity $I_A + I_T$ of refracted light. This usually requires an angle of incidence close to $\pi/2$ for water–glass interfaces. In this approximation, the intensity of the absorbed light can be expressed as

$$I_A = I_0 - I_T \tag{11}$$

For the derivation of the law that governs the absorption of light by molecules in a solution, consider an infinitesimal element of solution with a width dl and a concentration c of molecules (Fig. 4). For monochromatic light, each molecule also exhibits a specific cross-sectional area α. We will assume that the amount of light absorbed is proportional to the number of photon-molecule collisions,

$$-dI = \alpha IC\, dl \tag{12}$$

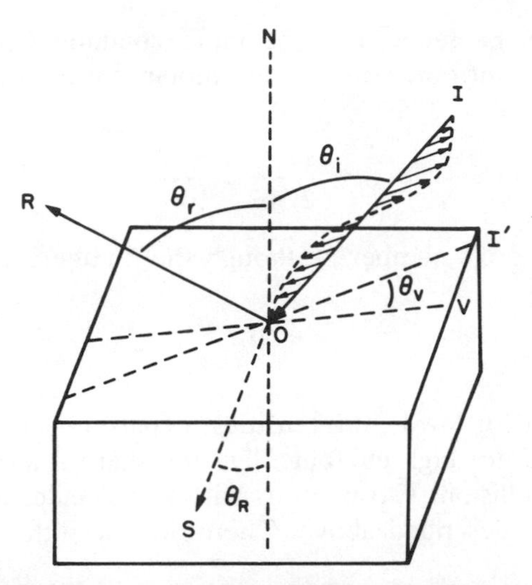

Figure 3. Reflection and refraction. A ray of linearly polarized light experiences reflection in a direction $\overrightarrow{OR}$ and refraction in a direction $\overrightarrow{OS}$ at the interface of two media. The vectors perpendicular to the incident ray, $\overrightarrow{IO}$, denote vibrations in a plane (vibration plane $I'ON$) forming an angle θ_v with a plane containing the incident ray ION and the perpendicular to the interface, $\overrightarrow{ND}$.

where c is the number of molecules per cubic centimeter and α is the cross section per molecule in square centimeters. This equation can be easily integrated when the concentration of absorbent is independent of l and I. Such a condition arises when the number of photons absorbed is only an insignificant fraction of the concentration c. Hence the integrated equation is

$$I_l = I_0 \exp(-\alpha c l) \tag{13}$$

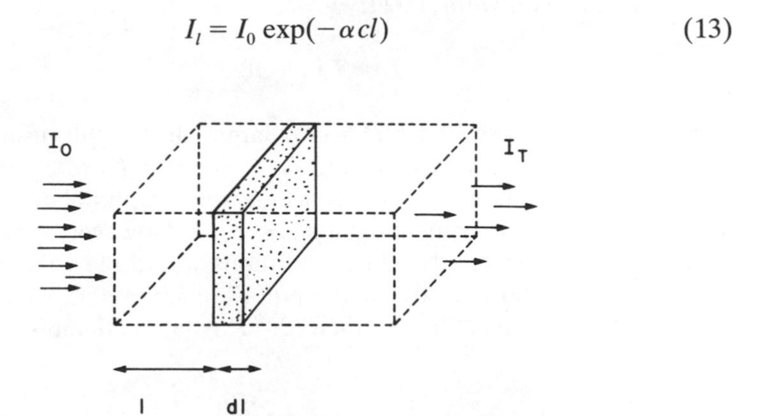

Figure 4. Absorption of light by material contained in an infinitesimal element of volume $dV = A\,dl$, where A is the cross section of the light beam.

This equation can be reduced to the more common Lambert–Beer relationship with the introduction of the molar concentration M and the extinction coefficient

$$\epsilon = \frac{N\alpha}{2.303 \times 10^3} \tag{14}$$

where N is Avogadro's number. Although the Lambert–Beer equation

$$\frac{I_T}{I_0} = 10^{-\epsilon l M} \tag{15}$$

is basically correct for low light intensities, departures are observed at high intensities. Indeed, for high photonic densities there is a certain probability of a three-body collision, two photons with a molecule, in addition to the two-body collision described above. Therefore the differential law can be expressed as

$$-dI = \epsilon_1 IM\, dl + \epsilon_2 I^2 M\, dl$$

which upon integration gives

$$\frac{\epsilon_1 + \epsilon_2 I_T}{I_T} = \frac{\epsilon_1 + \epsilon_2 I_0}{I_0} 10^{-\epsilon_1 Ml} \tag{16}$$

The departure from the Lambert–Beer law (eq. 13) caused by the dependence of the absorbent molecules, that is, c in eq. 12, on the light intensity will be discussed later in this chapter in connection with the photostationary regime.

In photochemical reactions it is important to know the amount of light absorbed by the solution of photolyte. This quantity can be estimated when eq. 15 is combined with eq. 11:

$$I_A = I_0(1 - 10^{-\epsilon l M}) \tag{17}$$

Hence, values of absorbance $D = \epsilon l M$ larger than 2 will insure a complete absorption of the incident light, for example, $I_A \approx I_0$. Many photochemical reactions are carried out with a concentration of photolyte and an optical path such that all the incident light is absorbed. However, in some cases it is convenient to use optically diluted solutions where the value of the photolyte absorbance allows us to expand eq. 17 in series and neglect terms with powers of absorbance larger than 2. This approximation leads to the equation

$$I_A = I_0 \epsilon l M \tag{18}$$

for absorbed light.

The quantities I_A, I_0, and I_T in the Lambert–Beer law (eqs. 13, 15, 17) are light intensities or light densities and can be expressed as number of photons per second or number of photons per second and square centimeter. It is also common to give the number of photons in einsteins (moles of photons) per unit time or unit time and unit area. In photochemical studies, some researchers use the total light intensity or the flux incident on the reaction cell whereas others use the average number of photons absorbed by the photolyte per unit volume and unit time I_a. These quantities are related by the expression

$$I_a = \frac{\int_A \mathbf{J} \cdot \check{n}\, dA}{\int_v dv} = \frac{I}{\int_v dv} \tag{19}$$

For a prismatic cell illuminated by a beam with constant photonic density, these expressions can be reduced to

$$I_a = \frac{|\mathbf{J}|s}{v} = \frac{I}{v} \tag{20}$$

where s is the area illuminated and v the volume of the solution. It is therefore irrelevant which quantity is used, provided the authors give all of the additional experimental information required for the complete definition of the photochemical conditions. For prismatic cells the total light intensity absorbed by the photolyte should be paired with the area illuminated and the average absorbed intensity per volume should be paired with the illuminated volume.

1-3 PHOTOCHEMICAL LAWS

Photochemical reactions are ways by which chemical systems degrade energy that has been supplied to them in the form of light. In a sense, this is the first law of photochemistry[10] (also called the law of Grotthus and Draper), which establishes that *the only radiations that are effective in inducing photochemical changes are those absorbed by the system*. Although this law is necessary for the definition of photochemical changes, it is not sufficient for the quantitative description of photochemical processes. Our understanding of the quantitative aspects of photochemical reactions followed the establishment of the corpuscular nature of light. The second law of photochemistry was then deduced, which, in its original form, stated: *A molecule in a photochemical reaction absorbs one quanta of the radiation that induces the reaction.*

Once the photonic energy has been absorbed by the molecule, this energy can be degraded in a variety of processes that induce chemical or physical changes. These processes have been named *primary photochemical processes* when they comprise the series of events starting with the absorption of light by a molecule and ending with the disappearance of the molecule of its conversion to a state such that its reactivity is statistically no greater than that of similar molecules in thermal equilibrium with their surroundings. We can define the quantum yield of a primary process as the number of molecules that experience a given transformation in such a process divided by the number of absorbed quanta. For the primary process

$$\mathrm{Co(NH_3)_4Br^{2+}} + h\nu \rightarrow \mathrm{Co^{2+}_{aq}} + 5\mathrm{NH_3} + \mathrm{Br}\cdot \tag{21}$$

the quantum yield is

$$\Psi_i = \frac{\text{mol of Co}^{2+}\text{ produced}}{\text{einstein of light absorbed}} = \frac{d[\mathrm{Co}^{2+}]/dt}{I_\mathrm{a}} = \frac{\text{mol of Co}^{2+}\text{ produced/dm}^3\text{ s}}{\text{einstein/dm}^3\text{ s}} \tag{22}$$

Equation 21 shows that the quantum yield Ψ_i carries no units when the light intensity and the rate of reaction are expressed in compatible units.

It is easy to demonstrate with the definitions given above that the quantum yield of the primary processes must be always less than or equal to unity. This allows us to restate the second photochemical law: *The absorption of light by a molecule is a one-quantum process, so that the sum of the totality of the primary processes quantum yields ϕ_i must be unity*:

$$\sum_{i=1}^{n} \Psi_i = 1 \tag{23}$$

The second photochemical law was developed for moderate and low light intensities and does not contemplate the existence of multiphotonic processes. For example, two types of biphotonic processes have been reported. One consists of the simultaneous absorption of two photons by a molecule in a three-body collision

$$\mathrm{A} + 2h\nu \rightarrow \text{products} \tag{24}$$

Another corresponds to the sequential absorption of two photons

$$\mathrm{A} + h\nu \rightarrow {}^*\mathrm{A} \xrightarrow{h\nu} \text{products} \tag{25}$$

where *A is an excited state of A. Insofar as these examples are observed at very high photonic densities, they can be considered special cases of the second law of photochemistry.

In photochemical experiments we measure the concentration of products, and the quantum yields obtained from these measurements are the overall quantum yields

$$\phi_x = \frac{1}{I_a}\frac{d[x]}{dt} = \frac{\text{mol of product formed/dm}^3\ \text{s}}{\text{einstein absorbed/dm}^3\ \text{s}} \tag{26}$$

Although the overall quantum yields provide information on the reaction mechanism, primary yields cannot always be derived from the overall yields.

1-4 PHOTOSTATIONARY STATE AND THE RATE LAW

The absorption of light on an element of an absorbing solution is given by the product of the concentration of photons and absorbing molecules (eq. 12). Insofar as the absorption of light is appreciable over the cell, the absorbed intensity depends on the distance to the front of the cell. This determines that the rate of the photochemical reaction, a rate that is proportional to the absorbed light, must change from one position to another in the cell. The measured rate therefore represents an average of the local rates. For a prismatic cell this rate is

$$\left[\frac{d[\mathrm{P}]}{dt}\right]_{\text{measured}} = \frac{1}{l}\int_0^l \left[\frac{d[\mathrm{P}]}{dt}\right]_{\text{local}} dl \tag{27}$$

where l is the distance to the front of the cell and $[P]$ is the product concentration. In a system with a complex mechanism, the local rate might have a complex dependence on the absorbed light intensity. For example

$$\left[\frac{d[\mathrm{P}]}{dt}\right]_{\text{local}} = \phi I_a^m \tag{28}$$

where I_a is the light absorbed at the point of evaluation in the cell. Assume now that the Lambert–Beer law for the determination of thc absorbed light intensity (eq. 17) is valid. In this approximation, the light absorbed between the distances l and $l + dl$ to the front of the cell (Fig. 4) is given by

$$I_a = 2.303\epsilon c I_0 \exp(-2.303\epsilon l c) \tag{29}$$

The insertion of eqs. 28 and 29 in eq. 27 gives an explicit expression for the observed rate of photolysis

$$\left[\frac{d[\mathrm{P}]}{dt}\right]_{\text{measured}} = \frac{\phi I_0(1 - \exp(-2.303 m\epsilon c l))}{ml(2.303\epsilon c)^{1-m}} \tag{30}$$

This expression can be simplified under two experimental conditions, either $m = 1$ and $2.303\epsilon c l \ll 0.1$ or $m = 1$ and $\epsilon l c > 2$. A solution of an absorbent that responds to the first set of conditions is usually described as being optically diluted. For such conditions expansion in series of the exponential and elimination of high-order terms in eq. 30 gives an observed photolysis rate:

$$\left[\frac{d[\mathrm{P}]}{dt}\right]_{\text{measured}} \approx \phi I_0 2.303\epsilon c \tag{31}$$

The second set of conditions, $m = 1$ and $\epsilon lc > 2$, correspond, in contrast to the optically diluted absorbent, to an optically concentrated absorbent that undergoes photolysis with a rate

$$\left[\frac{d[\mathrm{P}]}{dt}\right]_{\text{measured}} \approx \phi \frac{I_0}{l} \tag{32}$$

However, the average absorbed intensity is given by

$$\langle I_a \rangle = \frac{\int_0^l 2.303\epsilon c I_0 \exp(-2.303\epsilon cl)\, dl}{\int_0^l dl}$$

$$= \frac{I_0}{l}(1 - \exp(-2.303\epsilon cl)) \tag{33}$$

For an optically concentrated absorbent this equation can be reduced to the form

$$\langle I_a \rangle = \frac{I_0}{l} \tag{34}$$

Comparison of eqs. 32 and 34 shows that the observed rate of photolysis can now be expressed:

$$\left[\frac{d[\mathrm{P}]}{dt}\right]_{\text{measured}} = \phi \langle I_a \rangle \tag{35}$$

The average absorbed light intensity $\langle I_a \rangle$ can also be used in eq. 31, since for an optically diluted absorbent it is $\langle I_a \rangle \approx 2.303\epsilon c I_0$. Therefore eq. 35 can be regarded as a general rate equation for any reaction mechanism with $m = 1$. The use of optically diluted or concentrated solutions is advantageous from an experimental standpoint. This is discussed later in relationship with the experimental determination of the light intensity.

In the photostationary state approximation it is assumed that the reactive intermediates, species that do not appear as products, will reach a limiting concentration if the photochemical reaction is induced with a constant light intensity. Hence, the rate of formation equals the rate of disappearance of the intermediate under the steady state approximation. For example, consider the photodissociation of the μ-oxo dimer[11]

$$(\mathrm{OH_2})\mathrm{Mn^{III}(tspc)\text{–}O\text{–}Mn^{III}(txpc)(OH)^{9-}} \xrightarrow[h\nu]{\phi \simeq 10^{-2}} \mathrm{Mn^{II}(tspc)^{4-}}$$

$$+ \mathrm{Mn^{III}(tspc)(\dot{O})^{3-}} \tag{36}$$

$$\mathrm{Mn^{III}(tspc)(\dot{O})^{4-}} \xrightarrow[k_i = 10^6\ \mathrm{s^{-1}}]{} \mathrm{Mn^{III}(tspc)(OH_2)(OH)^{3-}} \tag{37}$$

$$Mn^{II}(tspc)^{4-} + Mn^{III}(ts\dot{p}c)(OH_2)(OH)^{3-} \xrightarrow[k_r = 10^2\ M^{-1}\ s^{-1}]{\text{slow}}$$

$$(OH_2)Mn^{III}(tspc){-}O{-}Mn^{III}(tspc)(OH)^{9-} + 2H^+ \tag{38}$$

In this sequence of reactions, the recombination reaction, eq. 38, is slow compared with the other reactions. Moreover the rate of $Mn^{III}(tspc)(\dot{O})^{4-}$ formation is

$$\frac{d[Mn^{III}(tspc)(\dot{O})^{4-}]}{dt} = \phi I - k_i[Mn^{III}(ts\dot{p}c)(\dot{O})^{4-}] \tag{39}$$

Under the steady state approximation

$$\frac{d[Mn^{III}(tspc)(\dot{O})^{4-}]}{dt} = 0 \tag{40}$$

the concentration of intermediate is

$$[Mn^{III}(tspc)(\dot{O})^{4-}]_{ps} = \frac{\phi I}{k_i} \tag{41}$$

The large value of the rate constant k_i reduces the steady state concentration (eq. 41) to a negligible value compared with the concentration of $Mn^{III}(ts\dot{p}c)(OH_2)(OH)^{4-}$. In this approximation, the concentrations of $Mn^{II}(tspc)^{4-}$ and $Mn^{III}(ts\dot{p}c)(OH_2)(OH)^{3-}$ are nearly the same and the rate of $Mn^{II}(tspc)^{4-}$ formation is

$$\frac{d[Mn^{II}(tpsc)^{4-}]}{dt} = \phi I - k_r[Mn^{II}(tspc)^{4-}]^2 \tag{42}$$

The integration of eq. 42 with the initial condition

$$[Mn^{II}(tspc)^{4-}] = 0 \qquad \text{for } t = 0$$

gives

$$[Mn^{II}(tspc)^{4-}] = \left(\frac{\phi I}{k_r}\right)^{1/2} \frac{(1 + \exp(1 - 2\phi It))}{(1 + \exp(1 - 2\phi It))} \tag{43}$$

Hence, the concentration of $Mn^{II}(tspc)^{4-}$ reaches a limit at very long times

$$\lim_{t \to \infty}[Mn^{II}(tspc)^{4-}] = \left(\frac{\phi I}{k_r}\right)^{1/2}$$

When the concentration of $Mn^{II}(tspc)^{4-}$ reaches the limiting value $(\phi I/k_r)^{1/2}$, this product is also under a photostationary state. In other words, when the rate of formation of $Mn^{II}(tspc)^{4-}$ equals its rate of disappearance

$$\frac{d[\mathrm{Mn^{II}(tspc)^{4-}}]}{dt} = 0$$

and according to eq. 42

$$[\mathrm{Mn^{II}(tspc)^{4-}}]_{\mathrm{ps}} = \left(\frac{\phi I}{k_{\mathrm{r}}}\right)^{1/2}$$

In real experiments the concentration of $Mn^{II}(tspc)^{4-}$ increases with irradiation time in a nonlinear manner and the limiting concentration that is reached is proportional to the square root of the light intensity (Fig. 5).

In the derivation of the Lambert–Beer law (eq. 13), we have assumed a photostationary state where the concentration of absorbent remains constant along the cell. One failure of this approximation arises when the light intensity is sufficiently large and the depletion of the ground state is significant. Consider that ground state molecules are consumed and replenished by the following processes,

$$\mathrm{A} + h\nu \rightarrow {}^*\mathrm{A} \tag{44}$$

$${}^*\mathrm{A} \xrightarrow{k} \mathrm{A} \tag{45}$$

and k is assumed to be independent of the light intensity. A photostationary

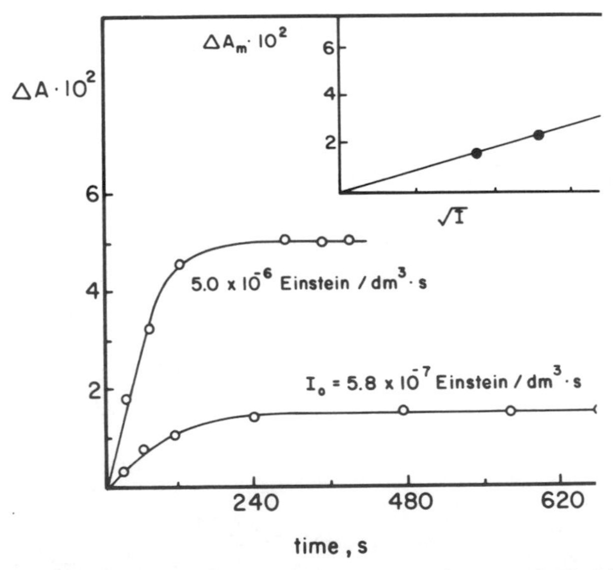

Figure 5. Nonlinear dependence of the product concentration, represented by a change of optical density ΔA on irradiation time. The insert shows that the maximum concentration of product produced with a given light intensity ΔA_m depends linearly on the square root of the light intensity.

regime on the concentration of the excited state gives

$$[^*A]_{ps} = I/k$$

Hence, a mass balance on the concentration of A gives

$$C_0 = [A] + [^*A] \simeq [A] + I/k$$

and the elemental form of the Lambert–Beer law (eq. 12) can be expressed as

$$-dI = \alpha I(C_0 - I/k)\, dl \tag{46}$$

For an optically diluted solution, one where the absorbent is uniformly distributed across the cell, the integration of eq. 46 leads to

$$\frac{(C_0 - I/k)}{I} = \frac{(C_0 - I_0/k)}{I_0} \exp(-\alpha l C_0)$$

where I_0 is the intensity of the incident light and I is the intensity after the light has traveled a distance l inside the solution. Hence, for very low light intensities, for example $C_0 > I_0/k$, such an expression reduces to the simple form of the Lambert–Beer law (eq. 12). It is easy to see that in the case of ground state depletion, the expression for the Lambert–Beer law will be determined by the set of processes that cause the ground state depletion.

1-5 CONTINUOUS PHOTOLYSIS, QUANTUM YIELDS, AND MEASUREMENT OF LIGHT INTENSITY

The irradiation of a sample with monochromatic light of a constant intensity is generically known as steady state or continuous photolysis. In these photochemical experiments, we analyze the total concentration of product that is formed as a function of the irradiation time; and from plots of the product concentration versus irradiation time we obtain information on the photochemical mechanism. For example, often the product concentration versus time plot is not linear. This situation arises when the product quantum yield depends on the irradiation time, the conversion of photolyte to product, or both (Fig. 6). To study systems where product concentration exhibits a nonlinear dependence on time, we can define the *instantaneous quantum yield* (eq. 47) as a ratio of the instantaneous rate of product formation, that is, $(d[P]/dt)_t$ at a time t in Figure 7, to the absorbed light intensity

$$\phi_i(t) = \frac{(d[P]/dt)_t}{I_a} \tag{47}$$

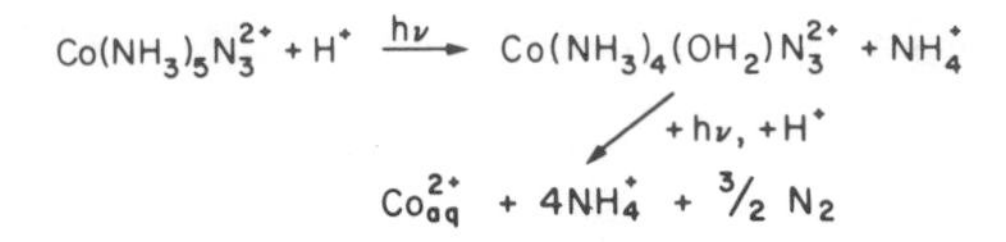

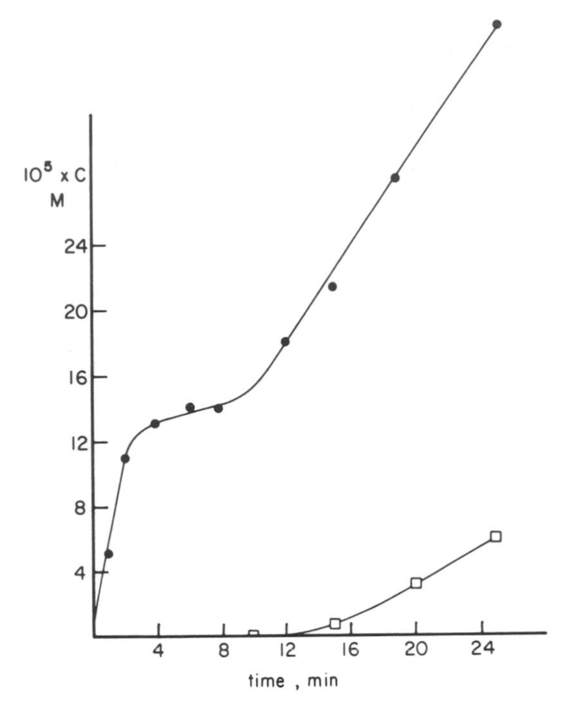

Figure 6. Secondary photolysis in the steady state photolysis of $Co(NH_3)_5N_3^{2+}$, a process shown in the reactions at the top of the figure. The solid circles correspond to changes in the $Co(NH_3)_5N_3^{2+}$ concentration calculated from spectra determined at various irradiation times under the assumption that the final products, Co_{aq}^{2+}, NH_3, and N_2, were formed in the primary photolysis. The open squares correspond to the concentration of Co_{aq}^{2+}. Data after G. Ferraudi, J. F. Endicott, and J. R. Barber, *J. Am. Chem. Soc.* **1975, 97,** 6406.

A linear dependence of the product concentration on irradiation time corresponds to a constant *overall quantum yield* $\phi_0(t)$, namely, the quantum yield defined as the ratio of the total amount of product to the total amount of light absorbed. The overall quantum yield and the instantaneous quantum yield have the same value when the concentration of product increases linearly with irradiation time (Fig. 7). Insofar as we can describe the product concentration versus time by a power series

$$[P] = \sum_n a_n t^n$$

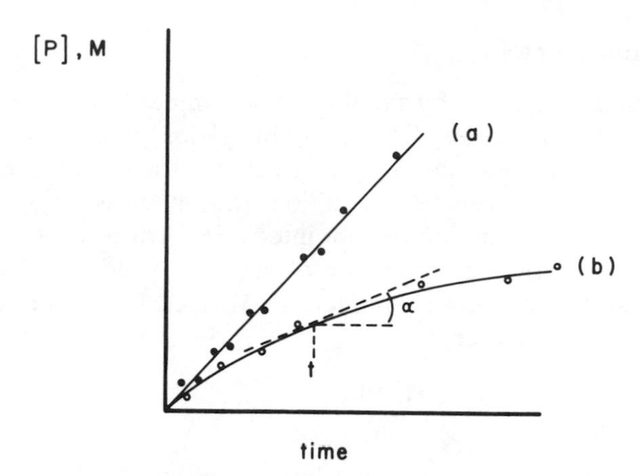

Figure 7. Linear (*a*), and nonlinear (*b*) dependences of the product concentration [*P*] on irradiation time. The slope α measures the instantaneous rate of reaction that must be used to determine an instantaneous quantum yield, that is, the quantum yield at time *t*.

it is easy to demonstrate that, in passing to the limit of a zero irradiation time, both yields must have the same value, namely

$$\lim_{t \to 0} \phi_0(t) = \lim_{t \to 0} \phi_i(t) \tag{48}$$

It should be noticed that the equality in eq. 48 is valid if in the time domain of continuous photolysis the product seems to be formed instantaneously. For example, if the product is formed a few microseconds after the absorption of the photon and we measure the concentration at 1-s intervals, eq. 48 is perfectly valid. However, if we sample the concentration of product at intervals equal to or shorter than a microsecond, there will be a period in which the product has not reached the steady state concentration and the equality in eq. 48 is no longer valid.

Thus, in the determination of quantum yields, the points discussed above show that it is convenient to limit the conversion to products to the initial portion in the plot of product concentration versus irradiation time, for example, when there is a linear dependence of the concentration on the irradiation time. When experimental restrictions such as the resolution of the analytical techniques do not allow us to reach the linear region, we must resort to pass to the limit for an irradiation period approaching zero. However, quantum yields determined by passing to the limit can be dependent on other experimental parameters, such as the concentration of the photolyte.

The other problem associated with determining quantum yields is evaluating the absorbed light intensity. Several techniques are briefly discussed below.

Chemical Actinometers

Chemical actinometers are by far the most commonly used instruments in determining light intensities. The quantum yields of a number of photoprocesses have been measured with great accuracy and are sufficiently independent of the experimental conditions that they can be used as primary standards in the determination of light intensities. This process is the reverse of that followed for the determination of quantum yields, in that the rate of product formation from the actinometer is divided by the known quantum yield ϕ_a of the actinometer

$$\frac{1}{\phi_a}\left(\frac{d[P]}{dt}\right)_{\text{actinometer}} = I_a \tag{49}$$

Since the concentration of product depends linearly on irradiation in most actinometers, eq. 49 can be reduced to

$$\frac{[P]}{\phi_a t} = I_a$$

For I_a to be an accurate measurement of the absorbed light intensity that must be used in determining the quantum yield, the conditions for photolysis of the actinometer must be as close as possible to those used in the irradiation of the photolyte. Hence, the reaction cells must have the same geometry and we must use the same volume for both solutions, which, in addition, must have the same solvent and the same optical density.

The aqueous solutions of three coordination complexes, namely $Fe(C_2O_4)_3^{3-}$,[11-14] UO_2-oxalate[15-19] and $Cr(NH_3)_2(SCN)_4^-$,[20] are frequently used as primary actinometers even though not all the details of the corresponding photochemical mechanisms are known. Monochromatic irradiations with UV or near-UV light of trisoxalateferrate(III) solutions form Fe(II) with quantum yields that are independent or exhibit a very low dependence on light intensity, Fe(III) complex, and Fe(II) product concentrations (Table 1). In order to increase the accuracy when measuring light intensities, the concentration of the product Fe_{aq}^{2+} has been determined by different analytical procedures. In uranyl oxalate, another popular actinometer, the irradiation induces the decomposition of oxalate ions. Although the reaction can be described as a photosensitized decomposition of oxalate in CO_2, CO, HCO_2H, and U_{aq}^{4+}, the process does not have a simple stoichiometry. For solutions with $0.01\,M$ UO_2SO_4 and $0.05\,M$ $H_2C_2O_4$ the quantum yield for the consumption of oxalate is nearly independent of temperature and light intensity. Moreover, the yield for oxalate loss doesn't exhibit much dependence on complex concentration. Trisoxalateferrate(III) and uranyl oxalate are actinometers useful at wavelengths shorter than 570 or 450 nm, respectively. In this context, neutral or moderately acidic solutions of $Cr(NH_3)_2(NCS)_4^-$ exhibit a quantum yield for

TABLE 1 Quantum Yields for the Measurement of Light Intensities with Chemical Actinometers

λ_{exc}	ϕ	Chemical and Optical Conditions
		$Fe(C_2O_4)_3^{3-}$
378 ± 1	0.013[c]	[Fe] = 0.15 *M*[a], *f* = 0.118[b]
546	0.15	[Fe] = 0.15 *M*, *f* = 0.061
509	0.86	[Fe] = 0.15 *M*, *f* = 0.132
480	0.94	[Fe] = 0.15 *M*, *f* = 0.578
468	0.93	[Fe] = 0.15 *M*, *f* = 0.850
436	1.01	[Fe] = 0.15 *M*, *f* = 0.997
	1.11	[Fe] = 0.006 *M*, *f* = 0.615
405	1.14	[Fe] = 0.006 *M*, *f* = 0.962
366	1.21	[Fe] = 0.006 *M*, *f* = 1.00
	1.15	[Fe] = 0.15 *M*, *f* = 1.00
334	1.23	[Fe] = 0.006 *M*, *f* = 1.00
313	1.24	[Fe] = 0.006 *M*, *f* = 1.00
300 ± 3	1.24	[Fe] = 0.006 *M*, *f* = 1.00
253.7	1.25	[Fe] = 0.006 *M*, *f* = 1.00
		$UO_2SO_4/C_2O_4^{2-}$[d]
435.8	0.58	
405.0	0.56	
366.0	0.49	
313.0	0.53	
278.0	0.58	
365.0	0.58	
245.0	0.61	
208.0	0.48	
		$Cr(NH_3)_2(NCS)_4^-$[e]
316	0.291	[Cr] = 1.1×10^{-3} *M*[a]
350	0.388	[Cr] = 3.0×10^{-3} *M*
392	0.316	[Cr] = 5.0×10^{-3} *M*
416	0.310	[Cr] = 8.0×10^{-3} *M*
452	0.311	[Cr] = 1.0×10^{-2} *M*
504	0.299	[Cr] = 5.0×10^{-3} *M*
520	0.289	[Cr] = 4.0×10^{-3} *M*
545	0.282	$6 \times 10^{-3} \geq$ [Cr] $\geq 5.0 \times 10^{-3}$ *M*
585	0.270	[Cr] = 1.0×10^{-2} *M*
600	0.276	[Cr] = 2.5×10^{-2} *M*
676	0.271	[Cr] = 4.5×10^{-2} *M*
713	0.284	[Cr] = 4.6×10^{-2} *m*
735	0.302	[Cr] = 4.5×10^{-2} *M*
750	0.273	[Cr] = 4.7×10^{-2} *M*

[a]Concentration of complex.
[b]Fraction of light absorbed for a 1.5-cm optical path.
[c]Quantum yield of Fe_{aq}^{2+} formation.
[d]Actinometer prepared from 0.01 *m* UO_2SO_4 and 0.01 *m* $H_2C_2O_4$. The yields reported are for the photochemical loss of ion oxalate.
[e]Quantum yields for photorelease of thiocyanate from an actinometer prepared with $Cr(NH_3)_2(SCN)_4^-$ at 23°C and pH5.3.

SCN^- release, $\phi \sim 0.3$, which is independent of excitation wavelength in the range of 316–750 nm (Table 1). The photoprocess can be described as a photosubstitution reaction (eq. 50) and the subsequent hydrolysis of the product (eq. 51).

$$Cr(NH_3)_2(NCS)^{4-} \xrightarrow{h\nu} Cr(NH_3)_2(OH_2)(NCS)_3 + NCS^- \tag{50}$$

$$Cr(NH_3)_2(OH_2)(NCS)_3 \rightleftarrows Cr(NH_3)_2(OH)_2(NCS)_2^+ + NCS^- \tag{51}$$

The mechanism is more complex, however, for irradiation at wavelengths shorter than 400 nm.

A number of less popular actinometers, for example, the liberation of chloride in 254-nm photolysis of chloroacetic acid,[21] are used for only specific applications.

Thermopiles

Thermopiles combined with galvanometers or microvoltmeters are very useful for measuring light intensities in photochemical work. Although there are several designs of thermopiles, all are based on the conversion of radiant energy into heat on a blackened surface that acts as a receiver. The receiver is covered with thermocouples that have the cold junction in the dark (Fig. 8). The conversion of radiant energy into heat produces, therefore, a rise in the temperature and creates a thermal emf with respect to the cold junction. The combined emf of the thermocouples in the thermopile is detected as a potential with a sensitive detector such as a galvanometer or a microvoltmeter.

The thermopile behaves as a blackbody and reaches an equilibrium with the incident radiation. In this context, the emf ϵ_v is proportional to the incident energy flux $h\nu I_0$ and it is possible to write

$$\frac{\epsilon_v}{\nu I_0} = \frac{\epsilon_v \lambda}{c I_0} = \alpha \tag{52}$$

where α is a constant, and ν is the frequency and λ is the wavelength of the incident radiation. Although eq. 52 can be applied directly to obtain relative values of the light intensity, the determination of absolute values requires calibrating the thermopile. The calibration is usually carried out with standard lamps that can be acquired at the National Bureau of Standards. It is advisable to use several lamps and several operators for calibrating each thermopile to avoid errors. Because the procedure is cumbersome and often inaccurate, even when thermopiles are perfectly calibrated, they have lost popularity as primary actinometers in photochemical work.

Photophysical Devices

Devices based on photophysical effects are very popular because they combine low noise and high sensitivity with practically no maintenance.

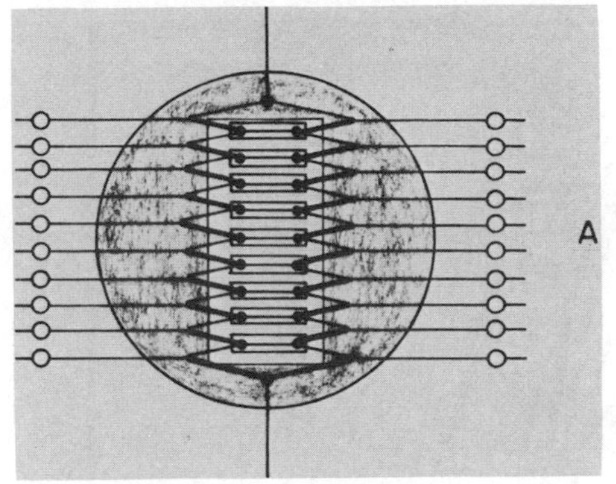

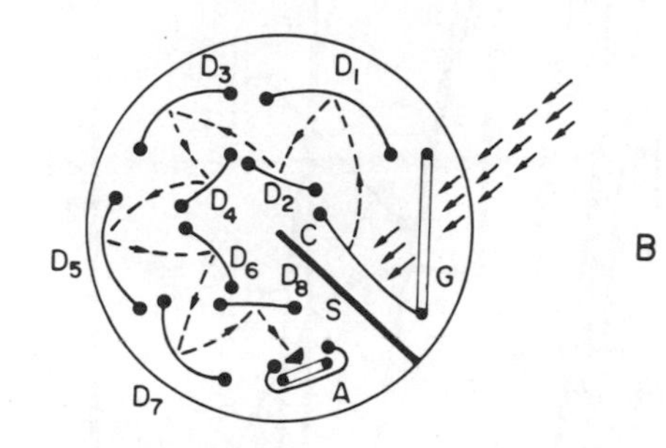

Figure 8. Light sensitive devices: the thermopile (*A*) and the photomultiplier (*B*). In the thermopile the central receivers are the hot junctions of thermocouples and are in the illuminated area, while two outside columns remain in the dark. In a photomultiplier the opaque photocathode (*C*) is part of the electron multiplier or dynode-chain structure D_i (where $i = 1-9$) with the anode or collector (A) at the end. Other elements in the photomultiplier are the shield (S) and the grill (G). The electron trajectory is indicated with a dashed line.

These devices are normally calibrated against primary actinometer sources, namely chemical actinometers or thermopiles, and are based on the photoelectric effect (e.g., phototubes and photomultipliers), the promotion of electrons to the conduction band of photodiodes (e.g., photoconductive cells), and the potential created between the layers of a solid or liquid material (e.g., photovoltaic or barrier cells).

In a phototube there is usually a single cathode or emitter and one anode or collector.[22] For work with high light intensities, the resistance to fatigue and the wide spectral response make the phototube a very convenient device. The photomultiplier intercalates several steps, namely plates or dynodes, between the emitter and the collector (Fig. 8).[22] The plates act as secondary emitters of electrons, namely each electron that strikes one of these plates will induce the release of a number of electrons. Thus the cascading effect from one plate to the next results in a multiplication of the number of electrons that dramatically increases the sensitivity of the photomultiplier with respect to the phototube. Figure 9 shows some of the spectral responses that characterize the photomultipliers.

Since photodiodes work on the principle that the number of charge carriers is proportional to the light intensity, the response is not necessarily linear with light intensity, nor does it exhibit a convenient sensitivity. Such drawbacks might not constitute a problem when we work in a narrow range of conditions, but they should be kept in mind when we are interested in

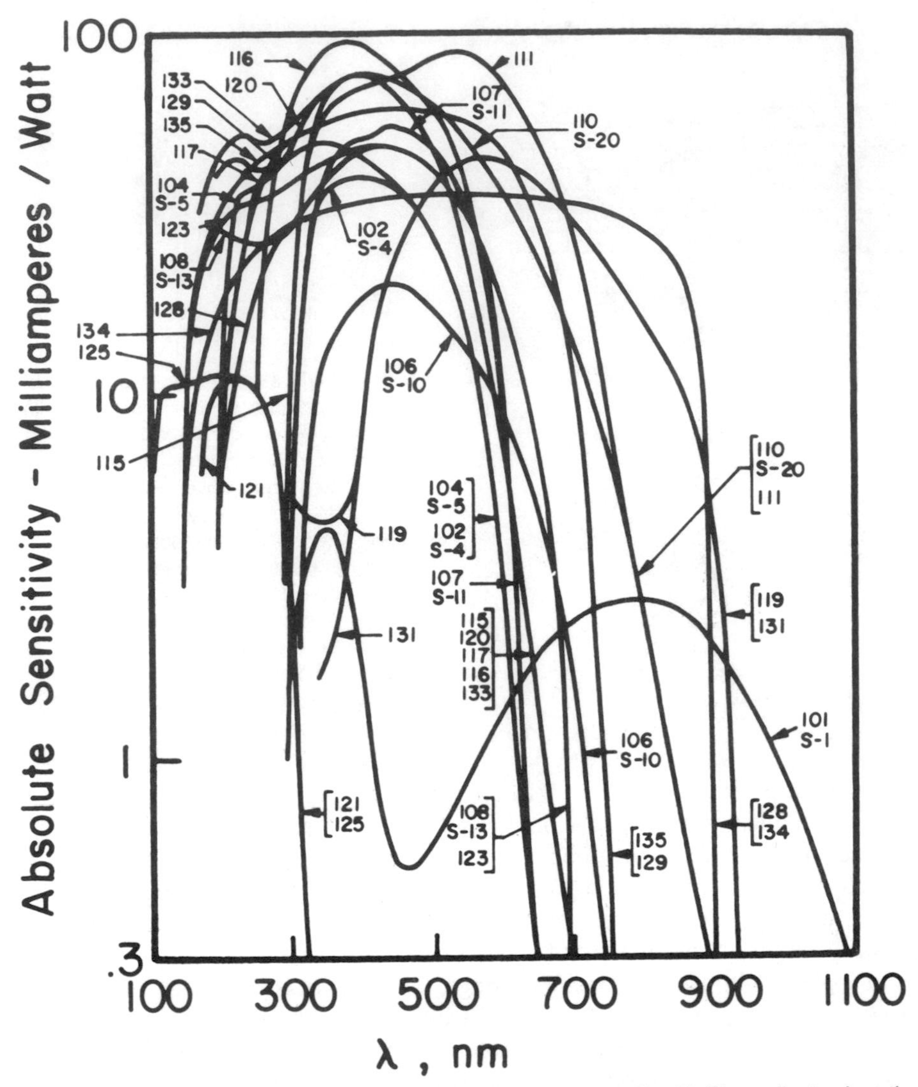

Figure 9. Absolute sensitivity of some photocathodes (see also Fig. 8). The ordinate gives the number of milliamperes per incident watt of monochromatic light. The numbers on the curves (e.g., 101) are used to identify the photocathode responses (e.g., S – 1).

determining accurate values of light intensity. A similar criticism can be raised with respect to the barrier cells.

Despite the different properties exhibited by various kinds of photocells, they have in common the relatively fast response to intensity changes, compared to that of chemical actinometers and thermopiles. Therefore phototubes and photomultipliers have a rather broad range of applications in either continuous photolysis or pulsed techniques.

1-6 PHOTOCHEMICAL REACTIONS IN CONDENSED MEDIA

A difference between gas phase and solution photochemistries results from complications introduced by solvent–molecule interactions in the liquid phase. For example, the so-called "cage effect," which is related to the encirclement of the reaction products by the solvent, is typical in solution photochemistry,[23,24] It is convenient to describe here the solution model that is usually used in the discussion of the effects encountered in solution photochemistry.[24,25] The density in the gas phase is very low, and changes in position by a molecule, known as diffusive movements, are important. On the other hand, the large density of liquids restricts the molecular movement in such a manner that diffusive movements have an activation energy, and diffusion coefficients in the liquid phase are several orders of magnitude smaller than coefficients in the gas phase. As a consequence of this large density in the liquid phase, molecules must spend a relatively large time oscillating around a position in the midst of solvent molecules, that is, colliding with nearest neighbors, and only occasionally experience a net displacement. According to this model, products generated by the thermal or photochemical dissociation of the same molecule, that is, a pair of radicals, are called geminate or primary species. It follows from this definition that geminate species are at a certain instant of the reaction in the same solvent cage. For example, two hydroxyl radicals generated in the photodissociation of hydrogen peroxide (eq. 53) are geminate or primary radicals if they have been originally produced by the splitting of the same molecule

$$\begin{gathered} H_2O_2 + h\nu \rightarrow 2HO^{\cdot} \\ (\phi \sim 0.5 \text{ for } \lambda_{ex} \sim 254\text{ nm})^{26} \end{gathered} \tag{53}$$

Photodissociation will produce the two primary hydroxyl radicals as nearest neighbors in a solvent cage as it is schematically shown in Figure 10. Collisions of the primary products within the solvent cage will lead to so-called primary recombination, a process that competes with diffusion out of the cage.

In this process the fragments may never attain a separation of as much as a molecular diameter and recombination takes place in a period that is longer than a vibration (10^{-13} s) and less than the time between diffusive displacements (10^{-11} s). When primary products undergo diffusive separation, that is, escape from the cage, there is still some probability that they will reencounter each other later, after a number of random displacements. Some authors have called the pair of primary products at this particular stage a solvent-separated pair and the recombination a diffusive geminate recombination. Products failing to undergo primary or geminate recombination must diffuse into the bulk of the solution and enter into steady state reactions, for example, reactions between unoriginal partners or other

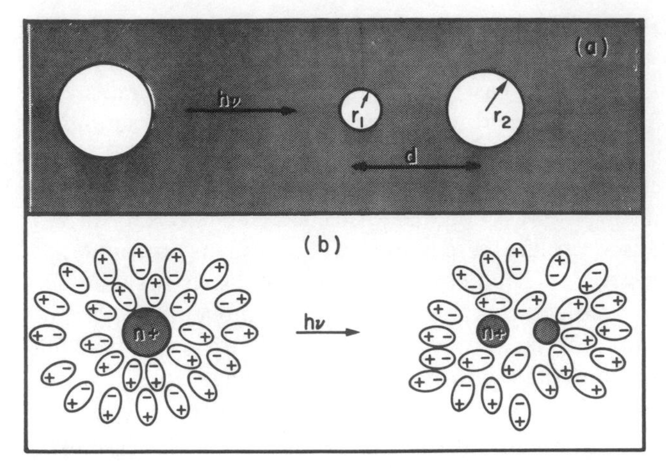

Figure 10. Photodissociation in condensed media. (*a*) The solvent is assumed to be a continuum; (*b*) the reactive species and products are solvated by dipolar molecules.

substrates. The events that take place during primary and geminate recombinations cannot be described by conventional kinetics since the probability of pair recombination at this time is proportional to $t^{-3/2}$.

There are several models that express the probability of geminate recombination in terms of various measurable parameters. Some models assume that the solvent is a continuum and that the diffusive displacements are randomly oriented. Moreover, long-range forces between radicals are neglected. These assumptions greatly diminish the mathematical complexity of the problem. In more recent treatments, the particulate nature of the solvent has been incorporated in a manner that requires the use of computers. Noyes' model takes the probability of geminate recombination as inversely proportional to the separation of the two primary products.[25,27] For example, let us represent the primary products by two spheres of radius r_1 and r_2 and masses m_1 and m_2 that are moving in opposite directions (Fig. 10). The distance between the centers of the spheres d is related to the corresponding distances x_1 and x_2 to the mass center by means of the well-known expressions of classical mechanics

$$d = x_1 + x_2 \tag{54}$$

$$m_1 x_1 = m_2 x_2 \tag{55}$$

Moreover, the movement of the spheres in a medium of viscosity η is hampered by the viscous drag of the solvent that imposes a deceleration given by the Stokes relationship

$$-\frac{\delta \dot{x}_i}{\delta t} = \frac{6\pi\eta r_i \dot{x}_i}{m_i} \qquad i = 1, 2 \tag{56}$$

where $\dot{x}_i = \delta x_i/\delta t$ with $i = 1, 2$. The double integration of eq. 56 leads to the distance separating the spheres at time t

$$d = \frac{m_i \dot{x}_1^0}{6\pi\eta r_1}\left(1 - \exp\frac{-6\pi\eta r_1}{m_1}\right) + \frac{m_2 \dot{x}_2^0}{6\pi\eta r_2}\left(1 - \exp\frac{-6\pi\nu r_2}{m_2}\right) \tag{57}$$

In eq. 57, $\dot{x}_1^0$ and $\dot{x}_2^0$ are the initial speeds of the spheres, that is, the speeds at the instant when the spheres start moving in opposite directions. The exponential terms decrease rapidly with time, and for $t = 10^{-12}$ s these terms are much less than unity. Hence, the simultaneous introduction of the reduced mass, $\mu = m_1 m_2/(m_1 + m_2)$, and the elimination of the exponential contributions in eq. 57 lead to

$$d = \frac{\mu \dot{d}^0}{6\pi\eta\rho} \tag{58}$$

where $\dot{d}^0 = \dot{x}_1 + \dot{x}_2^0$ and $\rho = r_1 r_2/(r_1 + r_2)$.

The probability that the pair of primary products recombine is

$$\beta = \frac{r_1 + r_2}{r_1 + r_2 + d} \tag{59}$$

and expressing d from eq. 59 in terms of eq. 58 gives

$$\beta = \frac{r_1 + r_2}{r_1 + r_2 + \mu \dot{d}^0/6\pi\eta\rho} \tag{60}$$

It is possible to relate the initial speed $\dot{d}^0$ to the photochemical or thermal energy ΔE^0 that propels the products apart in the initial instant of the reaction

$$\mu(\dot{d}^0)^2 = \Delta E^0 \tag{61}$$

Therefore eq. 60 can be recast as

$$\frac{1}{\beta} = 1 + \frac{(\Delta E^0 \mu)^{1/2}}{6\pi\eta\rho(r_1 + r_2)} \tag{62}$$

The parameter β, however, can be associated with the probability that a pair of products that have collided will collide again when the probability α of reaction in a collision is not unity. In this context, the probability of recombination, β', is given by a sum of the probabilities for successive collisions

$$\beta' = \alpha\beta + \alpha(1-\alpha)\beta^2 + \alpha(1-\alpha)^2\beta^3 + \cdots = \alpha\beta/(1 - \beta + \alpha\beta) \tag{63}$$

Moreover, the probability of the first collision may not be unity at the contact distance if the distance between the centers at the instant of the formation is larger than the contact distance $r_1 + r_2$ at the time of the encounter. The probability of reaction is, therefore, reduced by a factor $\sigma = (r_1 + r_2)/(\text{distance at the instant of formation})$

$$\beta' = \sigma\alpha\beta/(1 - \beta + \alpha\beta) \tag{64}$$

The explicit expression of β (eq. 62) can be introduced into eq. 64 to obtain

$$\beta' = \sigma\alpha \Bigg/ \left(\frac{(\Delta E^0_\mu)^{1/2}}{6\pi\eta\rho(r_1 + r_2)} + \alpha \right) \tag{65}$$

For primary products that separate with a thermal energy

$$\Delta E^0 = \frac{n}{2} \cdot kT \,,$$

(where k is the Boltzmann constant and n is the number of translation coordinates) and that experience recombination at each encounter $\alpha = 1$, the expression of β' is

$$\beta' = \sigma \Bigg/ \left(\frac{(nkT\mu/72)^{1/2}}{\pi\eta\rho(r_1 + r_2)} + 1 \right) \tag{66}$$

A value $n = 3$ in eq. 66 is used for diatomic molecules.

Insofar as the fraction of molecules that experience geminate recombination is $1 - \phi$, for a photodissociation in primary products with yield ϕ, this fraction can be equated to the fraction of molecules that undergoes geminate recombination, $\beta' = 1 - \phi$. It is assumed in Noyes' model that the separation between primary products increases with the excess energy, that is, with the difference between the bond energy and the energy of the quanta. This is equivalent to assuming that all of the excess energy appears as kinetic energy of the products, a hypothesis that is not always justified. However, the expression of β' in eq. 65 gives the yield of photodissociation $\phi = 1 - \beta'$ for photonic energies that are close to the bond dissociation energy D. The increase of the quantum yield with photonic energy is considered to arise from a decrease in the value of β', a parameter dependent on $(h\nu - D)$, that is,

$$1 - \phi = \sigma\alpha \Bigg/ \left(1 + \frac{(\mu(h\nu - D))^{1/2}}{6\pi\eta\rho(r_1 + r_2)} \right) \tag{67}$$

The functional dependences of the yield ϕ on the photonic energy $h\nu$ and the viscosity η that were found with a number of systems do not obey eq. 67. Several justifications have been advanced to explain the departure of the experimental data from theoretical predictions. According to Noyes,[24] the reason may be the treatment of the solvent as a continuum at distances of molecular dimensions. Moreover, the photonic energy of quanta absorbed by the molecule do not necessarily correlate with the energy that drives the reaction. For example, from the potential surfaces of the three electronic states in Figure 11, is clear that for excitations above point O the maximum energy that can propel the products apart is $h\nu_{eff} - E_D$. The energy $h\nu - h\nu_{eff}$ will be lost to the medium as thermal energy. The reason that $h\nu_{eff} - E_D$ is an upper limit of the kinetic energy of the products stems from the fact that this energy can be partitioned between various vibrational modes instead of being uniquely used as kinetic energy of the products.

The dependence of the cage effect can be rationalized by means of the rates of dissociation (eq. 68) and recombination (eq. 69) of caged geminate products (P_1, P_2)

$$P_1 - P_2 \xrightarrow{h\nu} (P_1, P_2) \xrightarrow{k_d} P_1 + P_2 \tag{68}$$

$$(P_1, P_2) \xrightarrow{k_c} P_1 - P_2 \tag{69}$$

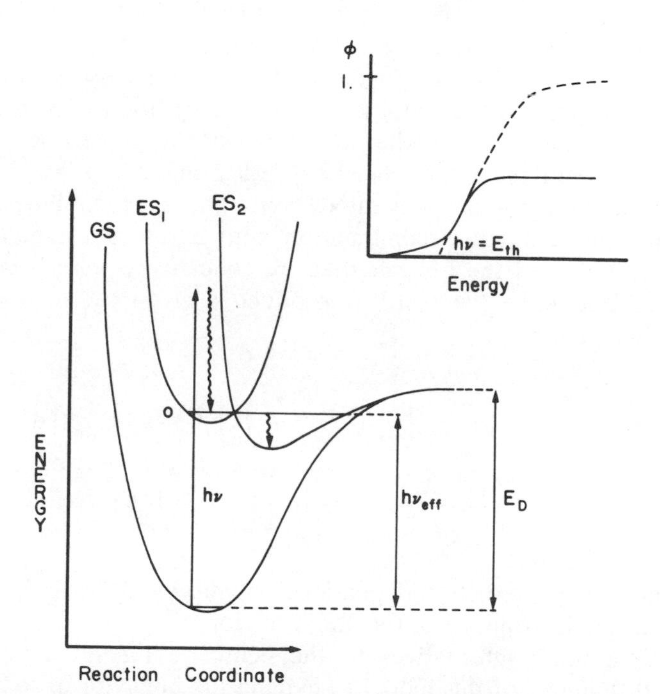

Figure 11. Potential energy diagram showing the relationship between the ground state (GS), the excited state produced by the absorption of light (ES_1), and the photoreactive excited state (ES_2). Notice that the photonic energy is much larger than the dissociation energy E_D. However, the useful energy for the photochemical reaction is only $h\nu_{eff}$. The solid line in the insert corresponds to a case where the yield of conversion of ES_1 to ES_2 is less than unity.

Most workers plot the reciprocal of the cage effect, $1/f = (k_c + k_d)/k_c$, as a function of the viscosity. In the model of Braun, Rajbenbach, and Eirich the reciprocal cage effect, $1/f$, depends on the square root of the fluidity, $(1/\eta)^{1/2}$, which is in agreement with experimental observations.[28,29] It is assumed that the geminate pairs diffuse apart with microscopic diffusion coefficients D. This treatment uses the diffusion equation, in one or three dimensions, and for cylindrical symmetry the solution shows the dependence on the square root of the viscosity

$$\chi(x, t) = \frac{A}{4a}\left[\operatorname{erf}\frac{(a/2)+x}{(4Dt)^{1/2}} + \operatorname{erf}\frac{(a/2)+x}{(4Dt)^{1/2}}\right] \tag{70}$$

The probability $\chi(x, t)$ that the particles are a distance x apart at a time t depends on the collision diameter a and the effective diffusion coefficient of the pair, D which can be equated to the viscosity of the medium by means of the Stokes–Einstein equation

$$D_i = kT/6\pi\eta r_i \tag{71}$$

This relationship expresses the diffusion coefficient D_i of a spherical particle with a radius r_i as a function of thermal energy.

To evaluate the yield of the caged products it is necessary to obtain the probabilities of geminate recombination. The integration of $\chi(x, t)$ eq. 70) with the limits $x = \pm a/2$ gives the probability of the pair to be caged. The integration between the limits $-\infty < x < -1/2$ and $\infty > x > a/2$ gives the probability of a diffusively separated pair. These probabilities must be modified by the geminate recombination with a rate constant k_c and by secondary reactions of the radicals that are collectively described by a rate constant k_s. Therefore the yield of product is given by an integral that combines these effects

$$\phi = \int_0^\infty k_c - \left(1 - 2\int_{a/2}^\infty \chi(x, t)\, dx\right) \exp[-(k_c + k_s)]\, \mathrm{dt}$$
$$+ 1\Big/\left(\frac{(k_c + k_s)}{k_c} + \sqrt{\frac{D}{k_c}}\, a^2\left[1 + \left(\frac{k_s}{k_c}\right)\right] b\right) \tag{72}$$

The Koenig model predicts a dependence of the yield on $(1/\eta)^{1/2}$ and an intercept different from zero for the function $(1/f) - 1$ when geminate products have other alternatives to the geminate recombination, that is when $k_s \neq 0$. Failures of this model to explain the behavior of some systems are attributed to the same reasons that have been already discussed in connection with Noyes' model.

The cage effect has been investigated with several inorganic reactions. For example, the photoredox decomposition of acidopentaammine cobalt-

(III) complexes (similar to the one shown in eq. 21) exhibits a pronounced dependence on the medium viscosity that can be equated with cage effects. Indeed, an increasing viscosity results in smaller quantum yields. These observations are discussed in Chapter 5 in relationship with charge transfer photochemistry. These studies are usually carried out with mixed-solvent solutions, for example, water–glycerol mixtures, of various compositions that result in different viscosities. In another experimental approach, net solvents with appropriate viscosities are used, for example, water, ethylene glycol, glycerol, instead of the mixed-solvent solutions. It must be pointed out, however, that neither approach is free of experimental artifacts. In mixed solvents, preferential solvation, that is, the solvation of the solute by one of the two solvents, is always possible. Such solvation results in an equilibrium between species with different solvation spheres that do not correlate with the bulk composition of the solvent. Moreover, with solutes that develop strong interactions with the solvent, the macroscopic viscosity may not give an appropriate description of the microscopic phenomena.

One further problem associated with the recombination reaction (eq. 76) is that it must obey the Witmer–Wigner spin conservation rule,[30,31] which states that *in elementary reactions the overall spin must be conserved*. The reason for this behavior is that for weak coupling certain degrees of freedom tend to be preserved, that is, experience little change, in collisional processes, and are therefore considered adiabatic. In this context, the spatial orientation of the magnetic field from a given electron can be changed only by a torque acting on the electron, that is, the torque created by the magnetic field associated with the spin of other electrons and the dynamic field created when electrons move rapidly in an electric field. Since mutual magnetic interactions depend on the sixth power of the distance between reactants, the intensity of the interaction decreases rapidly with distance and reaches a significant intensity only at a collision diameter, as in radical pairs, radical–ion pairs, or encounter complexes of energy transfer reactions. In this context, Wigner's conservation rule can be stated: *There must be at least one common value between the components of the spin angular momentum of the reactants and the products*. Hence, for the general reaction

$$^{(2S_A+1)}A + {}^{(2S_B+1)}B \rightarrow {}^{(2S_C+1)}C + {}^{(2S_D+1)}D \qquad (73)$$

where the superscripts indicate spin states, Wigner's rule requires the existence of at least a common value between the spin quantum numbers of the set of reactants

$$S_A + S_B\,, \qquad S_A + S_B - 1\,, \qquad \ldots, \qquad |S_A - S_B|$$

and the quantum numbers of the set of products

$$S_C + S_D\,, \qquad S_C + S_D - 1\,, \qquad \ldots, \qquad |S_C - S_D|$$

In the absence of a perturbation, such as a magnetic field or spin–orbit coupling, the microstates have the same energy and are equally populated. However, any transformation from reactants to products that involves a change of the spin quantum number represents a change in the spin state and requires that a certain activation energy be overcome (Fig. 12). For reactions proceeding only through potential surfaces that have no spin-induced activation energy, we can define a spin statistical factor g_s as the number of spin states of the reactants having common values with the products divided by the total number of spin microstates that are available to the reactants. For example, consider the reactions that Porter et al. have proposed as a mechanism for the energy transfer of triplet aromatic donors $^3M^*$ and ground state oxygen[32]

$$^3M^* + {}^3O_2 \rightarrow {}^1({}^3M^*, {}^3O_2) \rightarrow {}^1M + {}^1O_2^* \qquad g_s = \tfrac{1}{9} \tag{74}$$

$$^3M^* + {}^3O_2 \rightarrow {}^3({}^3M^*, {}^3O_2) \rightarrow {}^1M + {}^3O_2 \qquad g_s = \tfrac{1}{3} \tag{75}$$

$$^3M^* + {}^3O_2 \rightarrow {}^5({}^3M^*, {}^3O_2) \rightarrow {}^5(MO_2)^* \qquad g_s = \tfrac{5}{9} \tag{76}$$

The overall rate constant for each particular reaction if they are regarded as radiationless transitions, can be expressed as

$$k_0 = k g_e g_v g_s \tag{77}$$

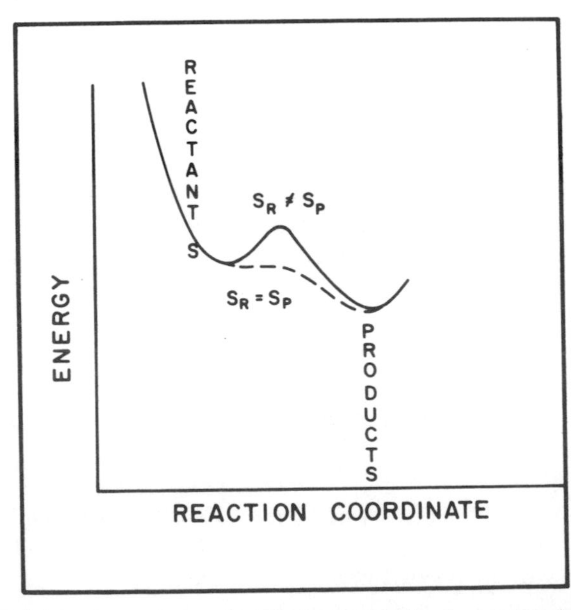

Figure 12. Potential energy surfaces showing the prohibition to crossing from the reactant to the product state when there is no spin conservation ($S_P \neq S_R$) in relationship to crossing when there is spin conservation ($S_P = S_R$).

where k represents the maximal rate and g_e, g_v, and g_s are prohibition factors introduced by electronic, vibrational and spin configuration changes in the reaction.[33] The prohibition factors are determined by the conservation laws for orbital symmetry and angular and spin momenta. Therefore, g_e can be evaluated by inspecting the "additive overlap" of the orbitals that are affected by the chemical transformation; g_v can be identified with Franck–Condon factors, which measure the prohibition to a change in nuclear configuration; and g_s can be identified with the spin statistical factor, which will in turn depend on the magnitude of the spin–orbit and hyperfine couplings. If the electronic and vibrational prohibition factors are the same for eqs. 74–76 and the processes are affected by a negligible spin–orbit coupling, their rate constants are expected to be in a 1:3:5 ratio. It is difficult to predict, however, how much of a prohibition factor must be applied when the spin–orbit perturbation is significantly strong. The perturbation will increase the probability of reaction in those surfaces that otherwise have a zero probability; a condition that can be associated with spin-statistical factors that are larger than the values calculated by applying Wigner's conservation rule.

The diffusion-controlled reactions, bimolecular reactions with activation energies that are largely determined by the activation energy of diffusive displacements, can be easily discussed in terms of Fick's first law[34]

$$\mathbf{J}_D = -D_i \nabla c \tag{78}$$

where $\mathbf{J}$ represents the flux of molecules with a diffusion coefficient D_i under a concentration gradient ∇c. For charged species under a coulombic potential U the flow of ions is

$$\mathbf{J}_e = -U_i c \nabla U \tag{79}$$

where U_i is the ionic mobility of the ions. It is possible to simplify the problem by using relative motions of the reactive species A and B, and placing A at the center of the coordinates. Hence, the total flux I of molecules B moving toward the center of coordinates that cross a sphere of radius r is

$$\left(\frac{I}{4\pi r^2}\right) = (D_A + D_B)\frac{d[B]}{dr} + (U_A + U_B)[B]\frac{dU}{dr} \tag{80}$$

where $(D_A + D_B)$ and $(U_A + U_B)$ are the sums of the diffusion coefficients and ionic mobilities, respectively. The rearrangement of eq. 80 into a linear differential equation (eq. 81) is achieved by means of the relationship

$$kt = D_i/u_i$$

between the thermal energy kT and the ratio D_i/u_i of the diffusion coefficient and the ionic mobility

$$d[\mathrm{B}] + [\mathrm{B}]\left(\frac{dU}{kT}\right) = \frac{I}{4\pi(D_{\mathrm{A}} + D_{\mathrm{B}})}\frac{dr}{r^2} \tag{81}$$

To solve eq. 81, we multiply both sides of the equality by the integrating factor $\exp(-U/kT)$ and express the coulombic potential

$$U = \frac{Z_{\mathrm{A}} Z_{\mathrm{B}} e^2}{\mathcal{K} r^2} \tag{82}$$

where r is the distance between ions and $\mathcal{K}$ is the dielectric constant of the solvent. Introduction of appropriate dimension factors leads to Smoluchowski's equation[35]

$$k_{\mathrm{d}} = \frac{4\pi R_{\mathrm{AB}}(D_{\mathrm{A}} + D_{\mathrm{B}})[Z_{\mathrm{A}} Z_{\mathrm{B}} e^2/\mathcal{K} kTR_{\mathrm{AB}}]N}{\{[\exp(Z_{\mathrm{A}} Z_{\mathrm{B}} e^2/\mathcal{K} kTR_{\mathrm{AB}})] - 1\} \times 10^3} \tag{83}$$

for the rate constant of a diffusion-controlled reaction. For uncharged species, $Z_{\mathrm{A}} = Z_{\mathrm{B}} = 0$, and eq. 83 can be reduced to

$$k_{\mathrm{d}} = \frac{4\pi R_{\mathrm{AB}}(D_{\mathrm{A}} + D_{\mathrm{B}})N}{10^3} \tag{84}$$

where N is Avogadro's number and R_{AB} is the encounter radius.

For elemental reactions where the activation energy can be expressed as a sum of contributions

$$k = A\exp\left(-\frac{E^{\ddagger}}{RT}\right); \qquad E^{\ddagger} = E^{\ddagger}_{\mathrm{e}} + E^{\ddagger}_{\mathrm{v}} + E^{\ddagger}_{\mathrm{s}} + \cdots \tag{85}$$

from the activation for reorganization of the electronic structure, $E^{\ddagger}_{\mathrm{e}}$, nuclear, $E^{\ddagger}_{\mathrm{v}}$, and spin, $E^{\ddagger}_{\mathrm{s}}$, configurations, there is a straightforward relationship between eq. 77 and eq. 85

$$\begin{aligned} k_0 &= A\left(\exp - \frac{E^{\ddagger}_{\mathrm{d}}}{RT}\right)\left(\exp - \frac{E^{\ddagger}_{\mathrm{e}}}{RT}\right)\left(\exp - \frac{E^{\ddagger}_{\mathrm{v}}}{RT}\right)\left(\exp - \frac{E^{\ddagger}_{\mathrm{s}}}{RT}\right) \\ &= kg_{\mathrm{e}} g_{\mathrm{v}} g_{\mathrm{s}} \end{aligned}$$

where $E^{\ddagger}_{\mathrm{d}}$ represents the activation energy for diffusive displacements. The diffusion-controlled reactions affected by Wigner's spin restriction are expected to have a rate constant determined also by a spin statistical factor

$$k = g_{\mathrm{s}} \frac{4\pi R_{\mathrm{AB}}(D_{\mathrm{A}} + D_{\mathrm{B}})(Z_{\mathrm{A}} Z_{\mathrm{B}} e^2/\mathcal{K} kTR_{\mathrm{AB}})N}{\{[\exp(Z_{\mathrm{A}} Z_{\mathrm{B}} e^2/\mathcal{K} kTR_{\mathrm{AB}})] - 1\}10^3} \tag{86}$$

if the condition $E^{\ddagger}_{\mathrm{v}}$, $E^{\ddagger}_{\mathrm{e}} < E^{\ddagger}_{\mathrm{d}} \approx E^{\ddagger}_{\mathrm{s}}$ holds.

2

DETECTION OF INTERMEDIATES

2-1 SCAVENGING OF REACTION INTERMEDIATES

One commonly used procedure for investigating a reaction mechanism consists of intercepting or scavenging reaction intermediates with appropriate reactants. The effect of these scavengers upon the reaction is determined from either changes of the product yields (in continuous photolysis) or lifetimes of the intermediates (in pulsed techniques) as a function of the scavenger concentration. For example, in the photolysis of aqueous triiodide one can use manganese(II) as a scavenger of the radicals I_2^- and $I^\cdot$

$$I_3^- \overset{h\nu}{\rightleftharpoons} I_2^- + I^\cdot \quad (1)$$

$$I^\cdot + I^- \rightleftharpoons I_2^- \quad (2)$$

$$I^\cdot + Mn(II) \longrightarrow I^- + Mn(III) \quad (3)$$

$$I_2^- + Mn(II) \longrightarrow 2I^- + Mn(III) \quad (4)$$

$$2I_2^- \longrightarrow I_3^- + I^- \quad (5)$$

The effect of the scavenger is to establish a competition between the reduction of the radicals to iodide (eqs. 3 and 4) and their disproportionation (eq. 5).

The selection of the scavenger is not dictated only by the nature of the intermediate that must be scavenged; the experimental conditions under

TABLE 2. Selected Properties of Substances Commonly Used as Scavengers

Typical Reaction[a] ($SH_2 + X^{\cdot} \rightarrow SH^{\cdot} + XH$)	k ($M^{-1}\,s^{-1}$)	Properties of $SH^{\cdot}$ and related Radicals[b]		
	Halides and Pseudohalides			
$SO_4^{\cdot-} + Cl^- \rightarrow SO_4^{2-} + Cl^{\cdot}$	1.9×10^8		$Cl^{\cdot}$	
$NO_3^{\cdot} + Cl^- \rightarrow NO_3^- + Cl^{\cdot}$	1.0×10^8	$Cl^{\cdot} + e^- \rightleftharpoons Cl^-$		$\epsilon^0 = 2.6\ V$
$Cl^{\cdot} + Cl^- \rightleftharpoons Cl_2^-$	$\{k_1 \geq 10^8$	$Cl^{\cdot} + Cl^{\cdot} \rightarrow Cl_2$		$k \sim 10^{10}\ M^{-1}\,s^{-1}$
	$\{k_{\bar{1}} = 1.1 \times 10^6\ s^{-1}$		Cl_2^-	
		$\lambda_{max} = 340\ nm$		$\epsilon_{max} = 1.2 \times 10^4\ M^{-1}\,cm^{-1}$
		$Cl_2^- + e^- \rightleftharpoons 2Cl^-$		$\epsilon^0 = 2.3\ V$
		$Cl_2 + e^- \rightleftharpoons Cl_2^-$		$\epsilon^0 = 0.6\ V$
		$Cl_2^- + Cl_2^- \rightarrow 2Cl^- + Cl_2$		$k = 4.0 \times 10^9\ M^{-1}\,s^{-1}$
$SO_4^{\cdot-} + Br^- \rightarrow SO_4^{2-} + Br^{\cdot}$	3.5×10^9		$Br^{\cdot}$	
$HPO_4^{\cdot-} + Br^- \rightarrow HPO_4^{2-} + Br^{\cdot}$	6.5×10^6	$Br^{\cdot} + e^- \rightleftharpoons Br^-$		$\epsilon^0 = 2.07\ V$
$Br^{\cdot} + Br^- \rightleftharpoons Br_2^-$	$\{k_1 = 5.6 \times 10^9$	$Br^{\cdot} + Br^{\cdot} \rightarrow Br_2$		$k \sim 10^{10}\ M^{-1}\,s^{-1}$
	$\{k_{\bar{1}} = 2.5 \times 10^4\ s^{-1}$		Br_2^-	
		$\lambda_{max} = 364\ nm$		$\epsilon_{max} = 7.8 \times 10^3\ M^{-1}\,cm^{-1}$
		$Br_2^- + e^- \rightleftharpoons 2Br^-$		$\epsilon^0 = 1.69\ V$
		$Br_2 + e^- \rightleftharpoons Br_2^-$		$\epsilon^0 = 0.51\ V$
		$Br_2^- + Br_2^- \rightleftharpoons 2Br^- + Br_2$		$k = 2.6 \times 10^9\ M^{-1}\,s^{-1}$
$CO_3^{\cdot-} + I^- \rightleftharpoons CO_3^{2-} + I^{\cdot}$	1.3×10^8		$I^{\cdot}$	
$PO_4^{2+} + I^- \rightarrow PO_4^{3-} + I^{\cdot}$	3.0×10^8	$I^{\cdot} + e^- \rightleftharpoons I^-$		$\epsilon^0 = 1.42\ V$
$I^{\cdot} + I^- \rightleftharpoons I_2^-$	$\{k_1 = 6.8 \times 10^9$	$I^{\cdot} + I^{\cdot} \rightarrow I_2$		$k = 10^{10}\ M^{-1}\,s^{-1}$
	$\{k_{\bar{1}} = 6.0 \times 10^4\ s^{-1}$		I_2^-	
		$\lambda_{max} = 380\ nm$		$\epsilon_{max} = 1.4 \times 10^4\ M^{-1}\,cm^{-1}$
		$I_2^- + e^- \rightleftharpoons 2I^-$		$\epsilon^0 = 1.13\ V$
		$I_2 + e^- \rightleftharpoons I_2^-$		$\epsilon^0 = 0.11\ V$
		$I_2^- + I_2^- \rightarrow 2I^- + I_2$		$k = 9 \times 10^9\ M^{-1}\,s^{-1}$
$CO_3^{\cdot-} + SCN^- \rightarrow CO_3^{2-} + SCN^{\cdot}$	8.0×10^5		$SCN^{\cdot}$	
$SO_4^{\cdot-} + SCN^- \rightarrow SO_4^{2-} + SCN^{\cdot}$	5.2×10^9	$SCN^{\cdot} + e^- \rightleftharpoons SCN^-$		$\epsilon^0 = 1.8\ V$
$Cl_2^- + SCN^- \rightarrow 2Cl^- + SCN^{\cdot}$	2.9×10^9		$BrSCN^-$	
$SCN^{\cdot} + SCN^- \rightleftharpoons (SCN)_2^-$	$\{k_1 = 6.8 \times 10^9$	$SCN^{\cdot} + Br^- \rightleftharpoons BrSCN^-$		$K_a = 1.7 \times 10^3$
	$\{k_{\bar{1}} = 3.4 \times 10^4\ s^{-1}$	$SCN^- + Br^{\cdot} \rightleftharpoons BrSCN^-$		$K_a = 2.2 \times 10^8$
$Br_2^- + SCN^- \rightarrow 2Br^- + SCN^{\cdot}$		$\lambda_{max} = 400\ nm$		$\epsilon_{max} = 2.3 \times 10^3\ M^{-1}\,cm^{-1}$
or			$ISCN^-$	
$\rightarrow Br^- + BrSCN^{\cdot}$	1.9×10^9	$SCN^{\cdot} + I^- \rightleftharpoons ISCN^-$		$K_a = 7.9 \times 10^7$
		$SCN^- + I^{\cdot} \rightleftharpoons ISCN^-$		$K_a = 2.1 \times 10^3$
		$\lambda_{max} = 420\ nm$		$\epsilon_{max} = 9.2 \times 10^3\ M^{-1}\,cm^{-1}$
			$(SCN)_2^-$	
		$SCN^- + SCN^{\cdot} \rightleftharpoons (SCN)_2^-$		$K_a = 2.0 \times 10^5$
		$(SCN)_2^- + e^- \rightleftharpoons 2SCN^-$		$\epsilon^0 = 1.5\ V$
		$(SCN)_2^- + (SCN)_2^- \rightarrow 2SCN^- + (SCN)_2$		$k = 2.1 \times 10^9\ M^{-1}\,s^{-1}$

Reaction	Rate constant	Radical properties	Value
$SO_4^{\cdot-} + N_3^- \rightleftharpoons SO_4^{2-} + N_3^{\cdot}$	7.4×10^8	$N_3^{\cdot}$	
$HPO_4^{\cdot-} + N_3^- \rightarrow HPO_4^{2-} + N_3^{\cdot}$	1.1×10^8	$\lambda_{max} = 275$ nm	$\epsilon_{max} = 1.1 \times 10^3\ M^{-1}\ cm^{-1}$
$Cl_2^{-} + N_3^- \rightarrow ClN_3^-(?) + Cl^-$	1.2×10^9	$N_3^{\cdot} + e^- \rightleftharpoons N^-3$	$\epsilon^0 = 1.9$ V
		$N_3^{\cdot} + N_3^{\cdot} \rightarrow 3N_2$	$k = 6.8 \times 10^9\ M^{-1}\ s^{-1}$
Alcohols			
$CH_3^{\cdot} + CH_3OH \rightarrow CH_4 + \cdot CH_2OH$	2.2×10^2	$\cdot CH_2OH$	
$CF_3^{\cdot} + CH_3OH \rightarrow HCF_3 + \cdot CH_2OH$	8.1×10^3	$\lambda max < 210$ nm	
$CO_3^{\cdot-} + CH_3OH \rightarrow CO_3^{2-} + H^+ + \cdot CH_2OH$	5.0×10^3	$\cdot CH_2OH + e^- \rightleftharpoons CH_3OH$	$\epsilon^0 = 1.29$ V
$SO_4^{\cdot-} + CH_3OH \rightarrow HSO_4^- + \cdot CH_2OH$	2.5×10^7	$CH_2O + e^- \rightleftharpoons \cdot CH_2OH$	$\epsilon^0 = 0.92$ V
$PO_4^{2-} + CH_3OH \rightarrow HPO_4^{2-} + \cdot CH_2OH$	1.0×10^7	$\cdot CH_2OH \rightleftharpoons \cdot CH_2O^- + H^+$	$K_a = 2.0 \times 10^{-11}$
$HPO_4^- + CH_3OH \rightarrow H_2PO_4^- + \cdot CH_2OH$	1.0×10^7	$\cdot CH_2OH + \cdot CH_2OH \rightarrow CH_2O + CH_3OH$	$k = 9.6 \times 10^7\ M^{-1}\ s^{-1}$
$Cl_2^- + CH_3OH \rightarrow 2Cl^- + H^+ + \cdot CH_2OH$	3.5×10^3	$\cdot CH_2OH + \cdot CH_2OH \rightarrow (CH_2OH)_2$	$k = 2.3 \times 10^9\ M^{-1}\ s^{-1}$
		$\cdot CH_2O^-$	
		$\lambda_{max} < 220$ nm	
		$CH_2O + e^- \rightleftharpoons \cdot CH_2O^-$	$\epsilon^0 = -1.49$ V
		$\cdot CH_2O^- + \cdot CH_2O^- \rightarrow CH_2O + Ch_3O^-$	$k = 9 \times 10^8\ M^{-1}\ s^{-1}$
$CH_3^{\cdot} + CH_3CH_2OH \rightarrow CH_4 + CH_3CHOH^{\cdot}$	5.9×10^2	$C_2H_4OH^{\cdot}$	
$CF_3^{\cdot} + CH_3CH_2OH \rightarrow HCF_3 + C_2H_4OH^{\cdot}$	4.6×10^4	$\lambda_{max} < 210$ nm	
$CO_3^{\cdot-} + CH_3CH_2OH \rightarrow CO_2 + CH_3\dot{C}HOH$	1.6×10^4	$CH_3\dot{C}HOH + e^- \rightleftharpoons CH_3CH_2OH$	$\epsilon^0 \sim 0.95$ V
$SO_4^{\cdot-} + CH_3CH_2OH \xrightarrow{H^+} HSO_4^- + CH_3\dot{C}HOH$	7.7×10^7 (pH4.8)	$CH_3CHO + e^- \rightleftharpoons CH_3\dot{C}HOH$	$\epsilon^0 = -0.97$ V
$PO_4^{2-} + CH_3CH_2OH \rightarrow HPO_4^{2-} + CH_3\dot{C}HOH$	1.9×10^7	$CH_3\dot{C}HOH \rightleftharpoons CH_3\dot{C}HO^- + H^+$	$K_a = 2.5 \times 10^{-12}$
$HPO_4^- + CH_3CH_2OH \rightarrow H_2PO_4^- + CH_3\dot{C}HOH$	4.0×10^7	$CH_3\dot{C}HOH + CH_3\dot{C}HOH \rightarrow CH_3CH_2OH + CH_3CHO$	$k = 2.3 \times 10^9$
$Cl_2^- + CH_3CH_2OH \rightarrow 2Cl^- + H^+ + C_2H_4OH^{\cdot}$	4.5×10^4	$CH_3\dot{C}HO^-$	
		$\lambda_{max} < 220$ nm	
		$Ch_3CHO + e^- \rightleftharpoons CH_3\dot{C}HO^-$	$\epsilon^0 = -1.65$ V
		$CH_3\dot{C}HO^- + CH_3\dot{C}HO^- \rightarrow$ [c]	$k = 5 \times 10^8\ M^{-1}\ s^{-1}$
$CH_3^{\cdot} + (CH_3)_2CHOH \rightarrow CH_4 + (CH_3)_2\dot{C}OH$	3.4×10^3	$(CH_3)_2\dot{C}OH$	
$CF_3^{\cdot} + (CH_3)_2CHOH \rightarrow HCF_3 + (CH_3)_2\dot{C}OH$	9.2×10^4	$\lambda_{max} < 230$ nm	
$CO_3^{\cdot-} + (CH_3)_2CHOH \rightarrow HCO_3^- + (CH_3)_2\dot{C}OH$	3.9×10^4	$(CH_3)_2\dot{C}OH + e^- \rightleftharpoons (CH_3)_2CHOH$	$\epsilon^0 = -1.2$ V
$SO_4^{\cdot-} + (CH_3)_2CHOH \rightarrow HSO_4^- + (CH_3)_2\dot{C}OH$	8.5×10^7	$(CH_3)_2\dot{C}OH \rightleftharpoons (CH_3)_2\dot{C}O^- + H^+$	$K_a = 6.1 \times 10^{-13}$
$PO_4^{2-} + (CH_3)_2CHOH \rightarrow HPO_4^{2-} + (CH_3)_2\dot{C}OH$	1.8×10^7	$(CH_3)_2\dot{C}OH + (CH_3)_2\dot{C}OH \rightarrow (CH_3)CHOH + (CH_3)_2CO$	$k = 1.4 \times 10^9\ M^{-1}\ s^{-1}$
$HPO_4^- + (CH_3)_2CHOH \rightarrow H_2PO_4^- + (CH_3)_2\dot{C}OH$	4.0×10^7		
$Cl_2^- + (CH_3)_2CHOH \rightarrow 2Cl^- + H^+(CH_3)_2\dot{C}OH$	1.2×10^5		
Olefinic Compounds			
$CH_3^{\cdot} + CH_2 = CH_2 \rightarrow \cdot CH_2CH_2CH_3$	4.9×10^3		
$CF_3^{\cdot} + CH_2 = CH_2 \rightarrow \cdot CH_2CH_2CF_3$	4.0×10^7		
$SO_4^{\cdot-} + CH_2 = CHCONH_2 \rightarrow {}^-O_3SOCH_2CHCONH_2^{\cdot}$	1.6×10^8		

TABLE 2 (Continued)

Typical Reaction[a] ($SH_2 + X^{\cdot} \rightarrow SH^{\cdot} + XH$)	k ($M^{-1}\,s^{-1}$)	Properties of $SH^{\cdot}$ and related Radicals[b]	
$H_2PO_4 + CH_2 = CHCONH_2 \rightarrow H_2O_3POCH_2\dot{C}HCONH_2$	2.2×10^8 (pH 4)		
$SO_4^{\dot{-}} + CH_2 = CHCO_2^- \rightarrow {}^-O_3SOCH_2CHCO_2^{\dot{-}}$	1.1×10^8		
$HPO_4 + CH_2 = CHCO_2^- \rightarrow {}^-HO_3POCH_2CHCO_2^{\dot{-}}$	6.2×10^6		
$H_2PO_4 + CH_2 = CHCO_2^- \rightarrow H_2O_3POCH_2\dot{C}HCO_2^-$	1.6×10^8		
$Cl_2^{\dot{-}} + CH_2 = CHCO_2^- \rightarrow Cl^- + ClCH_2\dot{C}HCO_2^-$	1.9×10^7		
	Coordination Complexes		
		$Co(NH_3)_5Cl^{2/1+}$ [c]	
$CO_3^{\dot{-}} + Co(NH_3)_5Cl^{2+} \xrightarrow[H^+]{} CO_2 + Co^{2+}_{aq} + 5NH_4^+ + Cl^-$	2.0×10^6	$Co(NH_3)_5Cl^{2+} + e^- \rightleftharpoons Co(NH_3)_5Cl^+$	$\epsilon^0 \sim 0.15$[d]
$\cdot CH_2OH + Co(NH_3)_5Cl^{2+} \xrightarrow[H^+]{} CH_2O + CO^{2+}_{aq} + 5NH_4^+ + Cl^-$	3.0×10^6	$Co(NH_3)_5Cl^{2+} + Co(NH_3)_5Cl^+ \rightarrow Co(NH_3)_5Cl^+ + Co(NH_3)_5Cl^{2+}$	$k \sim 10^{-7}\,M^{-1}\,s^{-1}$ [e]
$(CH_3)_2\dot{C}OH + Co(NH_3)_5Cl^{2+} \xrightarrow[H^+]{} (CH_3)_2CO + Co^{2+}_{aq} + 5NH_4^+ + Cl^-$	4.0×10^7		
$CO_2^{\dot{-}} + Co(NH_3)_5Cl^{2+} \xrightarrow[H^+]{} CO_2 + CO^{2+}_{aq} + 5NH_4^+ + Cl^-$	1.5×10^8		
$\cdot CH_2OH + Co(NH_3)_5OH_2^{3+} \xrightarrow[H^+]{} CH_2O + Co^{2+}_{aq} + 5NH_4^+$	1.5×10^6		
$CO_2^{\dot{-}} + Co(NH_3)_5OH_2^{3+} \xrightarrow[H^+]{} CO_2 + Co^{2+}_{aq} + 5NH_4^+$	1.7×10^8		
$\cdot CH_2OH + Co(NH_3)_5F^{2+} \xrightarrow[H^+]{} CH_2O + Co^{2+}_{aq} + 5NH_4^+ + F^-$	5.5×10^5		
		$Co(bipy)_3^{3/2+}$	
$\cdot CH_2OH + Co(bipy)_3^{3+} \rightarrow CH_2O + H^+ + Co(bipy)_3^{2+}$	2.0×10^8	$Co(bipy)_3^{3+} + e^- \rightleftharpoons Co(bipy)_3^{2+}$	$\epsilon^0 = 0.315$ V
$(CH_3)_2\dot{C}OH + Co(bipy)_3^{3+} \rightarrow (CH_3)_2CO + Co(bipy)_3^{2+}$	2.5×10^9	$Co(bipy)_3^{3+} + Co(bipy)_3^{2+} \rightarrow Co(bipy)_3^{2+} + Co(bipy)_3^{3+}$	$k_{ex} = 20\,M^{-1}\,s^{-1}$
$CO_2^{\dot{-}} + Co(bipy)_3^{3+} \rightarrow Co_2 + Co(bipy)_3^{2+}$	7.6×10^9		
$(CH_3)_2\dot{C}OH + Co(phen)_3^{3+} \rightarrow (CH_3)_2CO + H^+ + Co(phen)_3^{2+}$	4.6×10^9		
$\cdot CH_2OH + Co(5,6\text{-}Me_2phen)_3^{3+} \rightarrow CH_2O + H^+ + Co(5,6\text{-}Me_2phen)_3^{2+}$	4.9×10^8		
$CH_3\dot{C}HOH + Co(5,6\text{-}Me_2phen)_3^{3+} \rightarrow CH_3CHO + H^+ + Co(5,6\text{-}Me_2phen)_3^{2+}$	3.1×10^9		

Reaction	k	Couple / reaction	Value
$\cdot CH_2Oh + Cr^{2+}_{aq} \rightarrow CrCH_2OH^{2+}$	1.6×10^8	$Cr^{3/2+}_{aq}$	
$CH_3\dot{C}HOH + Cr^{2+}_{aq} \rightarrow CrCHOHCH_3^{2+}$	7.9×10^7	$Cr^{3+}_{aq} + e^- \rightleftharpoons Cr^{2+}_{aq}$	$\epsilon^0 = -0.41$ V
$(CH_3)_2\dot{C}OH + Cr^{2+}_{aq} \rightarrow CrCOH(CH_3)_2^{2+}$	5.1×10^7	$Cr^{3+}_{aq} + {}^1Cr^{2+}_{aq} \rightarrow Cr^{2+}_{aq} + {}^1Cr^{3+}_{aq}$	$k_{ex} \sim 1.2 \times 10^{-4}\ M^{-1}\ s^{-1}$
$\cdot CH_2C(CH_3)_2CO_2H + Cr^{2+}_{aq} \rightarrow CrCH_2C(CH_3)_2CO_2H$	1.1×10^8		
$CO_2^{\dot{-}} + Cr^{2+}_{aq} \rightarrow CrCO_2^+$	1.1×10^9		
$I_2^- + Cr^{2+}_{aq} \rightarrow CrI^{2+} + I^-$	1.5×10^9		
$Br_2^- + Cr^{2+}_{aq} \rightarrow CrBr^{2+} + Br^-$	1.9×10^9		
$SO_4^{\dot{-}} + Cr^{2+}_{aq} \rightarrow Cr^{3+}_{aq} + SO_4^{2-}$	$>10^9$		
		$Rh(bipy)_3^{3/2+}$	
$\cdot CH_2OH + Rh(bipy)_3^{3+} \rightarrow CH_2O + H^+ + Rh(bipy)_3^{2+}$	2.2×10^8	$Rh(bipy)_3^{3+} + e^- \rightleftharpoons Rh(bipy)_3^{2+}$	$\epsilon^0 = -0.67$ V
$CO_2^{\dot{-}} + Rh(bipy)_3^{3+} \rightarrow CO_2 + Rh(bipy)_3^{2+}$	6.2×10^9	$Rh(bipy)_3^{3+} + Rh(bipy)_3^{2+} \rightarrow Rh(bipy)^2 + 3 + Rh(bipy)_3^{2+}$	$k_{ex} = 1.7 \times 10^7\ M^{-1}\ s^{-1}$
$CH_3^{\cdot} + Cu^{2+}_{aq} \rightarrow CuCH_3^{2+}$	7.4×10^5	$Cu^{2/1+}_{aq}$	
$\cdot CH_2CH_2OH + Cu^{2+}_{aq} \rightarrow CuCH_2CH_2OH \xrightarrow{fast} Cu^+_{aq} + \text{products}$	1.9×10^7	$Cu^{2+}_{aq} + e^- \rightleftharpoons Cu^+_{aq}$	$\epsilon^0 = 0.158$ V
$\cdot CH_2OH + Cu^{2+}_{aq} \rightarrow Cu^+_{aq} + CH_2O + H^+$	1.1×10^8	$Cu^{2+}_{aq} + Cu^+_{aq} \rightarrow Cu^+_{aq} + Cu^{2+}_{aq}$	$k_{ex} = 10^{-5}\ M^{-1}\ s^{-1}$
$\cdot CH_2CO_2^- + Cu^{2+}_{aq} \rightarrow Cu(CH_2CO_2)^+ \rightarrow Cu^+_{aq} + \text{products}$	6.4×10^8		
$CO_2^{\dot{-}} + Cu^{2+}_{aq} \rightarrow CO_2 + Cu^+_{aq}$	1.5×10^8		
$CH_3^{\cdot} + IrCl_6^{2-} \rightarrow {}^c$	1.2×10^9		
$CCl_3^{\cdot} + IrCl_6^{2-} \rightarrow {}^c$	2.8×10^7		
$CO_2^{\dot{-}} + Ni(CN)_4^{2-} \rightarrow CO_2 + Ni(CN)_4^{3-}$	1.2×10^9		
$CO_2^{\dot{-}} + Fe(CN)_6^{3-} \rightarrow CO_2 + Fe(CN)_6^{4-}$	1.1×10^9	$Fe(CN)_6^{3/4-}$	
$\cdot Ch_3 + Fe(CN)_6^{3-} \rightarrow {}^c$	5.0×10^6	$Fe(CN)_6^{3-} + e^- \rightleftharpoons Fe(CN)_6^{4-}$	$\epsilon^0 \sim -0.69$ V
$\cdot CH_2OH + Fe(CN)_6^{3-} \rightarrow Ch_2O + H^+ + Fe(CN)_6^{4-}$	4.2×10^9	$Fe(CN)_6^{3-} + Fe(CN)_6^{4-} \rightarrow Fe(CN)_6^{4-} + Fe(CN)_6^{3-}$	$k_{ex} \sim 28\ M^{-1}\ s^{-1}$ [f]
$CH_3\dot{C}HOH + Fe(CN)_6^{3-} \rightarrow CH_3CHO + H^+ + Fe(CN)_6^{4-}$	5.3×10^9		
$(CH_3)_2\dot{C}OH + Fe(CN)_6^{3-} \rightarrow (CH_3)_2CO + H^+ + Fe(CN)_6^{4-}$	4.7×10^9		
$\cdot Ch_2OH + Fe^{3+}_{aq} \rightarrow CH_2O + H^+ + Fe^{2+}_{aq}$	1.0×10^8		
$CH_3\dot{C}HOH + Fe^{3+}_{aq} \rightarrow CH_3CHO + H^+ + Fe^{2+}_{aq}$	2.9×10^6		
$(CH_3)_2\dot{C}OH + Fe^{3+}_{aq} \rightarrow (CH_3)_2CO + Fe^{2+}_{aq}$	4.5×10^8		

[a] Data from Refs. 36–45 and references therein.

[b] Abbreviations: ϵ^0, standard redox potential versus NHE; k, specific rate constant; ϵ_{max}, extinction coefficient at the wavelength λ_{max} of the maximum; K, equilibrium constant.

[c] The Co(III) pentaammine complexes lose one NH_3 with $t_{1/2} \leq 1\ \mu s$ upon one electron reduction in acid solutions.

[d] System not completely characterized.

[e] Rate constant for the self-exchange electron transfer reaction.

[f] At 0.1°C with 2.5 M EDTA. The rate constant depends on the cations present in the reaction medium.

which the reaction must run may also introduce limitations to the type of scavengers that we can use. Primary and secondary alcohols, R_2CHOH, are useful hydrogen donors in hydrogen abstraction reactions (eq. 6) and electron donors in electron transfer reactions (eq. 7)

$$R_2CHOH + R\cdot \rightarrow R_2\dot{C}OH + RH \tag{6}$$

$$R_2CHOH + R\cdot \rightarrow R_2CH\dot{O}H^+ + R^- \tag{7}$$

In these equations, $R\cdot$ represents a radical that functions as a strong oxidant. Some amines, for example, triethylamine and triethanolamine, are used as electron donors in electron transfer reactions such as the reduction of $Ru(bipy)_3^{3+}$

$$Ru(bipy)_3^{3+} + TEOA \longrightarrow (^3CT)Ru(bipy)_3^{2+} + TEOA^{\cdot +} \tag{8}$$

$$(TEOA = \text{triethanolamine})\,.$$

The properties of a number of compounds, often used as scavengers, are reported in Table 2.

The scavenging of the reaction intermediates is a complex process that comprises several steps.[24,25] In terms of the models that describe the reactions in condensed media (Chapter 1), the intermediates that escaped primary and geminate recombinations can diffuse to the bulk of the solvent where the recombination of unoriginal partners and the scavenging processes obey conventional kinetics, that is, the steady state approximation is valid. The scavenging of the intermediates at this step, known as steady state scavenging, takes place at low scavenger concentrations, for example, $[\text{scavenger}] \leq 10^{-5}$ *M* in Figure 13. An example of this type of behavior is provided by the scavenging of solvated electrons in continuous photolysis of $Fe(CN)_6^{4-}$ in aqueous solutions[46,47]

$$Fe(CN)_6^{4-} \xrightarrow{h\nu,\phi} Fe(CN)_6^{3-} + e_{aq}^- \qquad (\lambda_{excit} \sim 254\ \text{nm}) \tag{9}$$

$$Fe(CN)_6^{3-} + e_{aq}^- \longrightarrow Fe(CN)_6^{4-} \tag{10}$$

$$e_{aq}^- + N_2O \longrightarrow N_2 + HO^\cdot \tag{11}$$

$$HO^\cdot + R_1R_2CHOH \longrightarrow H_2O + R_1R_2\dot{C}OH \tag{12}$$

$$R_1R_2\dot{C}OH \longrightarrow (\text{recombination and disproportionation products}) \tag{13}$$

The solvated electron reacts with nitrous oxide (eq. 11) in a reaction that competes with the steady state reduction of the Fe(III) complex (eq. 10). Such reactions have values of the specific rate constants $k_{10} \sim 3.0 \times 10^9\ M^{-1}\ s^{-1}$ and $k_{11} \sim 8.7 \times 10^9\ M^{-1}\ s^{-1}$, respectively, which correspond to

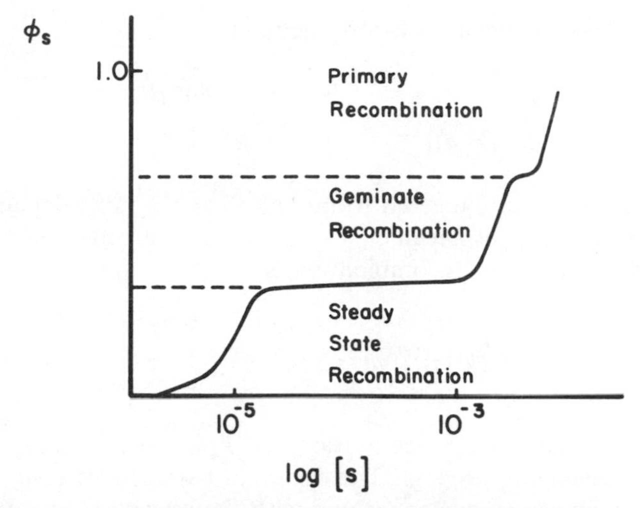

Figure 13. Steps (horizontal lines) in the scavenging of reaction intermediates. The scavenging process competes with steady state recombination, geminate recombination, and primary recombination.

processes with diffusion control. Alcohols (R_1R_2CHOH) are used to trap the hydroxyl radicals that are generated in the reduction of nitrous oxide (eq. 11). For the mathematical description of steady state scavenging of an electron with nitrous oxide, it is possible to apply the steady state approximation to the concentration of solvated electrons. This leads to an expression for the yield of Fe(III) as a function of the $Fe(CN)_6^{3-}$ and N_2O concentrations and the yield ϕ of products that escape primary and geminate recombination

$$\phi_{\mathrm{Fe(III)}} = \frac{1}{I_{\mathrm{ab}}} \frac{\partial[\mathrm{Fe(CN)}_6^{3-}]}{\partial t} = \phi\left(\frac{k_{11}[N_2O]}{k_{10}[\mathrm{Fe(CN)}_6^{3-}] + k_{11}[\mathrm{N_2O}]}\right) \tag{14}$$

For the scavenging process to compete with primary and geminate recombinations, the concentration of the scavenger has to be large. It is conceivable that molecules forming part of the wall of the cage may lead to scavenging events that compete with primary recombination. Nevertheless, these processes cannot be treated according to the rules of conventional kinetics since the probability of recombination of a pair of primary products will change with time.[48] The probability β' that the pairs will recombine can be expressed as an integral of the probability $h(t)$ that they will recombine in an interval between t and $t + dt$

$$\int_0^\infty h(t)\, dt = \beta' \tag{15}$$

The function $h(t)$ has been arbitrarily defined

$$h(t) = 0 \qquad \text{for } 0 < t < 4a^2/\beta'^2$$

$$h(t) = a/t^{3/2} \qquad \text{for } 4a^2/\beta'^2 \leq t \leq \infty$$

where the parameter a, expected to be less than $10^{-6}\ s^{1/2}$, depends on the frequency of diffusive displacements and on β'. The competition between scavenging and secondary recombination is given by

$$\int_0^\infty h(t)(1 - \exp(-2k_s[s]t))\, dt = 2a\sqrt{2\pi k_s[s]} - 8k_s a^2[s]/\beta' + \cdots \tag{16}$$

where $[s]$ is the scavenger concentration and k_s is the rate constant for the steady state scavenging process. The factor in parentheses gives the probability that the primary species reacted with the scavenger at a time t.

Equation 16 can be combined with expressions of β' that are functions of parameters such as the probability α that two primary species will react with each other, and the probability β that two primary species that have experienced a nonreactive encounter will encounter again. For example, eq. 17 can be obtained with the same arguments used for eq. 63 in Chapter 1

$$\beta' = \alpha\beta_0/(1 - \beta + \alpha\beta) \tag{17}$$

The limiting yield ϕ and the yield ϕ_s at a scavenger concentration $[s]$ are therefore related by

$$\ln(1 - \phi_s/\phi) = \ln \beta' - 2a/\beta'(\pi k_s[s])^{1/2} \tag{18}$$

The different steps for scavenging intermediates are shown in Figure 13. At very low scavenger concentrations, the scavenging reaction competes with other reactions of the intermediates in the bulk of the solvent. As the scavenger concentration is increased, all the bulk reactions are eliminated leading to the lower plateau in Figure 13. At higher concentrations of the scavenger, the scavenging process competes with primary and geminate recombinations and a new plateau is reached in Figure 13.

2-2 CONVENTIONAL FLASH PHOTOLYSIS AND SINGLE PHOTON COUNTING

The aim of these techniques is the investigation of those events in a photochemical reaction that are too fast for detection in the time domain of continuous photolysis.[49–51] They are based on irradiating the photolyte with very short and intense pulses of light from flash lamps or lasers (Fig. 14).

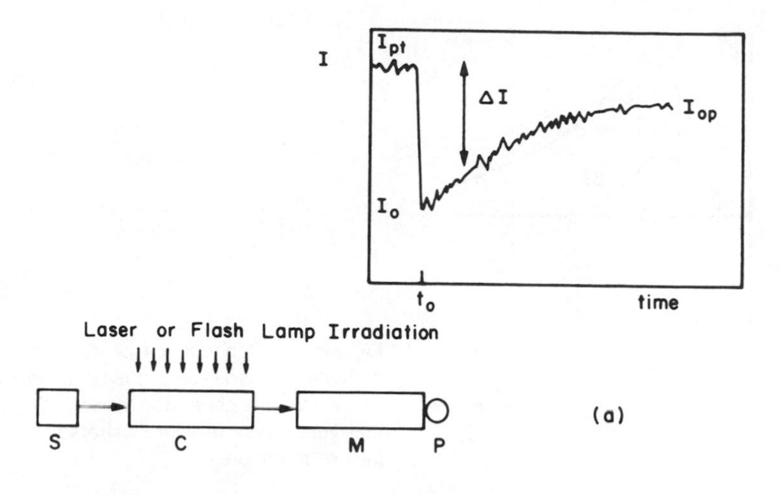

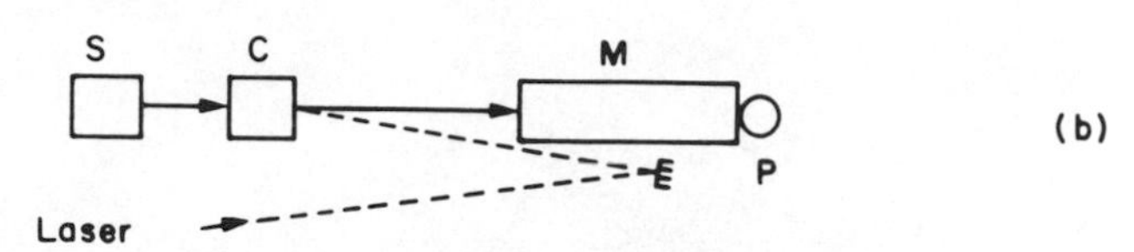

Figure 14. Geometrical configurations used in flash photolysis: (*a*) side irradiation; (*b*) front irradiation. Abbreviations: S, monitoring source; C, cell holder; M, monochromator; and P, phototube. The insert shows a typical trace corresponding to the time-resolved response of the phototube: I_{pt}, phototube pretriggering voltage; ΔI, drop in voltage (at time t) induced by the production of transient species absorbing some of the monitoring light. The flash irradiation takes place at t_0 causing the phototube voltage to drop to I_0.

Although this form of irradiation helps generate large concentrations of the reaction intermediates, real light pulses have a certain length, that is, they are not infinitesimally short, and this imposes a limit to the time resolution for the observation of a given process. The relationship between the length of the pulse and the evolution in time of a transient species can be illustrated by irradiating a sample with a triangular light pulse that has the time dependence

$$I_a = bt \qquad \text{for } 0 \leq t < I_a^0/b$$

$$I_a = 2I_a^0 - bt \qquad \text{for } I_a^0/b \leq t \leq 2I_a^0/b$$

$$I_a = 0 \qquad \text{for } t > 2I_a^0/b$$

where I_a^0 is the absorbed light intensity at a time t (Fig. 15). Let us also assume that the absorption of a photon by A produces an intermediate B (eq. 19) and that in the time domain of our observations B is formed

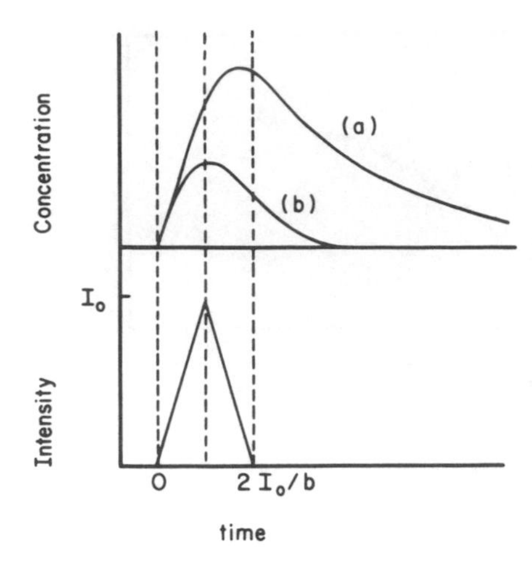

Figure 15. Dependence on time of an intermediate concentration, curves *a* and *b*. The intermediate is generated by irradiating with the triangular pulse of light illustrated in the lower part of the figure.

instantaneously. This intermediate B evolves into a product P by a first-order reaction

$$A + h\nu \xrightarrow{\phi} B \tag{19}$$

$$A \xrightarrow{k_{20}} P \tag{20}$$

The rate at which the concentration of B changes with time is given by

$$d[B]/dt = \phi bt - k_{20}[B] \qquad \text{for } 0 \leq t < I_a^0/b \tag{21}$$

$$d[B]/dt = \phi I_a^0 - \phi bt - k_{20}[B] \qquad \text{for } I_a^0/b \leq t < 2I_a^0/b \tag{22}$$

$$d[B]/dt = -k_{20}[B] \qquad \text{for } t \geq 2I_a^0/b \tag{23}$$

Integrating eq. 21 with the initial condition $[B] = 0$ for $t = 0$ gives the expression

$$[B] = \frac{\phi b}{k_{20}^2} [k_{20}t - 1 + \exp(-k_{20}t)] \tag{24}$$

for the concentration of B in the time interval $0 \leq t \leq I_a^0/b$. For a time $t = I_a^0/b$, the concentration of B reaches a value

$$\mathcal{B} = \frac{\phi b}{k_{20}^2} \left(\frac{k_{20}I_a^0}{b} - 1 + \exp\left(- \frac{k_{20}I_a^0}{b} \right) \right) \tag{25}$$

In the same manner, integrating eq. 22 with the initial condition $[B] = \mathcal{B}$ for $t = I_a^0/b$ leads to

$$[\mathrm{B}] = \frac{\phi I_a^0}{k_{20}} + \frac{\phi b}{k_{20}^2}(k_{20}t - 1) + \left(\beta - \frac{2\phi I_a^0}{k_{20}} + \frac{\phi b}{k_{20}^2}\right) \exp - \left[k_{20}t - \frac{k_{20}I_a^0}{b}\right] \tag{26}$$

The cumbersome expressions derived for the concentration of B (eq. 24, 26) reflect the variation of absorbed light with time and the consumption of the intermediate through its conversion to the product P (eq. 20). For the study of reaction kinetics it is convenient to use experimental conditions that separate contributions (to the concentration of the intermediate) related to the manner of irradiating the sample from those associated with the reaction mechanism. For example, the irradiation of the sample with a pulse having a half-height width $\delta = I_a^0/b$ one or more orders of magnitude shorter than thc lifetime of the intermediate is an experimental condition that reduces eq. 24 and 26 to

$$[\mathrm{B}] \simeq (\phi b/k_{20})t \qquad \text{for } I_a^0/b \leq 10^{-1}/k_{20} = 10^{-1}\tau \tag{27}$$

Moreover, observations at times $t > 2I_a^0/b$ will show the intermediate decaying with an integrated rate law

$$[\mathrm{B}] = \frac{2\phi I_a^0}{k_{20}} \exp(-k_{20}t) \tag{28}$$

It is possible to regard a time between three and five times the half-height width of the light pulse as the time resolution of a flash photolysis apparatus for the study of reaction kinetics. Deconvolution techniques can be used for improving the time resolution. However, the use of deconvolution techniques requires the exact knowledge of the pulse profile and the use of time-consuming calculations.

Further conditions for the observation of a transient are dictated by the type of detection used for the investigation of the intermediates. These conditions can be deduced by considering the sensitivity the detection system must have for us to observe the intermediate, and the speed of the response required to study the reaction. For example, the circuit in Figure 16 represents a typical way of wiring a phototube that must quickly respond to changes of the monitoring beam intensity in flash photolysis.[22] In the phototube circuit, the set of resistors R_d form the voltage divider, which causes a stepwise positive increase of the potential from one dynode to the next, starting from the photocathode. All the resistors R_d have the same value, which is much larger than the internal resistance of the phototube. For fast-pulse applications that develop high currents through the photomultiplier, capacitors are placed parallel to the resistors R_d. In the simple circuit of Figure 16, the photocurrent is detected by the potential drop across the load resistor R_L with an oscilloscope. The capacitor C_L parallel to the load resistor acts as a filter that removes high-frequency electronic noise.

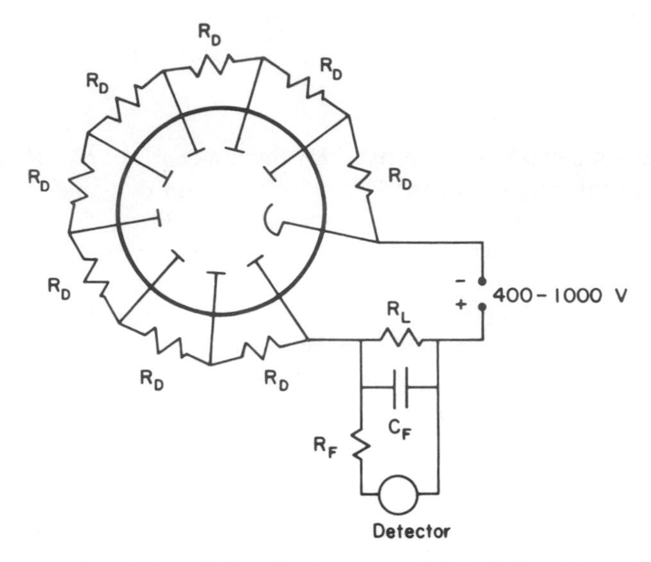

Figure 16. Wiring of a photomultiplier. Each of the photomultiplier stages, starting in the photocathode, are linked stepwise by R_D resistors (all of them with the same value). The photocurrent is read as a voltage drop across the load resistor R_L with a high-impedance detector (an oscilloscope). The intercalated $C_F R_F$ circuit is an appropriate electronic filter.

The formation and decay of a transient species in flash photolysis, as previously described in eq. 28, induce changes in the intensity of a monitoring light beam when the difference between the extinction coefficients of the transient species and the photolyte is as large or larger than required by the sensitivity of the detection system (Fig. 14). This also causes the photocurrent and the associated voltage drop across R_L to change in time, as shown in Figures 14 and 15. The voltage drop v_D registered by the oscilloscope is given by

$$v_D = R_L \theta I_\infty + \frac{\theta}{C_L} (I_0 - I_\infty) \frac{1}{(1/R_L C_L) - k_{20}} \exp(-k_{20} t) - \frac{\theta}{C_L} R_L C_L I_\infty + \frac{(I_0 - I_\infty)}{(1/R_L C_L) - k_{20}} \exp(-t/R_L C_L) \tag{29}$$

where I_0 and I_∞ are the light intensities of the monitoring beam at zero and infinite time, respectively (Fig. 14), k_{20} is the decay rate constant in eq. 28, and θ is the quantum efficiency of the photomultiplier. Hence, it is easy to see that the voltage drop v_D will have the same temporal dependence of the transient concentration (eq. 28) for $(1/R_L C_L) \geq 10k_{20}$. Under this condition, the exponential term $\exp(-t/R_L C_L)$ has a negligible value compared with the other exponential term $\exp(-k_{20}t)$, and the voltage drop is

$$v_D \simeq R_L \theta I_\infty + R_L \theta (I_0 - I_\infty) \exp(-k_{20} t) \tag{30}$$

Hence, the temporal part of the voltage drop is the same in eq. 28, and we can obtain information on the kinetics of the transient transformation by following the voltage in time.

It is possible to demonstrate the change in the optical density of the solution, caused by the generation of transient species by

$$\Delta \mathrm{OD} = \log \frac{I_{\mathrm{pt}}}{I_{\mathrm{pt}} - \Delta I} = \log \frac{v_{\mathrm{pt}}}{v_{\mathrm{pt}} - \Delta v} \tag{31}$$

For the change in optical density $\Delta \mathrm{OD}$, the parameters I_{pt} and v_{pt} are respectively the light intensity and the voltage drop before irradiation of the sample (Fig. 14). ΔI and Δv are the changes in light intensity and voltage evaluated with respect to the preirradiation values I_{pt} and v_{pt}. In this context, eq. 30 can be recast in the form

$$\Delta v = \Delta v_{\infty} + (\Delta v_0 - \Delta v_{\infty}) \exp(-k_{20} t) \tag{32}$$

where $\Delta v_{\infty} = R_{\mathrm{L}} \theta (I_{\infty} - I_{\mathrm{pt}})$ and $\Delta v_0 = R_{\mathrm{L}} \theta\ (I_0 - I_{\mathrm{pt}})$. The limitations imposed on the detection of an intermediate with a given lifetime (rate constant) by the electrical properties of the circuit are illustrated in Table 3.

The usual treatment of error propagation and eq. 31 allows us to express the error that is associated with the experimental determination of $\Delta \mathrm{OD}$ as a function of the errors $d(\Delta v)$ and $d(v_{\mathrm{pt}})$ of Δv and v_{pt}, respectively

$$d(\Delta \mathrm{OD}) = \pm 2.303 \frac{\Delta v}{(v_{\mathrm{pt}} - \Delta v)} \left[\left(\frac{dv_{\mathrm{pt}}}{v_{\mathrm{pt}}} \right)^2 + \left(\frac{d\Delta v}{\Delta v} \right)^2 \right]^{1/2} \tag{33}$$

For small values of Δv and $d(\Delta v)$ ($\Delta v < v_{\mathrm{pt}}$ and $d(\Delta v) < d(v_{\mathrm{pt}})$), the error (eq. 33) can be reduced to the form

$$d(\Delta \mathrm{OD}) \simeq \pm 2.303 \frac{\Delta v}{v_{\mathrm{pt}}} \left[\left(\frac{dv}{v_{\mathrm{pt}}} \right)^2 + \left(\frac{d\Delta v}{\Delta v} \right)^2 \right]^{1/2} \tag{34}$$

TABLE 3. Relationship between the Lifetime of an Intermediate and the Relaxation of the Detection Circuit[a]

	Delay $\delta t/\tau$ for a given deviation percent from τ[b]			
RC/τ	10%	5%	1%	0.1%
0.05	0.12	0.16	0.24	0.36
0.10	0.25	0.32	0.50	0.76
0.25	0.68	0.91	1.44	2.21
0.50	1.70	2.35	3.92	6.22
0.75	3.54	5.25	9.75	16.57
0.95	7.06	12.69	33.88	>50.00

[a]Values from J. N. Demas, *Excited State Lifetime Measurements*, Chap. 7, New York: Academic, **1983**, p. 112–126.

[b]Delays in units of τ measured from the peak of the trace.

For small signals with a large error ($\Delta v < v_{pt}$ and $d(\Delta v) \sim d(v_{pt})$) the error is

$$d(\Delta\text{OD}) \simeq \pm 2.303 \frac{d(v_{pt})}{v_{pt}} \tag{35}$$

In a typical experiment with $v_{pt} \sim 2$ V, $d(v_{pt}) = \pm 2 \times 10^{-3}$ V, $\Delta v \sim 5 \times 10^{-2}$ V, and $d(\Delta v) = \pm 5 \times 10^{-3}$ V, eq. 35 gives $d(\Delta\text{OD}) = 2.3 \times 10^{-3}$. This value compares very well with the value $d(\Delta\text{OD}) = \pm 6 \times 10^{-3}$, estimated with eq. 33. The sensitivity of the detection system can be improved by using an average of many determinations, a procedure called signal averaging that is discussed elsewhere in this chapter.

Flash fluorometers are based on the same principles described above. In some applications the input voltage applied to the phototube is pulsed to obtain large gain for a period corresponding to several lifetimes of the investigated light emission. Despite the large sensitivity of the detection system, the technique requires the use of signal averaging to increase the signal-to-noise ratio. Another technique for measuring emission lifetimes, the single-photon counting technique, is more popular than flash fluorometry.[52] In the single-photon counting technique, the input voltage to the photomultiplier is pulsed to obtain a moderate gain for less than one lifetime of the emission. The excitation source and the photomultiplier are triggered synchronously at a repetition frequency larger than a few kilohertz and the

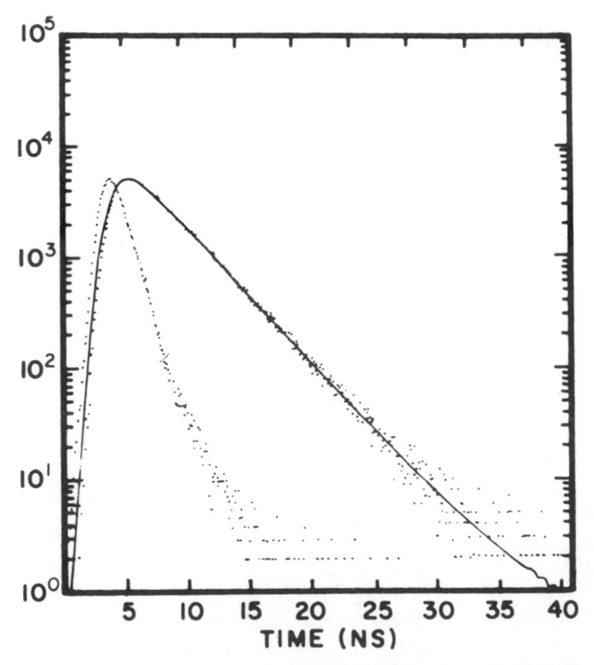

Figure 17. Typical trace recorded in single-photon counting techniques. The unfitted pulse corresponds to the lamp pulse profile.

decay curve of the emission is determined by changing the delay between the excitation and the observation. The output corresponds to the average of a large number of counts at each particular delay and these averages are plotted as a function of the delay period (Fig. 17).

2-3 THE ROTATING SECTOR TECHNIQUE

In this technique, the sample is intermittently illuminated with a train of light pulses.[53] This train is generated in some experiments through the use of a slotted wheel that is intercalated between the light source and the sample (Fig. 18). The wheel is rotated with a constant angular frequency so that the ratio of dark to light periods is a constant $r(w)$ for each angular frequency w. For long dark periods, the observed rate of product formation is the average of the rate under steady state illumination and a zero rate for the fraction of the time that the sample is in darkness. Let the rate of a product formation under steady state illumination be

$$\left(\frac{d[\mathrm{P}]}{dt}\right)_{\mathrm{s}} = f(I) \tag{36}$$

where $f(I)$ is a nonlinear function of the light intensity. Since $1/(1 + r(w))$ is

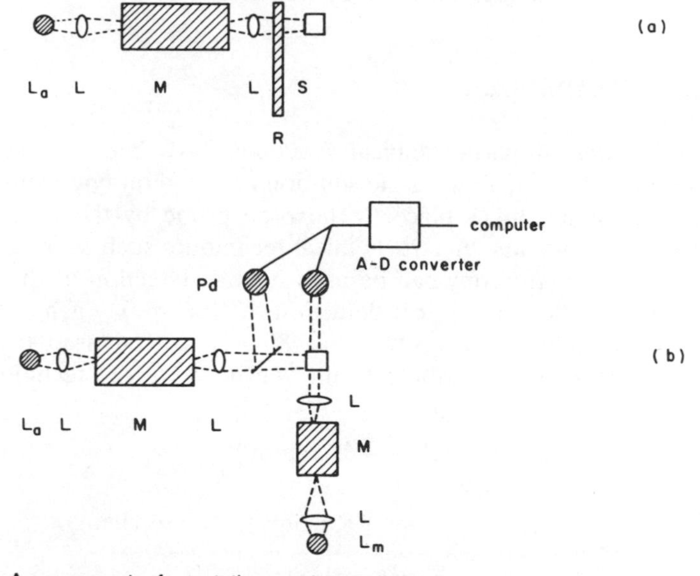

Figure 18. Arrangements for rotating sector experiments. (*a*) L_a, high intensity lamp; L, lenses; M, monochromator; R, slotted rotating wheel; and S, sample. (*b*) L_m, monitoring source; Pt, photodiode; A-D converter, analog-to-digital converter. Dashed lines represent light beams.

the fraction of time that the sample is illuminated, the rate with a slow sector is

$$\left(\frac{d[\mathrm{P}]}{dt}\right)_{\text{slow}} = \left(\frac{1}{1 + r(w)}\right) f(I) \tag{37}$$

When the dark time between the illumination periods is short, the effective light intensity is $I/(1 + r(w))$ and the rate of product formation is

$$\left(\frac{d[\mathrm{P}]}{dt}\right)_{\text{fast}} = f\left(\frac{1}{(1 - r(w))}\right) \tag{38}$$

The change from a slow to a fast regime corresponds to a period of time between light pulses that is of the same order of magnitude as the lifetime of the intermediates.

The rotating sector has been combined with various detection techniques, that is, the effect of the intermittent illumination on the reaction has been followed by means of changes in quantum yields, in the optical density of the solution and in the EPR spectrum. Furthermore, in some new instruments the rotating wheel has been replaced by a lamp that is driven by a modulated power supply (Fig. 18) to generate light pulses of various shapes and frequencies. This and the solution of the rate equations combined with the use of high-speed computers make the technique of the rotating sector a very powerful tool for the study of reaction mechanisms.

2-4 FLOW TECHNIQUES

The intermediates of photochemical reactions have been investigated in continuous photolysis by flowing the solution of the photolyte from the zone where the irradiation takes place to the zone probe by the detector (Fig. 19). In these experiments, electrochemical techniques such as polarography, ESR, or optical spectroscopy can be used for the detection of the transient species. For example, the optical detection of $\mathrm{Ru(bipy)_3^{3+}}$ generated when $\mathrm{Fe_{aq}^{3+}}$ quenches excited $\mathrm{Ru(bipy)_3^{2+}}$ (eqs. 39–42) can be used to establish some of the factor that contribute to the sensitivity of the technique

$$\mathrm{Ru(bipy)_3^{2+}} \xrightarrow{h\nu,\ \phi} {}^*\mathrm{Ru(bipy)_3^{2+}} \tag{39}$$

$${}^*\mathrm{Ru(bipy)_3^{2+}} \xrightarrow{k_{40}} \mathrm{Ru(bipy)_3^{2+}} + h\nu' \text{ (heat)} \tag{40}$$

$$\mathrm{Fe_{aq}^{3+}} + {}^*\mathrm{Ru(bipy)_3^{2+}} \xrightarrow{k_{41}} \mathrm{Fe_{aq}^{2+}} + \mathrm{Ru(bipy)_3^{3+}} \tag{41}$$

$$\mathrm{Fe_{aq}^{2+}} + \mathrm{Ru(bipy)_3^{3+}} \xrightarrow{k_{42}} \mathrm{Fe_{aq}^{3+}} + \mathrm{Ru(bipy)_3^{2+}} \tag{42}$$

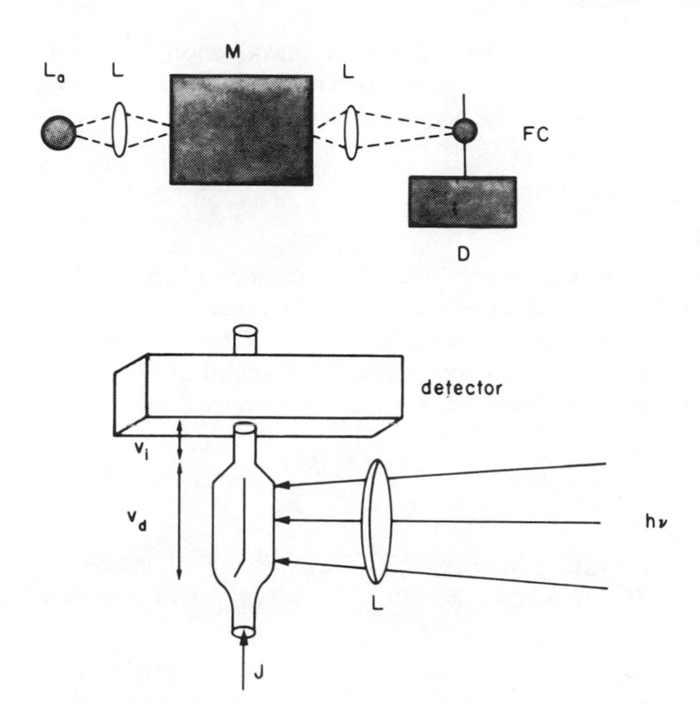

Figure 19. Flow techniques: (top) configuration used for the sample irradiation (L_a, photolyzing light source; L, lens; M, monochromator; FC, flow cell; D, detector) and (bottom) flow cell.

Since the rates of formation and decay of the excited state (eqs. 39–41) are fast in relation to the movement of the solution, it is possible to apply the steady state approximation to the concentration of the excited state. If $\mathcal{J}$ is the constant flow rate of the solution in cubic centimeters per second, $\mathcal{V}_i$ is the volume of the photolysis cell and $\mathcal{V}_d$ is the volume that the photolyzed solution must have swept to reach the detector, then an infinitesimal part of the solution moving along the photolysis cell will be exposed to the light for a time $t = \mathcal{V}_i/\mathcal{J}$. Therefore, the concentration of $Ru(bipy)_3^{3+}$ changes in time and reaches a maximum value

$$[Ru(bipy)_3^{3+}]_1 = \frac{a[1 - \exp(-2k_{42}a\mathcal{V}_i/\mathcal{J})]}{[1 + \exp(-2k_{42}a\mathcal{V}_i/\mathcal{J})]} \simeq \frac{\phi I \mathcal{V}_i}{\mathcal{J}} \tag{43}$$

where $a = [\phi I k_{41}[Fe^{3+}]/(k_{40} + k_{41}[Fe^{3+}])k_{42}]^{1/2}$. The same infinitesimal part of the photolyzed solution will be in the dark a time $t = \mathcal{V}_d/\mathcal{J}$ before reaching the detector. In this time, the back electron transfer reaction (eq. 42) reduces the concentration of $Ru(bipy)_3^{3+}$ to the final value

$$[Ru(bipy)_3^{3+}]_d = \frac{[Ru(bipy)_3^{3+}]_1}{1 + (k_{42}[Ru(bipy)_3^{3+}]_1 \mathcal{V}_d/\mathcal{J})} \tag{44}$$

that must be detected. Therefore, the introduction of the approximated form of eq. 43 in eq. 44 gives an explicit expression for $[Ru(bipy)_3^{3+}]_d$ (eq. 34) in terms of measurable experimental parameters

$$[Ru(bipy)_3^{3+}]_d = \frac{\phi I \mathcal{V}_i / \mathcal{J}}{1 + (k_{42} \phi I \mathcal{V}_i \mathcal{V}_d / \mathcal{J}^2)} \simeq \frac{\phi I \mathcal{V}_i}{\mathcal{J}} \tag{45}$$

Let δOD be the minimum value of the change in optical density that the detection system can detect $\Delta\epsilon$ the difference between the extinction coefficients of $Ru(bpy)_3^{3+}$ and $Ru(bpy)_3^{2+}$, and l the optical path of the detection cell. The experimental conditions required for the observation of the intermediate are established by the inequality

$$\frac{\delta \mathrm{OD}}{l\,\Delta\epsilon} \leq \frac{\phi I \mathcal{V}_i}{\mathcal{J}} \tag{46}$$

For example, with rather typical values for a photochemical reaction, $\delta\mathrm{OD} \sim 5 \times 10^{-3}$, $l \sim 1$ cm, $\phi \sim 10^{-1}$, $\mathcal{V}_i \sim 3\ \mathrm{cm}^3$, and $\mathcal{J} \sim 3\ \mathrm{cm}^3/\mathrm{s}$, eq. 46 leads to

$$\frac{5 \times 10^{-2}}{\Delta\epsilon} \leq I$$

and for an extinction coefficient $\Delta\epsilon \sim 300\ M^{-1}\ \mathrm{cm}^{-1}$, the required light intensity must be larger than 4.6×10^{-4} einstein/dm^3 s. The intensity of light required in these experiments can be reduced if we increase the sensitivity of the detector by using signal averaging. Indeed, the uncertainty in determining the optical density is principally caused by high-frequency noise in the detector. This noise is randomly distributed and in a large number of determinations any pair of values with errors ϵ and $-\epsilon$ are found with the same probability. The average of a large number of readings tends to cancel a large fraction of this error and the residual part decreases with the number of readings that are averaged. In a conventional spectrophotometer, the sensitivity of the detector limits the observation of changes in the optical density to values $\delta\mathrm{OD} \geq 2 \times 10^{-3}$. However, the introduction of signal averaging increases the sensitivity to such a level that changes in the optical density $\delta\mathrm{OD} \sim 10^{-4}$ can be easily detected. In the context of our example, the intensity required for the detection of $Ru(bpy)_3^{3+}$ is consequently two orders of magnitude smaller than the value determined above.

The flow technique has been used with great success in the detection of radicals by irradiation in the cavity of an EPR spectrometer. The sample can be irradiated on the same volume that is probed by the spectrometer, a condition that simplifies the equations that express the concentration of the intermediate.

2-5 PICOSECOND TRANSIENT KINETICS

The observation of intermediates with lifetimes in a nanosecond-to-second time domain can be accomplished by conventional flash photolysis where either flash lamps or lasers are used to excite the sample. However, detecting intermediates with lifetimes shorter than a nanosecond requires a different experimental approach with regard to excitation and detection. For these purposes mode-locked lasers acting as sources of intense picosecond pulses are used with appropriate optics. The mode-locking of a laser consists of overlapping longitudinal modes that are locked in phase.[54,55] Coupling a large number of modes results in narrow pulses whereby the laser can deliver a train of picosecond pulses with lengths of a few hundred nanoseconds. Since the pulses in a train vary in duration, spectral response, and intensity, it is convenient to single out one pulse from the train.[56] Initial pulses in the train can be perturbed by secondary processes that last more than the separation between pulses, that is, 5–20 ns. The separation is achieved with a Pockels cell placed between polarizers, and the pulse that is separated, one with picosecond width and high intensity, has a small photonic content and must be amplified.

The detection of intermediates is based on the constant speed of light, which allows us to establish a correlation between the space traveled by a given pulse and a corresponding time interval. A number of experimental arrangements based on this principle have been reported and some are illustrated in Figure 20. Two important features of these apparatuses are the picosecond shutter and an optical device that acts as a picosecond clock. The shutter consists of a cell containing CS_2 placed between two crossed polarizers.[57] The CS_2 behaves as a birefringent material under the action of the light's electric field, which orients the dipole of the molecules parallel to the field. The shutter therefore operates in a manner similar to the Pockels cell with an open time of a few picoseconds. The picosecond clock can be built around several optical devices. The stepped echelon, a block of quartz cut into a number of equal steps as shown in the insert to Figure 20, can be used for achieving time resolution with a single pulse.[58] Indeed, as a single pulse travels through or is reflected by the echelon, it experiences variable delays Δt that are a simple function

$$\Delta t = \frac{d}{p}\left([n^2 - \sin^2 \phi_i]^{1/2} - \cos \phi_i\right)$$

of the refraction index n of the echelon material, the thickness of the step d, and the angle of incidence ϕ_i.

The apparatus in Figure 20*a* combines the picosecond shutter with the echelon to time resolve the emission of a dye.[58] The pulse of light that constitutes the fluorescence from the dye is split by the echelon in an array

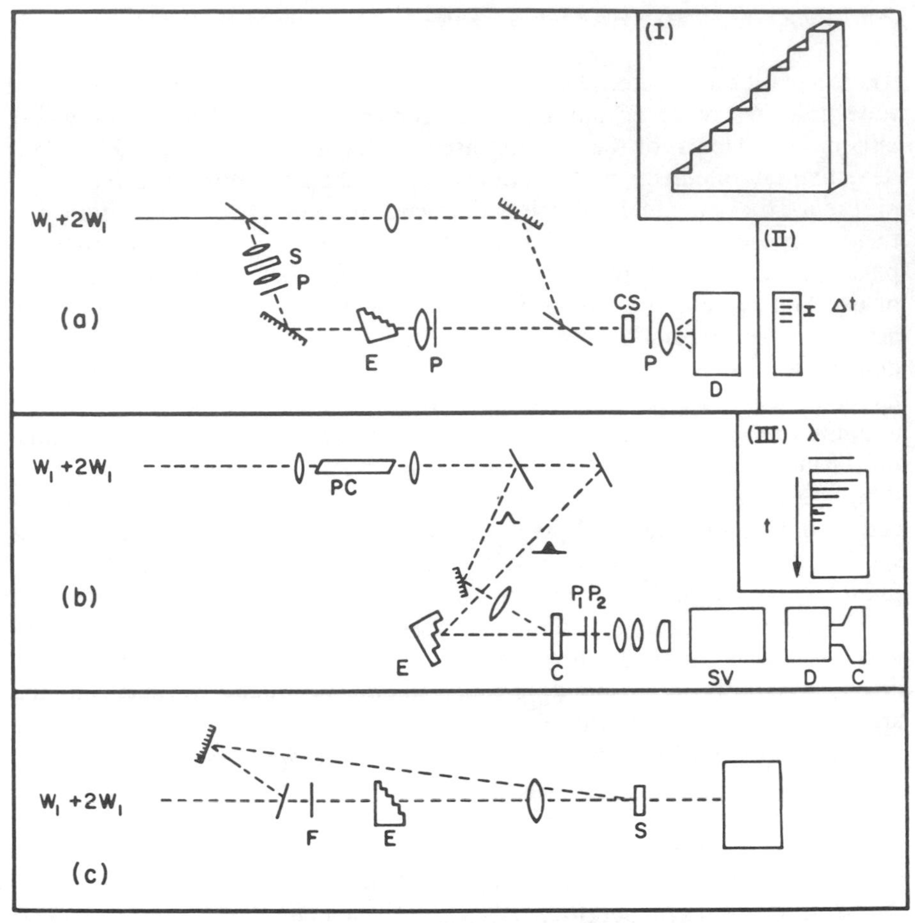

Figure 20. Optical arrangements used in picosecond flash photolysis. (*a*) Scheme for the shutter that allows the selection of one pulse from the set produced by the echelon (E) and insert I. The polarizers (P) and CS_2 cell (CS) set as a light gate as in a pockels cell. Emission of the dye in the cell S appears as an array of tracks of decreasing intensity and spaced by intervals Δt (insert II) in the spectrograph, D. (*b*) Experimental arrangement, based on the shutter described in (*a*), for the determination of transient spectra in a picosecond time domain. A cell with some liquid (PC) is used for the generation of light corresponding to a continuum of frequencies (▲). The light is sorted in a time–wavelength array for probing the photochemical event initiated by the laser pulse (Λ) in the sample C. The time-resolved spectra in the spectrograph picture is shown in insert III. (*c*) Picosecond flash photolysis apparatus used for the detection of transient absorptions in a picosecond time domain.

of identical pulses that are delayed a constant time. Part or all of the light corresponding to pulses that arrive when the picosecond shutter is closed will be cut off. Insofar as the lenses L_4 and L_5 form a telescopic system, the spatial arrangement created by the echelon is recovered in the detector. Thus, portions of light from the fluorescence arrive at the detector with a spatial organization that describes the change in light intensity with time.

For time-resolved absorption spectroscopy, it is necessary to have a probing beam of appropriate intensity and polychromatic nature. A continuum of several thousand wavenumbers and a duration of a few picoseconds can be generated by interaction of an intense laser pulse with simple liquids, for example, H_2O, D_2O, alcohols, benzene, and CCl_4. The picosecond continuum can be split and spatially organized by passing it through or reflecting it in the echelon as shown in Figure 20(*b*).[59] The arrival of the pulse used for sample excitation is adjusted to coincide with a delayed probing pulse. The use of a spectrographic system, for example, silicon-vidicom, cathode-ray tube and photographic camera, will give a time- and wavelength-resolved display.

A simpler arrangement can be used when picosecond transient kinetics is followed at a single wavelength (Figure 20*c*).[60,61] In this apparatus a high laser harmonic pulse is split in an array of pulses with various delays at the echelon. The array of pulses is used for probing the photolyzed sample.

3

ELEMENTS OF INORGANIC SPECTROSCOPY

3-1 BORN–OPPENHEIMER APPROXIMATION

A basic approximation that is regularly invoked in quantum mechanical treatments of molecules is the Born–Oppenheimer approximation, abbreviated BO, for separating various forms of motion.[62–64] It is possible to express the energy of the molecule as a sum of a number of contributions

$$E = E_{\text{translation}} + E_{\text{rotation}} + E_{\text{vibration}} + E_{\text{electronic}} + E_{\text{int}} + E_{\text{ext}} \qquad (1)$$

The term E_{int} represents those terms that account for the interaction between two different forms of motion, for example, vibration with rotation, and E_{ext} represents additional terms describing contributions from the interaction of the molecule with external fields. In the BO approximation, the contribution from E_{int} terms to the total energy E is considered to be small, and each form of motion can be treated independently of the rest. This introduces a great simplification in the mathematical treatment of the motion equations. Let us express the molecular Hamiltonian in a condensed form

$$\mathscr{H} = \hat{T}(q) + \hat{T}(Q) + P(q) + P(q, Q) \qquad (2)$$

with

$$\hat{T}(q) = -\sum_i \frac{\hbar^2}{2m} \nabla_i^2 = -\sum_i \frac{\hbar^2}{2m} \Delta_i$$

$$\hat{T}(Q) = -\sum_j \frac{\hbar^2}{2M_\mu} \nabla_\mu^2 = -\sum_j \frac{\hbar^2}{2M\mu} \Delta_\mu$$

where $P(q, Q)$ is the total coulombic potential energy, and $P(Q)$ is the potential energy of the nuclei.

It is also possible to define an electronic Hamiltonian,

$$\hat{\mathscr{H}}_{\mathrm{el}} = \hat{T}(q) + P(q, Q) + P(Q) = \hat{T}(q) + U(q, Q) \tag{3}$$

for a given nuclear configuration Q. The electronic Hamiltonian can be used to determine the electronic wave functions $\theta_n(q, Q)$, which are solutions of the equation for the electronic motion

$$(\hat{\mathscr{H}}_{\mathrm{el}} - E_n(Q))\theta_n(q, Q) = 0 \tag{4}$$

The set of eigenfunctions of eq. 4 can be used for the expansion of the complete wave function

$$\Psi(q, Q) = \sum_n X_n(Q)\theta_n(q, Q) \tag{5}$$

Thus the coefficients $X_n(Q)$ can be obtained by using the total Hamiltonian (eq. 2), the expression of the wave function (eq. 5), and the orthonormal properties of the set $\theta_n(q, Q)$. This gives an expression for the coefficients X_n that contains all the terms related to the coupling of the electronic and nuclear motions

$$[\hat{T}(Q) + E_n(Q) + \langle\theta_n|\hat{T}(Q)|\theta_n\rangle - W_n]X_n(Q) + \sum_{s\neq n}\left[\langle\theta_n|\hat{T}(Q)|\theta_s\rangle - \sum_k \frac{\hbar^2}{2M_k}\langle\theta_n|\frac{\delta}{\delta Q_k}|\theta_k\rangle\frac{\delta}{\delta Q_k}\right]X_s(Q) = 0 \tag{6}$$

The Born–Oppenheimer approximation is introduced here by ignoring all the coupling terms in eq. 6. For this purpose, let us define an operator $\hat{A}_{n,m}$

$$\hat{A}_{n,m} = \langle\theta_n|\hat{T}(Q)|\theta_m\rangle - 2\sum_k \frac{\hbar^2}{2M_k}\langle\theta_n|\frac{\delta}{\delta Q_k}|\theta_m\rangle\frac{\delta}{\delta Q_k} \tag{7}$$

and recast eq. 6 in the form

$$[\hat{T}(Q) + E_n(Q) - W_n - \hat{A}_{n,n}]X_n(Q) - \sum_{s\neq n}\hat{A}_{n,s}X_s(Q) = 0 \tag{8}$$

It can be demonstrated (see below) that the contributions from operators $\hat{A}_{n,n}$ and $\hat{A}_{n,s}$ can be neglected with the consequent simplification of the expansion coefficients

$$[\hat{T}(Q) + E_n(Q)]X_{n,v}(Q) = W_{n,v}X_{n,v}(Q) \tag{9}$$

Close inspection of this equation reveals that its solutions describe the motion of the nuclei in an effective potential field of electrons; that is $E_n(Q)$ is acting as the potential in eq. 9. Therefore we can, in practice, obtain the total wave function for the molecule by solving the equation of the electronic wave functions for all values of Q and using these functions as a basis set for the molecular wave function. The immediate result of ignoring contributions from operators $\hat{A}_{n,m}$, called adiabatic approximation,[65] is the reduction of the complete wave function (eq. 5) to a single term

$$\varphi_{n,v} = X_{n,v}(Q)\theta_n(q, Q) \tag{10}$$

For these wave functions, it is possible to obtain the normalization restrictions

$$\langle \theta_n | \theta_m \rangle = \delta_{n,m} \tag{11}$$

$$\langle X_{n,i} | X_{n,j} \rangle = \delta_{i,j} \tag{12}$$

$$\langle \varphi_{n,i} | \varphi_{m,j} \rangle = \delta_{n,m}\delta_{i,j} \tag{13}$$

Moreover, the integral of the expansion coefficients of two different electronic states is called a Franck–Condon overlap factor

$$\langle X_{n,i} | X_{m,j} \rangle = S(n, i \;\; m, j) \tag{14}$$

The normalization restrictions (eqs. 11–13) establish very stringent selection rules for the conversion of onc state into another. In fact the neglected terms of the total Hamiltonian, that is, terms corresponding to mixed motion modes, can be treated as perturbations that determine the mixing of different states.

The approximate nature of the solutions obtained above makes it necessary to establish the range of validity of the Born–Oppenheimer approximation. In this regard, we should note that the nuclear kinetic energy operator $\hat{T}(Q)$ in eq. 2 acts on the electronic and nuclear parts of the wave function obtained under the adiabatic approximation (eq. 10). Hence, deviations will be measured by the matrix elements of the nuclear kinetic energy operator and the nuclear momentum. The off-diagonal matrix elements between different vibronic states are

$$\langle \varphi_{n,i} | \hat{\mathcal{H}}_{\text{el}} | \varphi_{nj} \rangle = W_{ni}\delta_{ij}$$

$$\begin{aligned}\langle \varphi_{m,i} | \hat{\mathcal{H}}_{\text{el}} | \varphi_{n,j} \rangle &= \langle X_{m,i}(Q) | \hat{A}_{n,m} | X_{n,j}(Q) \rangle \\ &= \int X^*_{m,i}(Q) \langle \theta_m | \hat{T}(Q) | \theta_n \rangle X_{n,j}(Q)\, dQ \\ &\quad - 2\sum_{\mu} \frac{\hbar^2}{2M_\mu} \int X^*_{m,i}(Q) \left\langle \theta_m \left| \frac{\partial}{\partial Q_\mu} \right| \theta_n \right\rangle \frac{\partial}{\partial Q_\mu} X_{n,j}(Q)\, dQ\end{aligned}$$

Rejection of these elements (eqs. 8 and 9) make the BO approximation diagonal for an electronic state, while matrix elements, expected to be small, connect different electronic states. The following relationships describe the matrix elements in the off-diagonal terms

$$\langle \theta_m | \frac{\partial}{\partial Q_\mu} | \theta_n \rangle = \frac{\langle \theta_m | \partial U / \partial Q_\mu | \theta_n \rangle}{E_n(Q) - E_m(Q)}$$

$$\langle \theta_m | \frac{\partial^2}{\partial Q_\mu^2} | \theta_n \rangle = \sum_l \langle \theta_m | \frac{\partial}{\partial Q_\mu} | \theta_l \rangle \langle \theta_l | \frac{\partial}{\partial Q_\mu} | \theta_n \rangle + \frac{\partial}{\partial Q_\mu} \langle \theta_m | \frac{\partial}{\partial Q_\mu} | \theta_n \rangle$$

From these results, it is apparent that the adiabatic approximation is accurate when

$$E_n(Q) - E_m(Q) > \int X^*_{m,i}(Q) \langle \theta_m | \frac{\partial}{\partial Q_\mu} | \theta_n \rangle \frac{\partial}{\partial Q_\mu} X_{n,j}(Q)\, dQ$$

This inequality establishes that the energy difference between the zero-order electronic states θ_n and θ_m is larger than the nuclear matrix elements connecting these states. The Jahn–Teller effect, static or dynamic, is considered a manifestation of the Born–Oppenheimer approximation breakdown.

3-2 POTENTIAL SURFACES—THE NUCLEAR MOTIONS

Although the motion of a molecule with N atoms can be described by $3N$ Cartesian coordinates, $Q_1, Q_2, \ldots, Q_{3N}$, it is convenient to use coordinates that can be more easily associated with the vibrations.[31,66,67] In this regard, we can use the so-called *mass-weighted displacement coordinates* ρ_i, defined as the square root of the nucleus mass times the displacement from the position of equilibrium. The mass-weighted coordinates are

$$\tilde{\rho}_i = \sqrt{M_\mu}\,(Q_{\mu,q} - Q^0_{\mu,q}) \tag{15}$$

with $1 \le \mu \le N$, $1 \le i \le 3N$, and $q = x$, y, or z. The equilibrium position is determined by the condition

$$\left(\frac{\partial E_n(Q)}{\partial Q_\mu} \right)_{Q^0\mu} = 0 \qquad 1 \le \mu \le N \tag{16}$$

for all the nuclei. Hence, the expansion of the potential energy in a Taylor series is

$$E_n(S) = E_n^0 + \frac{1}{2} \sum_{k,l} \left(\frac{\partial^2 E_n(S)}{\partial \tilde{\rho}_k\, \partial \tilde{\rho}_l} \right) \rho_k \rho_l + \cdots \tag{17}$$

The coordinates ρ_k can be linearly combined according to

$$\Xi_n = \sum_m b_{m,n}\tilde{\rho}_m \tag{18}$$

These coordinates can be used to describe the nuclear motion and leave unchanged the center of mass of the molecules. If we impose upon this set of coordinates conditions that remove the rotation and translation of the molecule as a rigid body and eliminate cross-products $\rho_i\rho_j$ in the expression for the potential energy (eq. 17), the resulting $3N-6$ linear combinations ($3N-5$ for a linear molecule) are called *normal coordinates*. With these coordinates, the total wave function from the BO separation can be recast

$$\Psi_{n,v}(q, \Xi) = X_{n,v}(\Xi)\Psi_n^0(q) \tag{19}$$

where the electronic wave function $\Psi_n^0(q)$ was estimated at the nuclear equilibrium position.

The set of normal coordinates allows us to express the nuclear vibrations as $3N-6$ or $3N-5$ modes with their own fixed frequency. Each of the normal modes of vibration forms a basis for an irreducible representation in the point group of the molecule. Consider, for example, the *trans* isomer of ML_4X_2, a given molecule with D_{4h} symmetry. The nuclear motions are described by 21 Cartesian displacement vectors as indicated in Figure 21. The application of symmetry operations to the molecule in Figure 21 allows us to deduce the corresponding transformation matrices (Appendices II and III). For example, for the rotation around the C_4 axis we get a 21×21 matrix with a character (the sum over the diagonal elements) equal to 6. The characters of the transformation matrices in the D_{4h} point group are

	E	$2C_4$	C_2	$2C_2'$	$2C_2''$	i	$2S_4$	σ_n	$2\sigma_v$	$2\sigma_d$
Γ	21	3	-3	-3	-1	-3	-1	5	5	3

Decomposition of the reducible representation Γ in irreducible representations can be achieved by means of the relationship

$$a_i = \frac{1}{h_0}\sum_k n_k X(k)X^i(k)$$

where a_i is the number of times that the ith irreducible representation is contained in the reducible representation. The summation is carried over all the classes k (with an order n_k) of the h_0-dimensional point group. Hence, the reducible representation Γ is decomposed in the irreducible representations

$$\Gamma = 2A_{1g} + A_{2g} + B_{1g} + B_{2g} + 2E_g + 3A_{2u} + B_{2u} + 4E_u$$

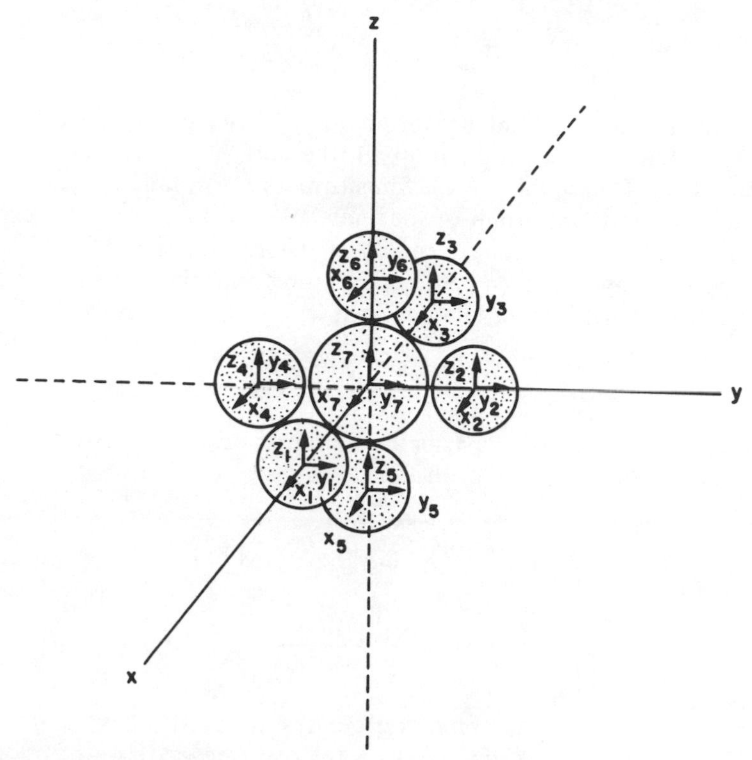

Figure 21. Cartesian coordinates used for the description of nuclear displacements in hexacoordinate complexes.

The redundant coordinates for the translation and rotation of the molecule span the irreducible representations

$$\Gamma(\text{rotation} + \text{translation}) = \mathrm{A_{2g} + E_g + A_{2u} + E_u}$$

Therefore the normal modes of vibration, $3N - 6 = 15$, are

$$\mathrm{2A_{1g} + B_{1g} + B_{2g} + E_g + 2A_{2u} + B_{2u} + 3E_u}$$

These results can be used to interpret the infrared spectrum of a molecule with D_{4h} symmetry. It is also possible to used the so-called projection operations of group theory (Appendix III) if we wish to express the normal coordinates in terms of Cartesian vectors, as defined in Figure 21.

In the introduction of the normal coordinates as a unitary transformation of the mass-weighted coordinates (eq. 18), the relationship between the normal coordinates and the electronic state has not yet been explored. For this purpose, let us expand the Hamiltonian around the equilibrium position Q_0 for the ground state

$$\hat{\mathcal{H}} = \hat{\mathcal{H}}^0 + \sum_{k=1}^{3N-6} \left(\frac{\partial U}{\partial \tilde{\rho}_k}\right)_0 \tilde{\rho}_k + \frac{1}{2} \sum_{i,j}^{3N-6} \left(\frac{\partial U}{\partial \tilde{\rho}_i \, \partial \tilde{\rho}_j}\right)_0 \tilde{\rho}_i \tilde{\rho}_j + \cdots \tag{20}$$

In eq. 20, the mass-weighted coordinates ρ_k from eq. 15 have been symmetry-adapted in the point group of the molecule and denoted by $\tilde{\rho}_k$. Insofar as the Hamiltonian must transform as the totally symmetric representation in the point group of the molecule, each term in the expansion must transform in the same form. This determines that $(\partial U/\partial \tilde{\rho}_k)_0$ transforms like ρ_k and that $(\partial^2 U/\partial \tilde{\rho}_i \, \partial \tilde{\rho}_j)_0$ transforms as $\rho_i \rho_j$. The potential energy for molecular vibrations, defined in eq. 9, is given by

$$V_n = \langle \Psi_n^0(q) | \hat{\mathcal{H}} | \Psi_n^0(q) \rangle \tag{21}$$

and the introduction of the Hamiltonian expansion (eq. 20) allows us to recast the potential energy in the form

$$V_n(\tilde{\rho}) = V_n^0 + \langle \Psi_n^0 | \sum_{k}^{3N-6} \left(\frac{\partial U}{\partial \tilde{\rho}_k}\right)_0 \tilde{\rho}_k | \Psi_n^0 \rangle + \frac{1}{2} \langle \Psi_n^0 | \sum_{i,j}^{3N-6} \left(\frac{\partial^2 U}{\partial \tilde{\rho}_i \, \partial \tilde{\rho}_j}\right)_0 \tilde{\rho}_i \tilde{\rho}_j | \Psi_n^0 \rangle + \cdots$$

Integration over the electronic coordinates demands that $(\delta U/\delta \tilde{\rho}_k)_0$ and $(\delta U/\delta \tilde{\rho}_i \, \delta \tilde{\rho}_j)_0$ must transform like the totally symmetric representation. So $\tilde{\rho}_k$ transforms like the totally symmetric coordinate $\tilde{\rho}_1$ and $\tilde{\rho}_i$ transforms like $\tilde{\rho}_j$. Two additional conditions are $V_0^0 = 0$ and $\langle \Psi_0 | (\partial U/\partial \tilde{\rho}_1)_0 | \Psi_0 \rangle = 0$ because the evaluation is made at a minimum of the electronic energy. Hence, the potential energy for the ground state takes a quadratic form known as the harmonic approximation[66,67]

$$V_0(\tilde{\rho}) = \frac{1}{2} \sum_{i,j}^{3N-6} \langle \Psi_0^0 | \left(\frac{\partial U}{\partial \tilde{\rho}_i \, \partial \tilde{\rho}_j}\right)_0 | \Psi_0^0 \rangle \, \tilde{\rho}_i \tilde{\rho}_j = \frac{1}{2} \sum_{i,j}^{3N-6} a_{ij} \tilde{\rho}_i \tilde{\rho}_j \tag{22}$$

The potential energy expression can be diagonalized by using normal coordinates

$$V_0(\epsilon) = \frac{1}{2} \sum_{i}^{3N-6} \alpha_i \Xi_i^2 \tag{23}$$

A similar expression can be applied to an excited state where $n \neq 0$ and $V_n^0 \neq 0$,

$$V_n(\epsilon) = V_n^0 + \frac{1}{2} \sum_{i-1}^{3N-6} \alpha_{i,n} \Xi_{i,n}^2 \tag{24}$$

The set of coefficients $\alpha_{i,n}$ will depend on the excited state, and the function $V_n(\epsilon)$ must respond to certain symmetry restrictions. Indeed, let $\hat{R}$ be a symmetry operator in the point group of the molecule which commutes with the Hamiltonian, that is, $[\hat{R}, \hat{\mathcal{H}}] = 0$. In this sense, the eigenvalue equation can be expressed

$$\hat{R}\hat{\mathcal{H}}\Psi_n^0 = \hat{\mathcal{H}}\hat{R}\Psi_n^0 = (\hat{R}V_n(\epsilon))(\hat{R}\Psi_n^0) \tag{25}$$

The commutation between the operator $\hat{\mathcal{H}}$ and $\hat{R}$ determines that Ψ_n^0 should also be an eigenfunction of $\hat{R}$

$$\hat{R}\Psi_n^0 = c\Psi_n^0 \tag{26}$$

Substituting eq. 26 into eq. 25 leads to the equality

$$\hat{\mathcal{H}}\Psi_n^0 = (\hat{R}V_n(\Xi))\Psi_n^0 = V_n(\Xi)\Psi_n^0 \tag{27}$$

which establishes that the potential energy remains invariant under the operations in the point group of the molecule.

The definition of potential energy (eq. 21) can be applied to any state where the Born–Oppenheimer (adiabatic) approximation is valid. For these states it is possible to define potential surfaces, also called adiabatic surfaces, that under certain conditions can intersect each other. This is the case with two states n and m, which are described by two orthonormal wave functions Ψ_n^0 and Ψ_m^0 (Fig. 22*a*). These two wave functions will give an appropriate description of the system when the corresponding potential energies $H_{nn} = \langle\Psi_n^0|\hat{\mathcal{H}}|\Psi_n^0\rangle$ and $H_{mm} = \langle\Psi_m^0|\hat{\mathcal{H}}|\Psi_m^0\rangle$ are very different. However, it is necessary to use linear combinations

$$\Psi = 2\Psi_n^0 + (1-\lambda^2)^{1/2}\Psi_m^0 \tag{28}$$

when the energies of the two states are very close (point T in Fig. 22). The energy of the molecule is given by operating with $\langle\Psi_n^0|\hat{\mathcal{H}}|$ and $\langle\Psi_m^0|\hat{\mathcal{H}}|$ on the linear combination given in eq. 28

$$\lambda(\langle\Psi_n^0|\hat{\mathcal{H}}|\Psi_n^0\rangle - E) + (1-\lambda^2)^{1/2}\langle\Psi_n^0|\hat{\mathcal{H}}|\Psi_m^0\rangle = 0 \tag{29}$$

$$\lambda(\langle\Psi_m^0|\hat{\mathcal{H}}|\Psi_n^0\rangle + (1-\lambda^2)^{1/2}(\langle\Psi_m^0|\hat{\mathcal{H}}|\Psi_m^0\rangle - E) = 0 \tag{30}$$

The solutions for the set of eqs. 29 and 30 can now be obtained by solving the determinant

$$\begin{pmatrix} H_{nn} - E & H_{nm} \\ H_{nm} & H_{mm} - E \end{pmatrix} = 0 \qquad \text{with } H_{ij} = \langle\Psi_i^0|\hat{\mathcal{H}}|\Psi_j^0\rangle \tag{31}$$

For points of the potential surfaces around T in Figure 22, the solutions in a first approximation are $E = H_{nn}$ and $E = H_{mm}$. Therefore, the solutions in a

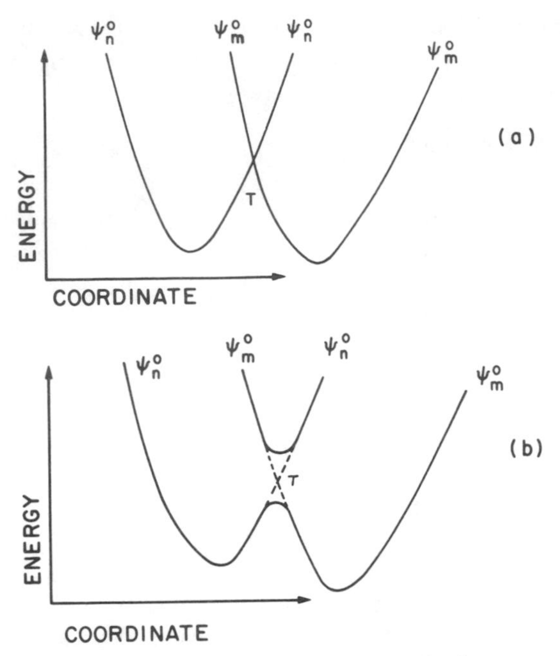

Figure 22. Potential surfaces crossing. Solutions obtained under the assumption of interaction energies (*a*) $H_{nm} = 0$ and (*b*) $H_{nm} > 0$ between the states described by zero-order eigenfunctions Ψ_n^0 and Ψ_m^0.

higher approximation are given by

$$E = H_{nn} - \frac{H_{nm}^2}{H_{mm} - H_{nn}} \tag{32}$$

$$E = H_{mm} + \frac{H_{nm}^2}{H_{mm} - H_{nn}} \tag{33}$$

These solutions show that as the curves approach the intersection, the upper and lower surfaces (Fig. 22*b*) are destabilized and stabilized, respectively, in an energy that depends on the interaction (resonance) energy H_{nm}. In this sense, the conditions $H_{nn} = H_{mm}$ and $H_{nm} = 0$ must be satisfied for the potential surfaces to cross as the internuclear distances are changed. The commutation of the Hamiltonian with other operators, for example, the square of the total angular momentum, gives place to additional conservation relationships such as the conservation of the total angular momentum $\langle \Psi_n^0 | \hat{J}^2 | \Psi_m^0 \rangle = 0$.[31]

Each of the conditions specified above, $H_{nn} = H_{mm}$ and $H_{nm} = 0$, will fix one variable, the internal coordinates, and two n-dimensional adiabatic surfaces will intersect along an $n - 2$ hyperline. Therefore, for two bidimensional surfaces (each a function of the internal coordinates Ξ_1 and Ξ_2) the line of intersection can be a parabola, an ellipse, or a circumference. Furthermore, in diatomic molecules with only one coordinate to vary, the condition $H_{nn} = H_{mm}$ can be satisfied by changing the internuclear distance and the condition $H_{nm} = 0$ can be satisfied only if the matrix element vanishes for all values of the internuclear distance. The second condition gives place, therefore, to the following symmetry rules: potential surfaces cannot cross if they correspond to states of the electronic Hamiltonian that span the same irreducible representation in the point group of the molecule. For polyatomic molecules with $3N-6$ or $3N-5$ internal coordinates, it is possible to change two different sets of internal coordinates independently of each other so as to fulfill both conditions. In other words, because of the many internal coordinates, H_{nm} does not need to vanish for all values of the internal coordinates as is the case if we apply the symmetry rule for diatomic molecules.

The definition of the potential surfaces (eq. 21) is based on (a) the adiabatic approximation that separates the electronic from the nuclear motions (these motions are not considered to interact or be mixed) and (b) the absence in the electronic Hamiltonian of terms that operate on the spin functions. This is correct in a zero-order approximation but operators for coupling the electronic movements with nuclear vibrations (vibronic coupling) and with the spin (spin–orbit coupling) must be introduced in a more refined treatment. These new operators can induce the mixing of zero-order wave functions, that is, those solutions of the electronic Hamiltonian that were obtained without the perturbing operators for spin–orbit and vibronic coupling. Thus, potential surfaces that were unable to cross (or mix) under zero-order approximation can cross under the effect of these perturbations. Furthermore, the matrix element H_{nm} is complex when an external magnetic field brings into action the spin perturbations. It is thus necessary to vanish the real and imaginary parts of H_{nm} if the condition $H_{nm} = 0$ is to be obeyed under an external magnetic field. The vanishing of the imaginary part fixes the value of one additional internal coordinate, and the intersection of two n-dimensional surfaces takes place along a line with $3n-3$ dimensions.

The potential surfaces described above are of a particular interest in the discussion of chemical systems that evolve in time such as chemical reactions. For a system in a state Ψ_n^0 and moving within the boundaries imposed by the potential surface $\langle \Psi_n^0 | \mathscr{H} | \Psi_n^0 \rangle$ (Fig. 22) the probability of changing to the state Ψ_m^0 depends on remaining on the lower surface when passing above the maximum energy, point T in Figure 22. The question of the probability x of crossing from one surface to another has been considered by Landau, Zener, and Stueckelberg who arrived at the expression[68–71]

$$x = \exp\left(- \frac{2\pi H_{nm}^2}{\hbar v |\lambda_m - \lambda_n|} \right)$$

where $|\lambda_m - \lambda_n|$ is the absolute value of the difference between the slopes of the zero-order surfaces and v is the velocity with which the system passes through the configuration corresponding to the least separation between the surfaces. This equation is valid when $H_{nm} \ll \frac{1}{2}\mu v^2$ where μ is the reduced mass of the system. In this case, the probability of remaining on the lower surface and change from the state Ψ_n^0 to the state Ψ_m^0 is given by

$$\rho = 1 - \exp\left(-\frac{4\pi H_{nm}^2}{\hbar v|\lambda_m - \lambda_n|}\right) \approx \frac{2\pi H_{nm}}{\hbar v|\lambda_m - \lambda_n|}$$

To visualize chemical transformations along potential surfaces, it is important to distinguish between nuclear motions described by *configuration coordinates* and *reaction coordinates*.[72] One configuration coordinate in octahedral complexes, $M(CO)_6$, is the symmetric stretch a_{1g} and the dissociation of the complex when it is energized along this coordinate can be described as an isotropic explosion

$$M(CO)_6 \rightarrow M + 6CO$$

The reaction mode

$$M(CO)_6 \rightarrow M(CO)_5 + CO$$

can be better described along the reaction coordinate that may involve a combination of normal modes (Chapter 6).

3-3 DISTORTION OF POTENTIAL SURFACES: THE JAHN–TELLER EFFECT

In coordination chemistry it is common to observe that certain nuclear symmetries are never reached in some molecules; for example, a perfect octahedral configuration cannot be obtained in copper(II) complexes.[73,74] For many of these arrangements the solutions of the electronic Hamiltonian are degenerate, that is, they have the same energy. Displacements of the nucleus that destroys the symmetry will also remove the degeneracy, that is, it will induce the splitting of the degenerate state. The expanded Hamiltonian (eq. 20) shows that this splitting is determined by the matrix elements

$$\sum_{i-1}^{3N-6} \langle \Psi_m^0 | \left(\frac{\partial v}{\partial \tilde{\rho}_i}\right)_0 | \Psi_n^0 \rangle \tilde{\rho}_i$$

A totally symmetric displacement cannot destroy the symmetry of the molecule. Hence, the non-totally symmetric displacement $\tilde{\rho}_i$ are the active ones for splitting the state. Ψ_n^0 and Ψ_m^0 will be coupled in consonance with the symmetry of the non-totally symmetric displacement $\tilde{\rho}_i$, and fulfill the condition

$$\Gamma_n \Gamma_i = \Gamma_m$$

where the subscripts signal the representation of the functions Ψ_n, Ψ_m and the coordinate $\tilde{\rho}_i$ in the point group of the molecule. These considerations are an expression of the so-called Jahn–Teller theorem, which establishes that the electronic degeneracy in a molecule forces departures from the symmetry on which the degeneracy was evaluated.[75–77] Therefore, for nonlinear molecules, a nuclear arrangement that will lead to electronic degeneracies will automatically have at least one non-totally symmetric vibrational coordinate. Consider an octahedral complex possessing a degenerated vibration that expands the irreducible representation E_g. Inspection of the O_h character table (Appendix II) shows that the two symmetry coordinates $\tilde{\rho}_1$ and $\tilde{\rho}_2$ must transform like the functions $2z^2 - x^2 - y^2$ and $x^2 - y^2$, respectively. Therefore, the expansion of the Hamiltonian (eq. 20) gives place to

$$\begin{aligned} \hat{\mathcal{H}} &= \hat{\mathcal{H}}^0 + \left(\frac{\partial \hat{\mathcal{H}}}{\partial \tilde{\rho}_i}\right)_0 \tilde{\rho}_1 + \left(\frac{\partial \hat{\mathcal{H}}}{\partial \tilde{\rho}_2}\right)_0 \tilde{\rho}_2 \\ &= \hat{\mathcal{H}}_0 + \frac{k_1}{\sqrt{3}} (3z^2 - r^2)\tilde{\rho}_1 + k_2(x^2 - y^2)\tilde{\rho}_2 \end{aligned} \tag{34}$$

The Hamiltonian can be more conveniently expressed by using the angular momentum operators $\hat{l}_z^2$, $\hat{l}_+ = \hat{l}_x + i\hat{l}_y$ and $\hat{l}_- = \hat{l}_x - i\hat{l}_y$

$$\hat{\mathcal{H}}^1 = \hat{\mathcal{H}} - \hat{\mathcal{H}}^0 = \frac{k_1}{\sqrt{3}} (\hat{l}_z^2 - l^2)\tilde{\rho}_1 + \frac{1}{2} k_2(\hat{l}_+^2 + \hat{l}_-^2)\tilde{\rho}_2 \tag{35}$$

Therefore, the use of the operator $\hat{\mathcal{H}}^1$ in first-order perturbation theory and the zero-order wave functions, that is d orbitals $\Psi^0_{z^2}$ and $\Psi^0_{x^2-y^2}$ corresponding to a degenerated electronic state that expands the irreducible representation E_g, gives the secular determinant

$$\begin{pmatrix} H_{11} - U(\tilde{\rho}_1 \tilde{\rho}_2) & H_{12} \\ H_{12} & H_{22} - U(\tilde{\rho}_1 \tilde{\rho}_2) \end{pmatrix} = 0 \tag{36}$$

where $H_{11} = \langle \Psi_{x^2-y^2} | \hat{\mathcal{H}}^1 | \Psi_{x^2-y^2} \rangle$, $H_{22} = \langle \Psi_{z^2} | \hat{\mathcal{H}}^1 | \Psi_{z^2} \rangle$ and $H_{12} = \langle \Psi_{z^2} | \hat{\mathcal{H}}^1 | \Psi_{x^2-y^2} \rangle$. Combining eqs. 35 and 36 gives an explicit expression for the Jahn–Teller perturbation energies

$$\Delta U = \pm \frac{c}{2} \sqrt{\tilde{\rho}_1^2 + \tilde{\rho}_2^2} \tag{37}$$

These solutions must be added to the potential defined by the zero-order electronic functions, a diagonalized harmonic expression similar to that given in eqs. 22 and 23,

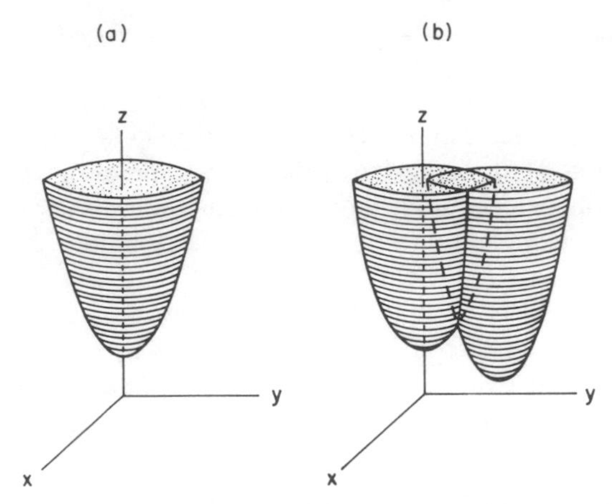

Figure 23. Jahn–Teller effect on a double degenerate state: (*a*) zero-order solution under octahedral symmetry and (*b*) splitting in two potential surfaces induced by distortion from octahedral symmetry.

$$U(\tilde{\rho}_1 \tilde{\rho}_2) = \frac{k}{2} (\tilde{\rho}_1^2 + \tilde{\rho}_2^2) \pm \frac{c}{2} \sqrt{\tilde{\rho}_1^2 + \tilde{\rho}_2^2} \tag{38}$$

It is possible to see that the zero-order potential surface is split in two (Fig. 23). If the energy separation between the first-order surfaces is large compared to the zero-point vibrational energy, the Born–Oppenheimer approximation is valid and the system remains in the lower potential surface. However, if the energy separation is close to the zero-point vibrational energy, the motion is not confined to the lower surface and a coupling between the electronic and nuclear motions arises. This corresponds to a case where more than one distortion can stabilize the system, which can then oscillate between various configurations. This effect is described as a *dynamic* Jahn–Teller and the one discussed above is known as the *static* Jahn–Teller effect.

3-4 VIBRONIC COUPLING

The electronic wave functions obtained under the Born–Oppenheimer separation respond to the assumption that the coupling of electronic and nuclear motions (vibronic coupling) is negligible. Although that is a good zero-order approximation, the manifestation of vibronic coupling is seen in the coupling of zero-order solutions.[78–80] One way of introducing vibronic coupling is to expand the Hamiltonian in terms of the vibrational coordinates (eq. 20) and use the linear terms in a first-order perturbation theory

$$\hat{\mathcal{H}}^1 = \sum_{i=1}^{3N-6} \left(\frac{\partial \hat{\mathcal{H}}}{\partial \tilde{\rho}_i} \right) \tilde{\rho}_i$$

Consider two zero-order electronic wave functions Ψ_n^0 and Ψ_m^0 with different symmetries, $\langle\Psi_n^0|\mathcal{H}^0|\Psi_m^0\rangle = 0$. First-order perturbation theory allows us to mix the two wave functions

$$\Psi_n^1 = \Psi_n^0 + \frac{H_{nm}^1}{E_n - E_m}\Psi_m^0 \tag{39}$$

with

$$H_{n,m}^1 = \sum_{i=1}^{3N-6} \langle\Psi_n^0|\left(\frac{\partial\hat{\mathcal{H}}}{\partial\tilde{\rho}_i}\right)\Psi_m^0\rangle\,\tilde{\rho}_i$$

Since the symmetries of $(\delta U/\delta\tilde{\rho}_i)$ and $\tilde{\rho}_i$ are the same, the non-totally symmetric coordinates will couple functions according to the direct product of the irreducible representations $\Gamma(\Psi_n^0)\cdot\Gamma(\Psi_m^0) = \Gamma(\tilde{\rho}_i)$. Consider a case where there is a single vibrational mode Q and the harmonic wave functions are X_p and X_q. The vibronic coupling will mix into a zero-order wave function $\Psi_{n,g}$, where g denotes a certain parity, another zero-order wave function $\Psi_{n,u}$, with another parity u. Hence, the first-order wave function can be represented by

$$\Psi_{n,p} = \Psi_n X_p = (\Psi_{n,g}^0 + C_n Q\Psi_{n,u}^0)X_p$$

for a state $\Psi_{n,p}$ and

$$\Psi_{v,q} = \Psi_v X_g = (\Psi_{v,g}^0 + C_v Q\Psi_{v,u}^0)X_q$$

for another state $\Psi_{v,q}$. Thus, for an operator $\hat{R}$ with a zero transition value when it operates on the zero-order wave functions, that is, $\langle\Psi_n^0|\hat{R}|\Psi_v^0\rangle = 0$, the introduction of the vibronic coupling leads to

$$\begin{aligned} R_{np,vq} &= \langle\Psi_{n,p}|\hat{R}|\Psi_{v,q}\rangle \\ &= (C_n\langle\Psi_{n,u}^0|\hat{R}|\Psi_{v,g}^0\rangle + C_v\langle\Psi_{n,g}^0|\hat{R}|\Psi_{v,u}^0\rangle)\langle x_p|Q|x_q\rangle \end{aligned} \tag{40}$$

Therefore, mixing functions with different symmetries by vibronic coupling removes symmetry-imposed selection rules. Consider the case of a low-spin d^6 metal ion in an octahedral environment. For such a complex the electronic ground state is $^1A_{1g}$ and one can assume that it is in the totally symmetric vibrational ground state. Two other singlet states placed at higher energies are $^1T_{1g}$ and $^1T_{2g}$. In transitions to these states ($^1T_{1g} \leftarrow {}^1A_{1g}$ and $^1T_{2g} \leftarrow {}^1A_{1g}$), an operator that forms a basis for the T_{1u} representation in O_h, for example, the electronic dipolar momentum $\mathbf{p} = \mathbf{r}\cdot e$ where e is the electronic charge and $\mathbf{r}$ the electron's position vector, will have zero transition values. It is possible to use symmetry to investigate vibronic coupling. The expansions of the direct products

$$T_{1g} \cdot T_{1u} = A_{1u} + E_u + T_{1u} + T_{2u}$$

$$T_{2g} \cdot T_{1u} = A_{2u} + E_u + T_{1u} + T_{2u}$$

establish the symmetries of the electronic states that can be mixed with $^1T_{1g}$ and $^1T_{2g}$ and determine the normal modes that can be activated. In a complex MX_6 the normal modes are A_{1g}, E_g, $2 \cdot T_{1u}$, T_{2g} and T_{2u}; with T_{1u} or T_{2u} being the modes that can be activated in the $^1T_{1g} \leftarrow {}^1A_{1g}$ or $^1T_{2g} \leftarrow {}^1A_{1g}$ transitions.

The previous treatment of vibronic interactions assumes that vibronic coupling is weak and the Born–Oppenheimer approximation is valid for the deduction of zero-order solutions. The vibronic effect has recently been reinvestigated for conditions such as strong interaction between vibrational and electronic motions, where we cannot apply the BO approximation.[81,82] The BO approximation breaks down when there is no conservation of the partial angular momenta (electronic and vibrational) and we are required to work with the resultant angular momentum, that is, the vectorial sum of partial moments.

3-5 SPIN–ORBIT COUPLING AND MAGNETIC FIELD PERTURBATIONS

The spin of the electron manifests itself as a magnetic dipole that interacts with external fields and the orbital motion. Although these interactions do not appear in the Hamiltonian (eq. 1) they can be introduced as perturbations with equations of the form[83]

$$\mathscr{H}^1 = \beta_e \mathbf{H}(k\hat{L} + 2\hat{S}) + \lambda \hat{L}\hat{S} \tag{41}$$

where $\hat{L}$ and $\hat{S}$ are the operators for the total angular momentum and total spin momentum, respectively, β_e is the Bohr magneton, and λ is the spin–orbit coupling constant. The first term of eq. 41 accounts for the effect of magnetic fields, for example, the Zeeman effect, and the second term accounts for spin–orbit coupling.

Discussion of the electronic spin requires the use of relativistic quantum mechanics. In this context, spin–orbit coupling is also a relativistic problem whose solutions are simple non relativistic equations. For a given electron interacting with its own orbit the spin–orbit coupling equation is given as

$$\mathscr{H}^1 = \sum_j \frac{1}{2m^2c^2} (\nabla U_j \times \hat{p}_j) \cdot \hat{s}_j \tag{42}$$

In eq. 42, ∇U_j is the potential gradient, $\hat{p}_j$ is the linear momentum, and $\hat{s}_j$ is the spin momentum operator for the electron j. One approximation that is usually introduced in eq. 42 consists of expanding the potential U_j in spherical coordinates (Fig. 24) and dividing the potential in a spherical component and a nonspherical contribution

$$U_j = U_j(r) + e^2 \int \frac{l(r)}{(r - r')}\, dr'$$

Therefore, the potential gradient is

$$\nabla U_j = \frac{1}{r} \frac{\partial U_j(r)}{\partial r} \mathbf{r} + T_j$$

Since $r \times p$ is the electronic angular momentum of the system, eq. 42 can be recast as

$$\hat{\mathcal{H}} = \sum_j \frac{1}{2m^2c^2} \frac{1}{r_j} \left(\frac{\partial U_j}{\partial r}\right) \hat{l}_j \cdot \hat{s}_j \cdot \hat{s}_j + \hat{P}_j$$

This expression shows that the orbital part of the spin–orbit coupling operator will transform like $\hat{l}_j$ in the point group of the molecule. For N centers in the molecule, the operator can be expressed in the form of N spherically symmetric centers

$$\hat{\mathcal{H}}^1 = \sum_{\mu=1}^{N} \sum_{j=1}^{n} \frac{1}{2m^2c^2} \left(\frac{1}{r_{\mu j}} \frac{\partial U_{\mu j}}{\partial r_{\mu j}}\right) \hat{l}_{\mu j} \cdot \hat{s}_j + \hat{P}_{\mu j}$$

The second part of the perturbation, $\hat{P}_{\mu j}$, is usually ignored and only the spherically symmetric contributions are retained

$$\begin{aligned} \hat{\mathcal{H}}^1 &= \sum_{\mu=1}^{N} \sum_{j=1}^{n} \frac{1}{2m^2c^2} \left(\frac{1}{r_{\mu j}} \frac{\partial U_{\mu j}}{\partial r_{\mu j}}\right) \hat{l}_{\mu j} \cdot \hat{s}_j \\ &= \sum_{\mu=1}^{N} \sum_{j=1}^{n} \xi(r_{\mu j}) \hat{l}_{\mu j} \hat{s}_j \end{aligned} \tag{43}$$

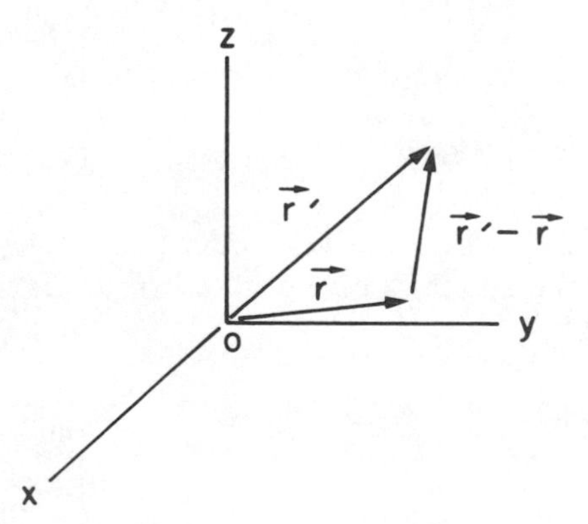

Figure 24. Spherical coordinates in the expansion of the potential. Arrows indicate vectors.

We must find here the relationship between the spin–orbit coupling constant λ in eq. 41 and the one-electron parameter $\xi_j(r_{\mu j})$. For this purpose we can use a determinantal wave function, (L, M_L, S, M_S), which is built of single orbitals

$$\Psi = \frac{1}{\sqrt{n!}} \begin{pmatrix} \Psi_1(1) & \Psi_2(1) & \cdots & \Psi_1(n) \\ & & \vdots & \\ \Psi_m(n) & \Psi_m(n) & \cdots & \Psi_m(n) \end{pmatrix}$$

So, for only one spherically symmetric contribution, $N = 1$ in eq. 43, the following equality can be built by using the diagonal element or each of the single orbitals of the determinantal wave function

$$\langle (L, M_L, S, M_S)|\mathscr{H}^1|(L, M_L, S, M_S)\rangle = \lambda M_L M_S = \sum_j \xi_j m_l^j m_S^j \quad (44)$$

This equality gives for the spin–orbit coupling constant

$$\lambda = \frac{1}{M_L M_S} \sum_j \xi_j m_l^j m_S^j \quad (45)$$

The expression of λ, however, is more complex, and for a wave function that involves several determinants with the same M_L and M_S

$$M_L M_S \sum_{\text{terms}} \lambda(L, S) = \sum_D \sum_j \xi_j m_l^j m_s^j \quad (46)$$

In eq. 46, the summation Σ_D contains contributions from all of these determinants.

The spin–orbit coupling operator is usually expressed in terms of the raising and lowering operators

$$\hat{L}_\pm = \hat{L}_x \pm i\hat{L}_y \qquad \hat{S}_\pm = \hat{S}_x \pm i\hat{S}_y \quad (47)$$

which can be combined with the operators $\hat{L}_z$ and $\hat{S}_z m$

$$\mathscr{H}^1 = \lambda(r)\hat{L} \cdot \hat{S} = \lambda(r)(\tfrac{1}{2}\hat{L}_+\hat{S}_- + \tfrac{1}{2}\hat{L}_-\hat{S}_+ + \hat{L}_z\hat{S}_z) \quad (48)$$

The use of this expression of the operator (eq. 48) and first-order perturbation theory leads to a wave function for the coupling of the state $\varphi_n\alpha$ to the state $\varphi_M\beta$

$$\Psi = \Psi_N\alpha + \frac{\lambda(r)\langle \Psi_N\alpha|\mathscr{H}^1|\Psi_M\beta\rangle}{E_N - E_M}\Psi_M\beta \quad (49)$$

Inspection of eq. 49 shows that the two states mix only under the following conditions

$$M_L^M = M_L^N, \qquad M_S^M = M_S^N$$
$$M_L^M = M_L^N \pm 1, \qquad M_S^M = M_S^N \pm 1$$

where the superscripts M and L refer to the two states coupled.

Although the spin–orbit coupling makes L and S bad quantum numbers for the characterization of the state, such a characterization can be achieved with an eigenvalue $J = L + S$, which responds to the eigenvalue equation

$$\hat{J}\Psi = (\hat{L} + \hat{S})\Psi = \hbar J\Psi = \hbar(L + S)\Psi$$

In this sense, the spin–orbit coupling perturbation is expressed as

$$\mathscr{H}^1 = \lambda(r)\hat{L} \cdot \hat{S} = \frac{\lambda(r)}{2}(\hat{J}^2 - \hat{L}^2 - \hat{S}^2) \tag{50}$$

Symmetry considerations show that the eigenvalues of the angular momentum operators $\hat{L}$ and $\hat{J}$ and the spin $\hat{S}$ can be used for the characterization of electronic states. So in a given symmetry group the rotation of an angle α gives characters for the representation spanned by the state wave functions. Such characters can be obtained with the help of the equation

$$\chi(\alpha) = \frac{\sin(M + \frac{1}{2})\alpha}{\sin \alpha/2} \qquad M = L, S, \text{ or } J \tag{51}$$

The characters $\chi(\alpha)$, obtained with integer values of $M = L$, S, or J in eq. 50, obey the equality

$$\chi(\alpha) = \chi(\alpha + 2\pi)$$

However, for the half integer values of $M = L$, S, or J

$$\chi(\alpha) = -\chi(\alpha + 2\pi)$$

So, in this case, we need to use Bethe's double groups (Appendix II). This can be seen more clearly with a couple of examples. First consider a free ion in a state with $L = 2$ and $S = \frac{1}{2}$, such as ^{2}D, which is subject to a spin–orbit coupling perturbation, and assume that this takes place under the symmetry of the point group D_4. A typical example of this class of perturbation is seen in crystal-field theory (Appendix IV) when a metal ion experiences a very large spin–orbit coupling, larger than the perturbations created by the environment. It is easy to see that $J = \frac{3}{2}$ and $\frac{5}{2}$ for such state. So the use of eq. 51, along with the symmetry operations of the double group D_4', gives the following characters

	E	R	$2C_4$	$2C_4R$	$2C_2$	$4C_2'$	$4C_2''$
$\frac{3}{2}$	4	-4	0	0	0	0	0
$\frac{5}{2}$	6	-6	$-\sqrt{2}$	$\sqrt{2}$	0	0	0

The orthonormal relationships allow us to find the number of times a_i that an irreducible representation Γ_i occurs in a reducible representation. Therefore the use of

$$a_i = \frac{1}{h_0} \sum_R \chi(R)\chi^i(R)$$

$$= \frac{1}{h_0} \sum_{\text{classes}} n_{\text{class}}\chi(\text{class})\chi^i(\text{class}) \tag{52}$$

where h_0 is the group order and the summations are over all the group elements Σ_R or classes Σ_{classes}, gives

$$\Gamma_{3/2} = \Gamma_6 + \Gamma_7 \qquad \text{and} \qquad \Gamma_{5/2} = \Gamma_6 + 2\Gamma_7$$

In this sense, the ^{2}D state is first split by the spin–orbit coupling into $^2D_{3/2}$ and $^2D_{5/2}$. The tetragonal field induces a further splitting of these states (Fig. 25). We can now consider the case where the spin–orbit coupling perturbation is sufficiently small that we can, therefore, work with symmetry-determined wave functions. It is easy to see from a character table for the D_{4h} point group that the tetragonal environment forces the splitting of this state in A_{1g}, B_{1g}, B_{2g}, and E_g. Of these state functions, those serving as bases for one-dimensional representations, for example, A_{1g}, B_{1g}, and B_{2g}, cannot suffer splitting by the spin–orbit coupling perturbation. These states correspond to the one-dimensional representations, Γ_1, Γ_3, and Γ_4, and the spin wave function spans Γ_6 in the double D_4'. Therefore, the direct products

$$\Gamma_1 \times \Gamma_6 = \Gamma_6$$

$$\Gamma_3 \times \Gamma_6 = \Gamma_7$$

$$\Gamma_4 \times \Gamma_6 = \Gamma_7$$

give the irreducible representations spanned by the perturbed wave functions. The two-dimensional representation E_g, that is, Γ_5 in D_4' will be split by a spin–orbit coupling perturbation

$$\Gamma_5 \times \Gamma_6 = \Gamma_6 + \Gamma_7$$

Figure 25 shows that the two different procedures followed for the splitting of the ^{2}D state give the same final result.

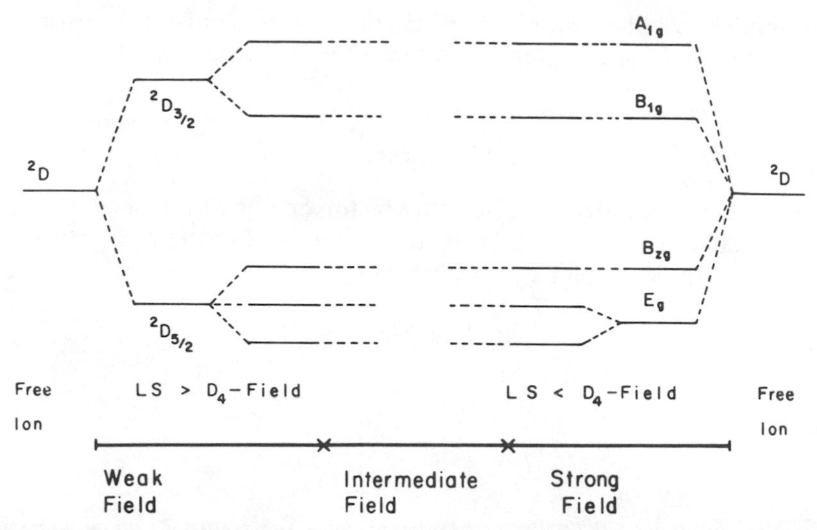

Figure 25. Splitting of a ^2D term by spin–orbit coupling and tetragonal distortions from an octahedral symmetry in the case of a weak distortion ($LS > D_4$ field) and strong distortion ($LS < D_4$ field), respectively.

The magnetic field induces perturbations (eq. 41) that are the result of a Zeeman effect, that is, the splitting of electronic levels and the mixing of electronic levels. The splitting of electronic levels can be detected, for example, in resonance experiments (e.g., ESR) or by the Faraday effect (e.g., the magnetic circular dichroism). Consider the case of Ni(II), in O_h, first using the approximation of the ligand field theory and second using the MO theory. The $(3d)^8$ ion can be treated as a $(3d)^2$ hole equivalent system, a simplification that gives for the orbital part of the ground state function

$$\Psi(A_{2g}) = (x^2 - y^2)_1(z^2)_2 - (x^2 - y^2)_2(z^2)_1$$

where the parentheses give the denomination of the d orbitals and the subscripts correspond to the two holes in the configuration. With this wave function, which spans the irreducible representation Γ_2 in the double group O', and the spin function for $S = 1$, which spans Γ_4, one can obtain

$$\Gamma_2 \times \Gamma_4 = \Gamma_5$$

The wave functions for Γ_5 are

$$\Psi_1 = \Psi(A_{2g})\alpha_1\alpha_2$$

$$\Psi_2 = \Psi(A_{2g})(\alpha_1\beta_2 - \alpha_2\beta_2)$$

$$\Psi_3 = \Psi(A_{2g})\beta_1\beta_2$$

This multiplet will be split by the magnetic field. It is possible to express the separations between components of the multiplet caused by a field H_k along the axis k

$$\hbar \nu = g_k \beta H_k \tag{53}$$

where g_k is the spectroscopic splitting factor or Landé factor along axis k, and β_e is the Bohr magneton. Using the first term of eq. 41 as a perturbation operator on the multiplet Γ_5 we obtain

$$\langle \Psi_1 | \hat{L}_z + 2\hat{S}_z | \Psi_1 \rangle = 2$$

$$\langle \Psi_2 | \hat{L}_z + 2\hat{S}_z | \Psi_2 \rangle = 0$$

$$\langle \Psi_3 | \hat{L}_z + 2\hat{S}_z | \Psi_3 \rangle = -2$$

Thus the spectroscopic splitting factor g takes the value 2. However, wave functions θ_i with $i = 1, 2, 3$ for a state ${}^3T_{2g}$ in O_h span Γ_5 in the double group O' and mix with $\Gamma_5({}^3A_{2g})$ under the spin–orbit coupling perturbation, eq. 49. The wave function for Γ_5 is

$$\Psi_i(\Gamma_5) = \Psi_i(\Gamma_5) + \frac{\langle \Gamma_5({}^3A_{2g}) | \lambda \hat{L}\hat{S} | \Gamma_5({}^3T_{2g}) \rangle}{E({}^3A_{2g}) - E({}^3T_{2g})} \theta_i(\Gamma_5) \qquad i = 1, 2, 3$$

with

$$\langle \Psi_i(\Gamma_5) | \lambda \hat{L}\hat{S} | \theta_i(\Gamma_5) \rangle = 2\sqrt{2}\lambda$$

and

$$E({}^3A_{2g}) - E({}^3T_{2g}) = 10D_q$$

The final expression for the perturbed wave function is

$$\Psi_i(\Gamma_5) = \Psi_i(\Gamma_5) - \frac{2\sqrt{2}\lambda}{10D_q} \theta_i(\Gamma_5)$$

Hence components of this multiplet give

$$\langle \Psi_1(\Gamma_5) | \hat{L}_z + 2\hat{S}_z | \Psi_1(\Gamma_5) \rangle = 2 - \frac{8\lambda}{10D_q}$$

$$\langle \Psi_2(\Gamma_5) | \hat{L}_z + 2\hat{S}_z | \Psi_2(\Gamma_5) \rangle = 0$$

$$\langle \Psi_3(\Gamma_5) | \hat{L}_z + 2\hat{S}_z | \Psi_3(\Gamma_5) \rangle = -\left(2 - \frac{8\lambda}{10D_q}\right)$$

which result in a spectroscopic splitting factor

$$g = 2 - \frac{8\lambda}{10D_q}$$

It is possible to arrive at a similar solution using the MO theory. The orbital part of the orbital functions for the $^3A_{2g}$ in the point group O_h are

$$\theta(x^2 - y^2) = \alpha(x^2 - y^2) - \frac{1}{2}\sqrt{1-\alpha^2}\sigma_1$$

$$\theta(z^2) = \alpha(z^2) - \frac{1}{2\sqrt{3}}\sqrt{1-\alpha^2}\sigma_2$$

where σ_1 and σ_2 denote symmetry-adapted linear combinations of the ligand orbitals. In a first approximation, the g factor must be obtained with the A_{2g} wave function

$$\Psi = \theta(x^2 - y^2)_1\theta_2(z^2) - \theta(x^2 - y^2)_2\theta_1(z^2)$$

where the subscripts still refer to the two holes in the electronic configuration. Hence, the g factor is

$$g = \langle\Psi| \sum_i \hat{l}_z + 2\hat{S}_z |\Psi\rangle$$

which has a value of 2 if one neglects the overlap terms between the metal atom and the ligands, for a truly ionic complex where the ligand field approximations are valid. The use of perturbation theory for mixing the ground state with the ^3T_2g excited state leads to

$$g = 2 - \frac{8\alpha^2\lambda}{10D_q}$$

which reduces to the g expression found above if the mixing coefficient α (the coefficient for the mixing of ligand and metal orbitals) is equal to one.

The spin–orbit coupling perturbation is small for the lighter main group and transition series. Despite this fact, spin–orbit coupling is an important effect in paramagnetic resonance experiments, especially in the theoretical calculation of g factors.[84–87] For the second and third transition series the perturbation is very strong and must be routinely introduced in any theoretical treatment of the electronic structure. For example, the strong field $(t_{2g})^4$ configuration of d^4 complexes gives rise to $^3T_{1g}$ (ground state) and $^1T_{2g}$, ^1E_g, $^1A_{1g}$ excited states. In Ru(IV), Os(IV), Ir(V), and Pt(VI) there is considerable spin–orbit mixing, which splits the ground state into four levels: A_{1g} (ground state), T_{1g}, T_{2g}, and E_g. The splitting manifests itself as transitions within the set of sublevels in Raman spectroscopy[88] and high-resolution emission spectroscopy.[89–93]

3-6 TIME-DEPENDENT PERTURBATIONS: ABSORPTION AND EMISSION OF RADIATION

The energy levels and wave functions used in previous sections depend only on the space coordinates. The problem of the variation of a wave function with time is directly addressed by quantum mechanics principles

$$\hat{\mathscr{H}}\Psi = i\,\frac{\partial \Psi}{\partial t} \tag{54}$$

To resolve eq. 54 for the emission and absorption of radiation, let us write the Hamiltonian in the form

$$\hat{\mathscr{H}} = \hat{\mathscr{H}}_0 + \hat{\mathscr{H}}^1$$

where $\hat{\mathscr{H}}$ (the Hamiltonian defined in eq. 1) is time independent, and $\hat{\mathscr{H}}^1$ is a time-dependent perturbation. The unperturbed wave functions Ψ^0 satisfy

$$\hat{\mathscr{H}}_0\Psi^0 = i\,\frac{\partial \Psi^0}{\partial t} \tag{55}$$

and have the general shape $\Psi_n^0(q,t) = \Psi_n^0(q)\exp(-iE_nt/\hbar)$. Therefore the solution to eq. 54 can be expanded in terms of the unperturbed wave functions[94]

$$\Psi(q,t) = \sum_n c_n(t)\Psi_n^0(q,t)$$

Substituting this wave function in eq. 54 gives

$$\sum_n c_n\hat{\mathscr{H}}_0\Psi_n^0 + \sum_n c_n\hat{\mathscr{H}}^1\Psi_n^0 = i\hbar\sum_n \frac{\partial c_n}{\partial t}\Psi_n^0 + i\hbar\sum_n c_n\frac{\partial \Psi_n^0}{\partial t}$$

Since the functions Ψ_n^0 satisfy eq. 55, this relationship reduces to

$$\sum_n c_n\,\hat{\mathscr{H}}^1\,\Psi_n^0 = i\hbar\sum_n \frac{\partial c_n}{\partial t}\Psi_n^0$$

Therefore, the projection over a component Ψ_m^0 of the set of zero-order solutions gives

$$\frac{\partial c_m}{\partial t} = \frac{1}{i\hbar}\sum_n c_n\langle\Psi_m^0|\hat{\mathscr{H}}^1|\Psi_n^0\rangle \tag{56}$$

Consider that the perturbation $\hat{\mathscr{H}}^1$ is a light wave that interacts with charged particles, that is, the electrons in the molecule. To derive the Hamiltonian for such perturbation we use the relationship between the force on a particle of mass m and charge e moving in an electromagnetic field with a velocity $\mathbf{v}$,

$$\mathbf{F} = e\left(\mathbf{E} + \frac{1}{c}\,[\mathbf{v} \wedge \mathbf{H}]\right)$$

where the square bracket is the vectorial product of the velocity with the field, c is the speed of light, and the fields are conveniently defined in terms of the electrostatic potential U and the vector potential $\mathbf{A}$

$$\mathbf{H} = \nabla \wedge \mathbf{A} \qquad \text{and} \qquad \mathbf{E} = -\left(\frac{1}{c}\,\frac{\partial \mathbf{A}}{\partial t} + \nabla \cdot U\right)$$

It is possible to express the problem in the form of a Hamiltonian

$$\hat{\mathscr{H}} = \frac{1}{2m}\left(-\hbar^2\nabla^2 + i\hbar\,\frac{e}{c}\,\nabla\mathbf{A} + 2i\hbar\,\frac{e}{c}\,\mathbf{A}\nabla + \frac{e^2}{c^2}\,|\mathbf{A}|^2\right) + eU$$

The electromagnetic field associated with a light wave obeys $\nabla \cdot \mathbf{A} = 0$ and $U = 0$, causing the second and last terms of this Hamiltonian to vanish. Moreover, the term on $|\mathbf{A}|^2$ is small and can be ignored unless the magnetic field is very strong. In this sense, we can express the perturbation Hamiltonian for a number of particles

$$\hat{\mathscr{H}}^1 = \sum_j \frac{ei\hbar}{m_j c}\,\mathbf{A}_j\nabla_j = \sum_j -\frac{e}{mjc}\,\mathbf{A}_j\mathbf{p}_j$$

This operator is usually simplified under the assumption that the vector potential is constant on a space equivalent to a molecular strength. For this approximation eq. 56 gives

$$\frac{\partial c_m}{\partial t} = -\frac{\mathbf{A}}{c\hbar^2}\,\langle \Psi_m^0 | e \sum_j \mathbf{r}_j | \Psi_0^0 \rangle (E_m - E_0) \exp\left(\frac{i(E_m - E_0)}{\hbar}\,t\right) \tag{57}$$

If the light has a frequency ν, the vector potential can be expressed as

$$\mathbf{A} = A_0 \exp(-i2\pi\nu t) + A_0^* \exp(i2\pi\nu t)$$

Therefore, replacing the expression of $\mathbf{A}$ in eq. 57 and integrating on t gives

$$c_m^* c_m = \frac{\pi^2\omega_{nm}^2}{c^2\hbar^2}\,\{|A_x^0|^2|X_{nm}|^2 + |A_y^0|^2|Y_{nm}|^2 + |A_z^0|^2|Z_{nm}|^2\}t$$

with

$$|A_x^0|^2 = |A_y^0|^2 = |A_z^0|^2 = \tfrac{1}{3}|A^0|^2$$

By following the principles of the electromagnetic theory, we can express the radiation density as

$$\rho(\nu_{nm}) = \frac{1}{4\pi} \overline{E^2(\nu_{nm})}$$

where $\overline{E^2(\nu_{nm})}$ is the average value of the square of the electric field associated with the light wave, namely

$$\mathbf{E}(\nu_{nm}) = \frac{2\pi\nu_{nm}}{c} \mathbf{A}^0(\nu_{nm}) \sin(2\pi\nu t) \tag{58}$$

Substitution of the radiation density in eq. 58 gives

$$c_m^* c_m = \frac{2\pi}{3h^2} |R_{nm}|^2 \rho(\nu_{nm}) t$$

$$|R_{nm}|^2 = |X_{nm}|^2 + |Y_{nm}|^2 + |Z_{nm}|^2 \tag{59}$$

Equation 59 states that the probability of a molecule being in state m at $t = 0$ is zero and is c_m^*, c_m at a time t. Therefore, the probability of transition from state n to state m in unit time is given as

$$B_{n \to m} \rho(\nu_{nm}) = \frac{2\pi}{3\hbar^2} |R_{nm}|^2 \rho(\nu_{nm}) \tag{60}$$

If the system is in state m at time $t = 0$, the probability of a transition to state n at time t, such as by the emission of light, is given as

$$B_{n \to m} \rho(\nu_{nm}) = B_{m \to n} \rho(\nu_{nm}) \tag{61}$$

The coefficients B_{nm} and B_{mn} are known as Einstein transition probability coefficients of induced absorption and emission. Although the Einstein coefficient for induced absorption is enough for characterizing the absorption of light, emission can take place from an excited state in a spontaneous form. The Einstein coefficient for spontaneous emission $C_{m \to n}$ can be derived by considering that the number of molecules in each state is controlled by Boltzmann's distribution law under equilibrium conditions, for example

$$\frac{Nm}{Nn} = \frac{g_m \exp(-Em/kT)}{\exp(-En/kT)} = \exp - \frac{h\nu_{nm}}{kT}$$

and for dynamical reasons by the rate equality

$$N_n B_{nm} \rho(\nu_{nm}) = N_m [B_{mn} \rho(\nu_{nm}) + C_{nm}]$$

Hence, we can obtain an expression for the coefficient

$$C_{nm} = g_m \frac{8\pi h\nu_{nm}^3}{c^3} B_{mn} = \frac{32\pi^3 \nu_{nm}^3}{3c^3\hbar} |R_{nm}|^2 g_m \tag{62}$$

which also depends on the matrix element for the transition electric dipole moment between the two states. In this sense, eqs. 60–62 show that any pair of states with $|R_{nm}| = 0$ cannot be coupled by the absorption or emission of radiation. Selection rules can be established in each point group by expanding the direct product of the representations of the state Ψ_m and the operator **R** in terms of the irreducible representations

$$\Gamma_m \times \Gamma_R = \sum_i \Gamma^i$$

Therefore the state Ψ_n must span one of the irreducible representations in the summation. The use of wave functions perturbed by vibronic coupling provides us with a mechanism for circumventing the symmetry-imposed selection rules. This point has already been discussed in connection with the vibronic mixing of two wave functions (Sections 3-2 and 3-3) as a general mechanism that gives values different from zero to the transition value of an operator $\hat{R}$ (eq. 40). Other mechanisms that help to make the transition probability different from zero result from the violation of the basic assumption that the magnetic field is constant over a molecular length, the assumption we used to derive eq. 57. Hence, the additional terms are the transition magnetic dipole moment

$$m = \langle \Psi_m | \frac{e}{2mc} \hat{R} \wedge \hat{p} | \Psi_n \rangle$$

and the electric quadripole moment

$$q = \langle \Psi_m | e\hat{R}\hat{R} | \Psi_n \rangle$$

By including all of the terms in the expression of the transition probability we obtain[95]

$$C_{nm} = \frac{32\pi^3 \nu_{nm}^3}{3C^2\hbar} g_m \left(|R_{nm}|^2 + |m|^2 + \frac{3}{10} \pi^3 \frac{\nu_{nm}^2}{C^2} |q|^2 \right)$$

where the orders of magnitude of these terms are

$$|R_{nm}|^2 \sim 6.5 \times 10^{-36} \text{cgs}$$

$$|m|^2 \sim 8.7 \times 10^{-41} \text{cgs}$$

$$\frac{3}{10} \pi^3 \frac{\nu_{nm}^2}{C^2} |q|^2 \sim 6.8 \times 10^{-43} \text{cgs}$$

Hence, the probability of an electronic transition with intensity acquired by a magnetic dipole or electric quadrupole mechanism can be negligible compared to the probability of electric dipole radiation.

The Einstein coefficient of spontaneous emission can be seen from a dynamical standpoint as the number of events per second in which an excited molecule emits light. Thus one can express the intensity of the emitted light, I_{em}, as

$$I_{em} = C_{mn} \cdot N_m$$

and the rate constant for the radiative relaxation of the excited state, that is, the specific rate constant for the elementary step

$$\text{excited state} \rightarrow \text{ground state} + \text{light}$$

is therefore given by

$$k_{rad} = \frac{1}{\tau_{rad}} = C_{mn} \tag{63}$$

The coefficients B_{nm}, B_{mn}, and C_{nm}, defined in eqs. 60, 61, and 62, depend on the transition electric dipole momentum between the two states. It is possible to expand the electric dipole moment by using the functions

$$\Psi_{m,p} = \Psi^0_m X_p(\tilde{\rho}) \qquad \text{and} \qquad \Psi_{n,q} = \Psi^0_n X_n(\tilde{\rho})$$

where Ψ^0_n and Ψ^0_m are the electronic wave functions at the nuclear equilibrium position of the ground state and $X_p(\tilde{\rho})$, $X_q(\tilde{\rho})$ are functions of $3N-6$ (N = number of atoms in the molecule) internal vibrations described by the symmetry-adapted coordinates $\tilde{\rho}$. Hence, the transition electronic dipole momentum is

$$R_{nm} = \langle \Psi^0_m | \sum_i \hat{r}_i | \Psi^0_n \rangle \langle X_p(\tilde{\rho}) | X_q(\tilde{\rho}) \rangle$$

Each of the $3N-6$ modes can be described by a given vibrational wave function, and the transition momentum can be recast as

$$R_{nm} = \langle \Psi^0_m | \sum_i \hat{r}_i | \Psi^0_n \rangle \left\langle \prod_{v=1}^{3N-6} X_p(\tilde{\rho}_v) \middle| \prod_{v=1}^{3N-6} X_q(\tilde{\rho}_v) \right\rangle \tag{64}$$

The factors in the square of the transition momentum that depend on the nuclear coordinates are known as Franck–Condon factors. Since the overlap between vibrational functions determines the intensity of the transition, the maximum intensity occurs for transitions between vibrational levels that have the same nuclear configuration. This lead to the Franck–Condon principle, which states that the time required for the absorption of a photon and the corresponding electronic change is so short ($t \sim 10^{-15}$ s) compared with the period of nuclear motions ($t \sim 10^{-13}$ s) that the nuclei cannot move. As a consequence, electronic transitions between two states can be repre-

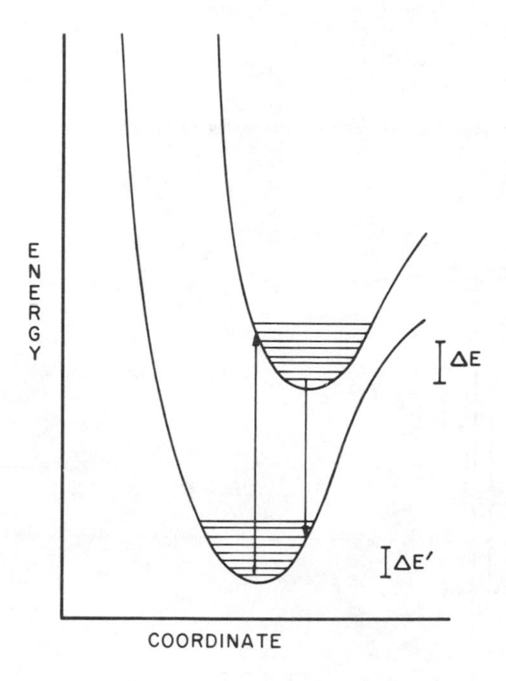

Figure 26. Vertical transitions for the absorption and emission of light. The difference in energy $\Delta E + \Delta E$ between absorption and emission is known as the Stokes shift.

sented by vertical lines in a potential diagram (Fig. 26). In this example, a transition from the vibrational level $v = 0$ of the ground state gives a maximum transition probability for $v = 7$, that is, for the excited state vibrational function that gives the best overlap with the vibrational function of the ground state.

3-7 INTENSITY OF ELECTRONIC TRANSITIONS (ABSORPTION BANDS)

It is a rather common procedure to classify electronic transitions according to the features of the orbitals engaged in the transition. In this context, the classification of molecular orbitals as bonding (σ or π), antibonding (σ^* or π^*), and nonbonding (n) leads to a description of the electronic transition, for example $\pi^* \leftarrow \pi$, $\sigma^* \leftarrow \sigma$, $\pi^* \leftarrow n$ or $\sigma^* \leftarrow n$, where the arrow indicates the direction of the transition. Many authors always write the excited state first and indicate the direction of the transition, for example, $\pi^* \leftarrow \pi$ for absorption and $\pi^* \rightarrow \pi$ for emission. Although this classification can also be applied to inorganic compounds, the assignment of the transitions as metal centered (d–d or f–f), charge transfer (CT), and ligand centered is more generally used. For example, the d–d metal centered transition $e_g \leftarrow t_{2g}$ in an ionic coordination complex with octahedral symmetry (Fig. 27) can also be described as $\sigma^* \leftarrow n$ while the ligand-to-metal charge transfer transition $e_g \leftarrow t1_u$ can be described as $\sigma^* \leftarrow \sigma$. Neither of the nomenclatures discussed

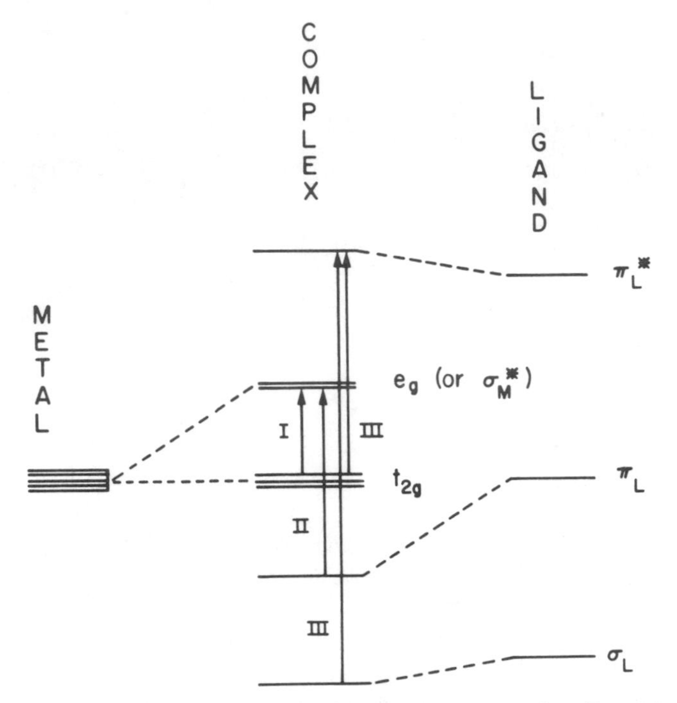

Figure 27. Electronic transitions in an octahedral coordination complex: (I) metal centered, (II) charge transfer ligand to metal, and (III) charge transfer metal to ligand.

above gives a complete description of the electronic transition. A combination of the two has been used in some instances. For example, the charge transfer transition in Figure 27 can be labeled $CT(\sigma_M^* \leftarrow \sigma_L)$, where the subscripts indicate the character of the orbitals (M, metal centered, and L, ligand centered).

It is more convenient to describe the electronic transitions in terms of the electronic states rather than the orbitals. For example, for $Cr(en)_3^{3+}$ an absorption band with $\lambda_{max} \sim 345$ nm can be assigned to an electronic transition ${}^4T_2 \leftarrow {}^4A_2$, and a second adsorption band with $\lambda_{max} \sim 345$ nm can be assigned to an electronic transition ${}^4T_1 \leftarrow {}^4A_2$. In this nomenclature, each excited state is labeled according to the corresponding irreducible representation that it spans in the point group of the molecule and the superscript gives the spin multiplicity of the state.

The mathematical framework developed in this chapter can now be applied to the interpretation of the band intensity and band shape. In this context, the Einstein transition probability coefficient for induced absorption (eq. 60) can be directly related to the extinction coefficient of Beer's law. However, the *oscillator strength* is more appropriate for the characterization of electronic transitions

$$f = 10^3 \left(\frac{mc^2}{N\pi e^2} \right) \ln 10 \int_{\nu_1}^{\nu_2} \epsilon(\nu)\, d\nu \approx 4.32 \times 10^{-9} \int_{\nu_1}^{\nu_2} \epsilon(\nu)\, d\nu \tag{65}$$

where e and m are the charge and mass of the electron, respectively, c is the speed of the light in vacuo, and N is Avogadro's number. It can also be related to the transition moment R_{nm} by using the corresponding Einstein transition probability coefficient

$$f = \left(\frac{8\pi mc}{3he^2} \right) g\nu |R_{nm}|^2 \approx 4.27 \times 10^{29} g\nu |R_{nm}|^2 \tag{66}$$

where $x = 1/g$ is the degeneracy of the states.

It is possible to see from eq. 66 that a zero oscillator strength leads to the so-called forbidden transitions. With molecular symmetries that incorporate an inversion center, the transition moment is odd with respect to the inversion and can couple only wave functions of different parities. This restriction is known as Laporte's rule for centrosymmetric complexes. One typical example are the electronic transitions involving d orbitals of an octahedral complex, for example, $Co(NH_3)_6^{3+}$, $Mn(OH_2)_6^{2+}$. Such transitions take place between orbitals that are even with respect to the inversion and, according to Laporte's rule, must have no intensity. The weak vibronic coupling, discussed in an earlier section, provides one of the mechanisms that increases the intensity of d–d (spin allowed–Laporte forbidden) transitions of centrosymmetric six-coordinated complexes. Transitions that are spin forbidden in addition to Laporte forbidden have smaller intensities than the spin allowed–Laporte forbidden transitions (Table 4). Octahedral complexes of d^5 ions, such as $Mn(OH_2)_6^{2+}$, exhibit these extremely weak transitions. In such a weak vibronic coupling the transition probability still depends on the transition electric dipole momentum. This is not the case in the strong vibronic coupling mechanism where the electronic transition, involving a charge displacement that leaves undisturbed the center of charge, is mapped into a mass displacement in an uncharged body. Such a motion is odd with regard to inversion in the symmetry center, for example, t_{1u} in *Oh*. The transition momentum operator, expanded in a multipole series, can then be reduced to the octupole term.[81,82]

For complexes exhibiting a high degree of covalency there is another mechanism, the intensity stealing mechanism, that increases the intensity of the forbidden transitions.[96] In this case, the Hamiltonian is expanded about an asymmetric vibration and the perturbation $\hat{\mathscr{H}}^1 = (\partial\hat{\mathscr{H}}/\partial Q) \cdot Q$ is used for coupling ground and excited states where we have incorporated the electronic density of the ligands. For example, for a d^1 metal ion, the ground state can be described by using the first-order perturbation theory to couple the zero-order determinant wave function

$$\Psi_0^0 = |(t_{1u})^6(t_{2g})|$$

TABLE 4 Intensities of Electronic Transitions

Electronic Transition	Extinction Coefficient (M^{-1} cm^{-1})
Spin forbidden–Laporte forbidden (tetrahedral and octahedral complexes of d^5 ions)	$10^{-3} \leq \epsilon \leq 1$
Spin forbidden–Laporte forbidden with intensity stealing Spin allowed–Laporte forbidden (ionic six-coordinate complexes)	$1 \leq \epsilon \leq 10$
Spin forbidden–Laporte forbidden (covalent tetrahedral complexes of d^5 metal ions, e.g., $FeBr_4^-$) Spin allowed–Laporte forbidden (some square complexes, e.g., $PdCl_4^{2-}$, copper complexes)	$10 \leq \epsilon \leq 10^2$
Spin allowed–Laporte forbidden (tetrahedral complexes, e.g., $NiCl_4^{2-}$. Six-coordinate complexes of low symmetry) Spin allowed–Laporte allowed (CT transitions to unsaturated ligands. Some forbidden CT transitions)	$10^2 \leq \epsilon \leq 10^3$
Spin allowed–Laporte forbidden (complexes lacking a symmetry center and covalent metal–ligand bonds)	$10^2 \leq \epsilon \leq 10^4$
Spin allowed–Laporte allowed (charge transfer transitions and electronically allowed transitions in organic molecules)	$10^3 \leq \epsilon \leq 10^6$

to determinant wave functions of excited states such as

$$\Psi_i^0 = |(t_{1u})^5(t_{2g})(e_g)|$$

In this context, the perturbed wave function for the ground state is

$$\Psi^0 = \Psi_0^0 + \sum_i \Psi_i^0(\langle \Psi_i^0|\hat{\mathscr{H}}^1|\Psi_0^0\rangle/(E_i^0 - E_0^0)) \tag{67}$$

A similar treatment for the excited state

$$\Psi_0^* = |(t_{1u})^6(e_g)|$$

leads to the perturbed excited state wave function

$$\Psi^* = \Psi_0^* + \sum_i \Psi_i^0(\langle \Psi_i^0|\hat{\mathscr{H}}^1|\Psi_0^*\rangle/(E_i^0 - E_0^*)) \tag{68}$$

Insofar as the transition moment $\langle \Psi_0^0|\hat{R}|\Psi_0^*\rangle$ is zero, the transition moment (eq. 64) expressed in terms of the perturbed functions (eqs. 67 and 68) is given by

$$\langle \Psi^0|\hat{R}|\Psi^*\rangle \approx \sum_i \langle \Psi^0|\hat{R}|\Psi_i^0\rangle \cdot \langle \Psi_i^0|\hat{\mathscr{H}}^1|\Psi_0^*\rangle/(\overset{0}{\underset{i}{E}} - E_0^*) \tag{69}$$

A further inspection of the $\Psi^* \leftarrow \Psi^0$ transition reveals that it induces a displacement of the electonic density, originally localized at the ligand, toward the metal ion. In this context, it is possible to describe the mechanism that enhances the intensity of a forbidden d–d transition as stealing intensity from an allowed charge transfer transition (Chapters 5 and 6).

One further point that should be discussed in relation to the absorption spectrum of inorganic compounds is the shape of absorption bands. We must now analyze how various factors contribute to the dependence of the transition probability with excitation energy. Insofar as the transition probability can be factored in the square of the transition moment and the square of the vibrational overlap integrals (Franck–Condon factors), in an electronically allowed transition the Franck–Condon factors determine the shape of the band. For a maximum overlap between vibrational functions of the ground and excited states one must consider the differences between the equilibrium nuclear configurations. Therefore, for an electronically allowed transition where the two states have the same equilibrium configurations, case (a) in Figure 28, the maximum overlap takes place between vibrational functions of the same parity, for example, $0^* \leftarrow 0$, $1^* \leftarrow 1, \ldots$. Insofar as the equilibrium configurations of the ground and excited states are different, case (b) in Figure 28, the transition from the vibrational level $v = 0$ must reach an excited vibronic state of the excited state. The vibrational overlap integral of the $0^* \leftarrow 0$ transition can be so small that the corresponding absorption does not appear in the spectrum. In a semiclassical model, the energies for the threshold and maximum absorptions can still be associated with the relative displacements of one surface with respect to the other. Therefore, an analysis of the vibrational progressions can give information

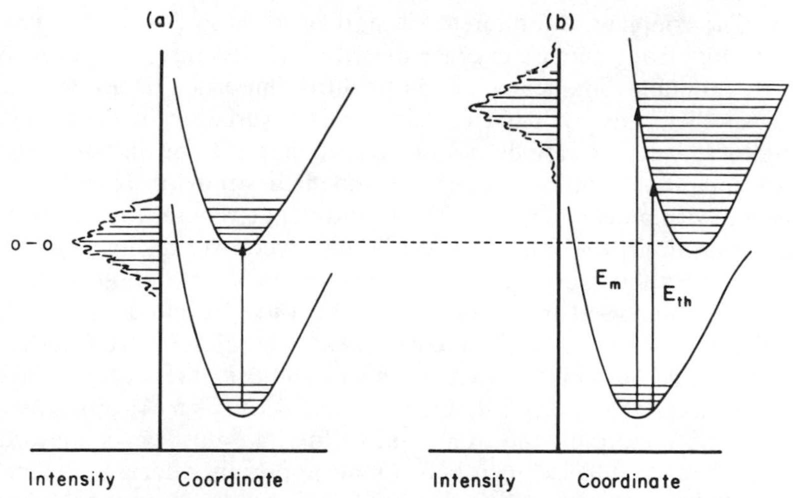

Figure 28. Relationship between the absorption spectrum (left side of (*a*) and (*b*)) and the relative position of the ground and excited state potential energy surfaces.

about the relative nuclear displacements. For such a correlation, however, we must consider that the vibrational frequencies in the ground and excited states can be different. Some examples of the vibronic structure in the absorption spectrum of coordination complexes are found in the low-temperature spectra of Mn(II) complexes. The spectrum of $((CH_3)_4N)_2MnBr_4$ shows that there is a significant spin–orbit coupling, and the vibrational structure is associated to the totally symmetric stretching M–Br mode (Fig. 29).[97] Vibrational structure is also considerable in the spectra of Pd(II) complexes, such as $PdCl_4^{2-}$. For such a compound, it has been possible to detect three different vibrational progressions: two associated with skeletal e_u vibrations and one with lattice e_u vibrations (Fig. 29).[98]

One important type of transition in coordination complexes is electronically forbidden but vibronically allowed. In such transitions the excited state can be coupled to the ground state only through a vibrational mode with appropriate symmetry. For example, for centrosymmetric complexes the coupling vibration must be odd in order to destroy the center of symmetry. Therefore the energy required for the observation of this type of transitions will be the Franck–Condon or vertical energy, plus the energy of the odd vibration. This threshold energy marks the beginning of vibronic transitions that contribute to the intensity distribution or, in other words to the band shape. Whenever more than one vibrational mode participates in the coupling of states, an equivalent number of vibrational progressions will contribute to the band shape. Furthermore, progressions for totally symmetric vibrations can also contribute to the band shape.

There are a number of mechanisms that cause broadening, and that in a limit may erase the vibrational structure in the absorption spectrum.[99] For example, every state is embedded in the vibrational continua of lower lying states. The coupling of a discrete vibrational to these continua is weak but the effect is strong enough to erase the vibronic structure. This broadening, called continuum broadening, is particularly important at room temperature. Another cause of band broadening is a variation in the ligand field strength. Indeed, vibrations will induce changes of the ligand–metal distances and in the intensity of the ligand field strength. Therefore, for a transition with a dependence on the ligand field strength, for example, d–d or CT transitions, the transition will be observed over a range of energies. In many cases this mechanism makes the largest contribution to the bandwidth of spin-allowed transitions. Very short-lived excited states will also broaden the sharp lines as a consequence of Heisenberg's uncertainty principle, which establishes that the uncertainty in the energy is inversely proportional to the uncertainty in time, that is, $\Delta E \geq \hbar / \Delta t$. For this mechanism to be important, the excited state lifetime must be of the order of magnitude of the Planck's constant. Other important effects in determining the band shape are the splittings of electronic states caused by spin–orbit coupling, the Jahn–Teller effect and contributions to the ligand field from low-symmetry components.

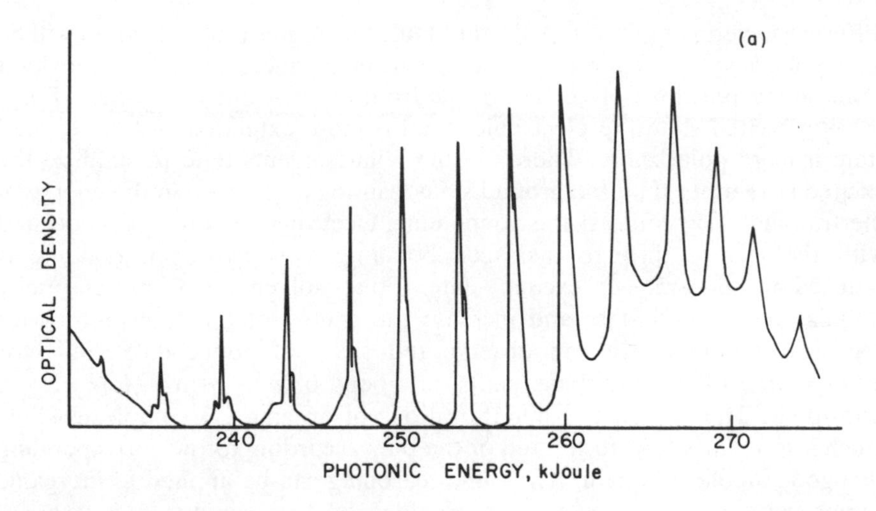

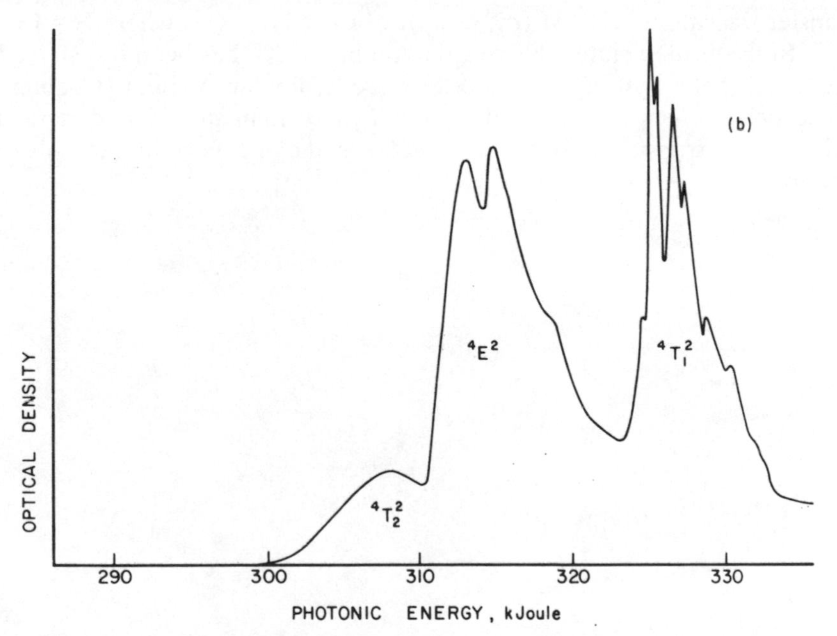

Figure 29. Fine structure in the absorption spectrum of coordination complexes. (*a*) Vibronic structure of the $^1A_{2g} \leftarrow {}^1A_{1g}$ transition in the single-crystal spectrum of K_2PdCl_4. (*b*) Single-crystal spectrum of $[(CH_3)_4N]_2MnBr_4$ at ~ 6 K.

For molecules in solution, the solvent plays a significant role in determining the position and the shape of the absorption bands. Indeed, the potential surfaces of molecules in condensed media, solid or liquid matrixes, do not represent pure intramolecular energies; they correspond instead to the energy of the molecule and its interaction with the solvent. The solvation energies of the ground and excited states are not necessarily the same in different solvents; the more polar the state, the larger its stabilization will be in a polar solvent. Therefore, the spectrum of a compound may exhibit shifts in the positions of certain bands from one solvent to another. For a $\pi^* \leftarrow \pi$ excited state the electronic cloud is more extended and the excited state is more polarizable. Therefore the polar solvents tend to stabilize the excited state more than the ground state, causing a decrease in the energy of the transition. By contrast the nonbonding electrons are strongly associated with the solvent through hydrogen bonding. This hydrogen bonding is reduced in the $\pi^* \leftarrow n$ excited state. Polar solvents will preferentially stabilize the ground state and increase the energy of the transition. With organic compounds we can observe red shifts of about 0.05 μm^{-1} for $\pi^* \leftarrow \pi$ transitions and blue shifts of about 0.08 is μm^{-1} for $\pi^* \leftarrow n$ transitions. The charge transfer transitions of organic CT-complexes show much larger shifts toward the red or the blue according to the corresponding change in dipolar moment. The same reasoning can be applied to inorganic compounds. A typical example is provided by the ligand-to-metal charge transfer transitions, CTTM ($\sigma_M^* \leftarrow \pi_L$), of $Co(NH_3)_5X^{2+}$, where $X = Cl^-$ or Br^-. In the excited state, charge from the ligand X^- has been transferred to the metal center with a significant decrease in dipolar moment (Chapter 5). Since polar solvents stabilize the ground more than the excited state, the shift of the transition to the blue follows the increase in the solvation energy.

4

KINETICS OF PHOTOLUMINESCENCE

4-1 THERMAL EFFECTS ON PHOTOLUMINESCENCE

Figure 30 shows a simplified scheme of two manifolds of states, Q_i and D_i, whose quantum spin numbers differ; for example Q_i can represent quartet states and D_i doublet states as in the case of Cr(III) complexes. The difference in spins is appropriate for a weak mixing of D_i and Q_i states through spin–orbit coupling, a mechanism that allows crossing from the D_i manifold into the Q_i manifold and vice versa. This radiationless crossing, called *intersystem crossing*, is indicated in Figure 30 with a specific rate constant k_{isc}. For a small energy gap between D_1 and Q_1 some of the molecules in D_1 can acquire enough thermal energy to return to the point of origin. This process is known as *back intersystem crossing* and has assigned a specific rate constant $k_{\overline{isc}}$ in Figure 30.

The radiative relaxation of the excited state $Q_1 \rightarrow Q_0$ takes place between states of the same spin multiplicity, and the associated emission of light is known as fluorescence. Phosphorescence is the emission of light associated with the radiative relaxation $D_1 \rightarrow Q_0$ between states with different spin multiplicities. The specific rate constants are respectively k_f and k_p.

Internal conversion is the radiationless process converting an excited state of a given spin multiplicity into a lower lying state of the same multiplicity, for example, $Q_2 \rightarrow Q_1$ in Figure 30. The internal conversion is generally very fast, which leads to rapid conversion of the upper excited states into the lowest excited state. This behavior is described by Kasha's rule: *Fluorescence always occurs from near the lowest vibrational level of the first excited state*. However, several exceptions to this rule are known and are discussed elsewhere in this chapter.

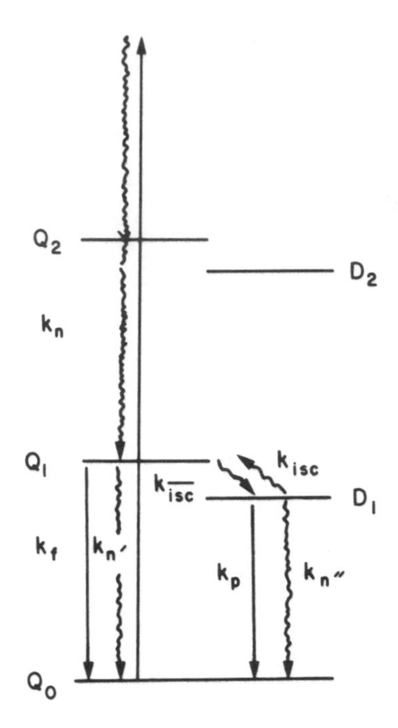

Figure 30. Radiative and nonradiative processes in manifolds of states (Q_i and D_i) whose quantum spin numbers differ by one.

The processes described above achieve a final constant speed when the sample is illuminated with a beam of exciting light of constant intensity. Under these conditions we can find the concentrations of molecules in each state by a steady state treatment

$$\frac{\partial[\mathrm{Q}_i]}{\partial t} = 0 \qquad \text{and} \qquad \frac{\partial[\mathrm{D}_1]}{\partial t} = 0$$

where the concentration of molecules in each particular state is considered constant. These conditions lead to the set of equations

$$0 = I_{\mathrm{a}} - k_{\mathrm{n}}[\mathrm{Q}_2] \tag{1}$$

$$0 = k_{\mathrm{n}}[\mathrm{Q}_2] - (k_{\mathrm{f}} + k_{\mathrm{n}'} + k_{\mathrm{isc}})[\mathrm{Q}_1] - k_{\overline{\mathrm{isc}}}[\mathrm{D}_1] \tag{2}$$

$$0 = k_{\mathrm{isc}}[\mathrm{Q}_1] - (k_{\mathrm{p}} + k_{\mathrm{n}''} + k_{\overline{\mathrm{isc}}})[\mathrm{D}_1] \tag{3}$$

The algebraic solution of eqs. 1–3 gives explicit expressions for the steady state concentrations of D_1 and Q_1

$$[\mathrm{Q}_1] = \frac{(k_{\mathrm{p}} + k_{\mathrm{n}'} + k_{\mathrm{isc}})}{(k_{\mathrm{f}} + k_{\mathrm{n}'})(k_{\mathrm{p}} + k_{\mathrm{n}''}) + (k_{\mathrm{f}} + k_{\mathrm{n}'})k_{\overline{\mathrm{isc}}} + (k_{\mathrm{p}} + k_{\mathrm{n}''})k_{\mathrm{isc}}}\, I_{\mathrm{a}} \tag{4}$$

$$[\mathrm{D_1}] = \frac{k_{\mathrm{isc}}}{(k_\mathrm{f} + k_{\mathrm{n}'})(k_\mathrm{p} + k_{\mathrm{n}''}) + (k_\mathrm{f} + k_{\mathrm{n}'})k_{\overline{\mathrm{isc}}} + (k_\mathrm{p} + k_{\mathrm{n}''})k_{\mathrm{isc}}} I_\mathrm{a} \tag{5}$$

The rate of decay of D_1 and Q_1 through the corresponding radiative paths can be associated with the intensities of fluorescence and phosphorescence

$$I_\mathrm{f} = k_\mathrm{f}[\mathrm{Q_1}] \tag{6}$$

$$I_\mathrm{p} = k_\mathrm{p}[\mathrm{D_1}] \tag{7}$$

It is now possible to define the fluorescence and phosphorescence quantum yields as a ratio of the emitted intensity, I_f or I_p, to the absorbed light intensity

$$\phi_\mathrm{f} = \frac{I_\mathrm{f}}{I_\mathrm{a}} = \frac{k_\mathrm{f}(k_\mathrm{p} + k_{\mathrm{n}''} + k_{\overline{\mathrm{isc}}})}{(k_\mathrm{f} + k_{\mathrm{n}'})(k_\mathrm{p} + k_{\mathrm{n}''}) + (k_\mathrm{f} + k_{\mathrm{n}'})k_{\overline{\mathrm{isc}}} + (k_\mathrm{p} + k_{\mathrm{n}''})k_{\mathrm{isc}}} \tag{8}$$

$$\phi_\mathrm{p} = \frac{I_\mathrm{p}}{I_\mathrm{a}} = \frac{k_\mathrm{p} k_{\mathrm{isc}}}{(k_\mathrm{f} + k_{\mathrm{n}'})(k_\mathrm{p} + k_{\mathrm{n}''}) + (k_\mathrm{f} + k_{\mathrm{n}'})k_{\overline{\mathrm{isc}}} + (k_\mathrm{p} + k_{\mathrm{n}''})k_{\mathrm{isc}}} \tag{9}$$

Back intersystem crossing has been proposed to explain certain features of the photoluminescence of a number of compounds, for example, dyes such as fluorescin or eosin and coordination complexes of Cr(III) and Ge(IV). In this context, it is interesting to compare eqs. 8 and 9 to those corresponding to the limiting case of a negligible back intersystem crossing. This can be achieved by passing eqs. 8 and 9 to the limit $k_{\overline{\mathrm{isc}}} \rightarrow 0$, procedure that leads to

$$\phi_\mathrm{f} = \frac{k_\mathrm{f}(k_\mathrm{p} + k_{\mathrm{n}''})}{(k_\mathrm{f} + k_{\mathrm{n}'})(k_\mathrm{p} + k_{\mathrm{n}''}) + (k_\mathrm{p} + k_{\mathrm{n}''})k_{\mathrm{isc}}} \tag{10}$$

$$\phi_\mathrm{p} = \frac{k_\mathrm{p} k_{\mathrm{isc}}}{(k_\mathrm{f} + k_{\mathrm{n}'})(k_\mathrm{p} + k_{\mathrm{n}''}) + (k_\mathrm{p} + k_{\mathrm{n}''})k_{\mathrm{isc}}} \tag{11}$$

From a comparison of eqs. 8–10 it is clear that the effect of the back intersystem crossing is to increase the yield of fluorescence and decrease the yield of phosphorescence. More important, we can explain the effect of the temperature on the yields of light emission in terms of an activation energy that determines the value of the rate constant for back intersystem crossing. This is illustrated more clearly in Figure 31, where the crossing of the Q_1 and D_1 potential surfaces arbitrarily corresponds to a case with zero activation energy for the forward intersystem crossing and an activation energy $E_{\overline{\mathrm{isc}}}$ for the back intersystem crossing. Therefore, the introduction of a rate constant $k_{\overline{\mathrm{isc}}} = k^0_{\overline{\mathrm{isc}}} \exp(-E_{\mathrm{isc}}/RT)$ in eqs. 8 and 9 leads to temperature-dependent quantum yields. This mechanism does not impose constraints on the rates of relaxation of the two excited states; that is, the populations of Q_1 and D_1 are not necessarily in a real equilibrium. For the

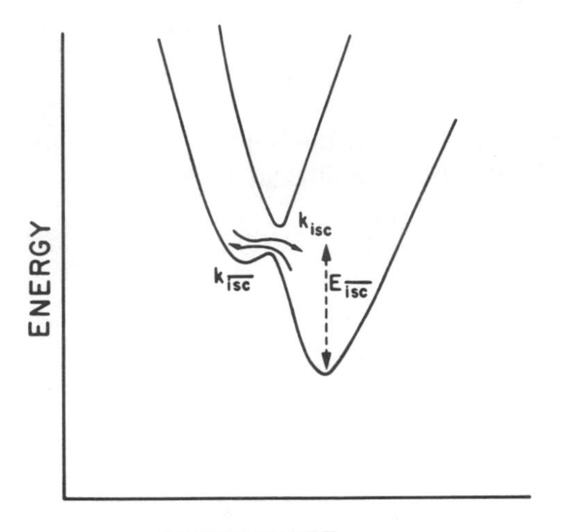

Figure 31. Potential surfaces for the crossing of two states, such as Q_1 and D_1 in Figure 30.

populations of Q_1 and D_1 to be in equilibrium, they must fulfill the condition

$$k_{isc}[Q_1] = k_{\overline{isc}}[D_1] \tag{12}$$

For eq. 12 to be valid, the rates of forward and back intersystem crossing must overwhelm the rates of relaxation of the excited states

$$(k_p + k_{n''}) \ll k_{\overline{isc}} \qquad \text{and} \qquad (k_f + k_{n'}) \ll k_{isc}$$

Therefore eqs. 8 and 9 can be reduced to

$$\phi_f = \frac{k_f K}{(k_f + k_{n'})K + (k_p + k_{n''})} \tag{13}$$

$$\phi_p = \frac{k_p}{(k_f + k_{n'})K + (k_p + k_{n''})} \tag{14}$$

where $K = k_{\overline{isc}}/k_{isc}$. A very useful relationship is obtained by taking the ratio of ϕ_f and ϕ_p

$$\frac{\phi_f}{\phi_p} = \frac{k_f}{k_p} K \tag{15}$$

From statistical thermodynamics the equilibrium constant can be expressed in terms of the degeneracies of Q_1 and D_1, g_Q and g_D, and the energy gap ΔE between these states

$$K = \frac{g_Q}{g_D} \exp\left(-\frac{\Delta E}{RT}\right) \tag{16}$$

A combination of the last two equations can be used for the experimental determination of ΔE and the ratio k_f/k_p

$$\ln \frac{\phi_f}{\phi_p} = \ln\left(\frac{k_f g_Q}{k_p g_D}\right) - \frac{\Delta E}{RT} \tag{17}$$

Equation 17 shows that increasing values of ΔE diminish the fluorescence yield as a consequence of decreasing contributions from back intersystem crossing. In $Cr(en)_3^{3+}$ and $Cr(NH_3)_6^{3+}$, for example, the energy gaps between the lowest quartet and doublet states are $\Delta E \sim 50$ kJ/mol and $\Delta E \sim 53$ kJ/mol, respectively. Therefore, back intersystem crossing can make significant contributions only at extremely high temperatures,[102–107] and thermal effects at low temperatures must be caused by variations of the rate constants for radiationless transformations.[108] The complexes $Cr(CH_3SO_2)_3$, $\Delta E \sim 14$ kJ/mol,[109–111] and $Cr(urea)_3^{3+}$, $\Delta E \sim 6$ kJ/mol,[112] are better candidates for the observation of back intersystem crossing. An interesting example of back intersystem crossing is provided by $Cs_2[MnF_6]$ (Fig. 32).[113–117] The emission spectrum of low temperatures consists of sharp lines assigned to phosphorescence, namely to the $^2E \rightarrow {}^4A_2$ electronic transition. The rise in temperature has several effects on the spectrum: the lines are broadened and new lines corresponding to higher vibrational levels are seen in the spectrum at moderate temperatures ($T < 400$ K), and a band at about 560 nm corresponding to delayed fluorescence appears at high temperatures ($T \geq 400 \cdot C$). The use of eq. 17 with the yields of fluorescence and phosphorescence of MnF_6^{2-} (Fig. 32) gives $\Delta E \simeq 47$ kJ/mol. This value compares poorly with the value $\Delta\epsilon \sim 55$ kJ/mol obtained from the analysis of the absorption and emission spectra.

Equally interesting is the relationship between the ^2E_g and $^2T_{1g}$ states in d^3 metal complexes with O_h or pseudo-O_h symmetry. In many cases these states are close enough (see Table 5 in Chap. 5) for the T_{1g} state to be in thermal equilibrium with the E_g states.

Although the examples represented above show that we can obtain some information on the back intersystem crossing from the dependence of the quantum yields on temperature, the interpretation of these results is open to controversy and is subject to the conclusions of further experimental tests.

Usually the emission spectrum is the mirror image of the first absorption band because the distribution of vibrational levels in the excited state, which determines the shape of the absorption band, is often similar to the distribution of vibrational levels in the ground state, which determines the shape of the emission spectrum. Therefore, the emission obeys the Stokes law, which states that the wavelength of the fluorescence is always longer than the wavelength of the excitation. When spectra exhibit a vibrational structure under conditions where thermal activation is significant, few molecules populate the low levels of the ground state and for excitations $0 \leftarrow n$ ($n > 0$) we can observe emissions corresponding to $(n-1) \rightarrow 0$ transitions (Fig. 33) that have more energy than the excitation.[118] Emission at

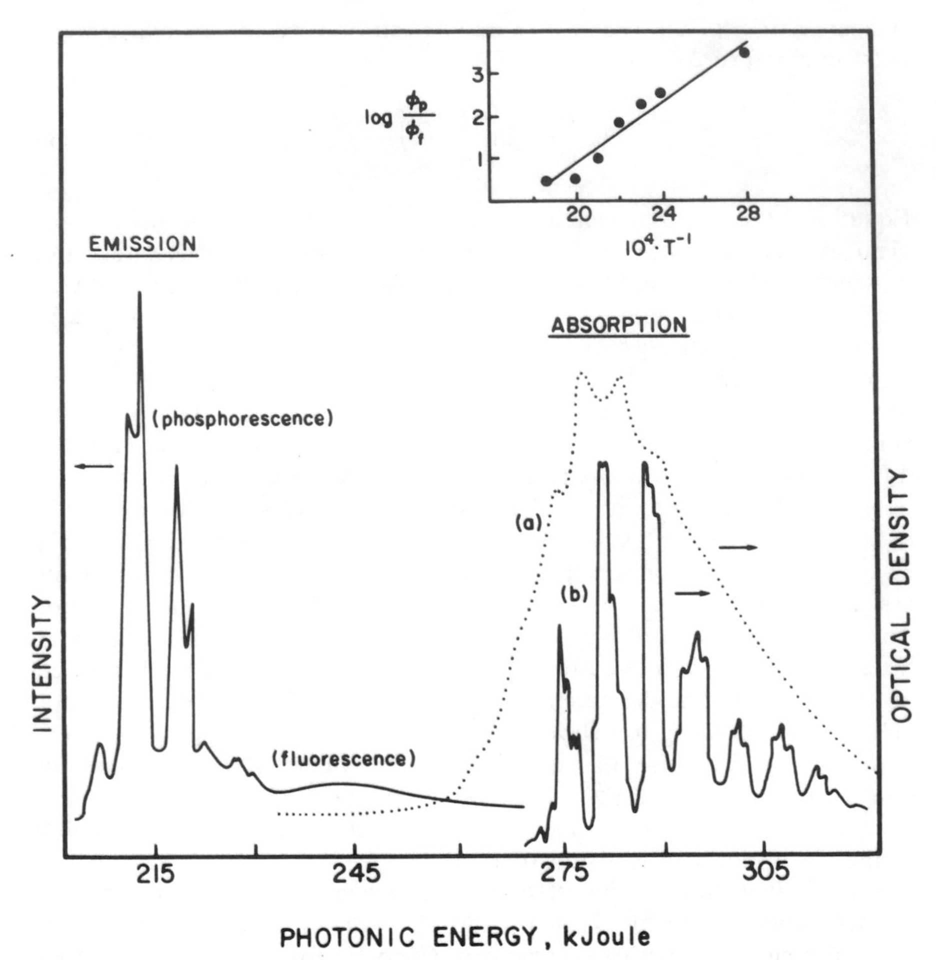

Figure 32. Optical properties of $Cs_2[MnF_6]$. Emission at 590 K showing phosphorescence and fluorescence. Absorption spectra at (*a*) 300 K and (*b*) 80 K for the transition $^4T_2 \leftarrow {}^4A_2$. The insert shows the effect of temperature on the yields of fluorescence ϕ_f and phosphorescence ϕ_p.

shorter wavelengths than the excitation is known as anti-Stokes fluorescence. The experimental observation of the abnormal emission requires monochromatic excitation at the lowest $0 \leftarrow n$ transition available. For such experiments, it is possible to take advantage of the monochromaticity and intensity of certain resonance lines from discharge lamps, such as low-pressure mercury lamps or lasers.

The effect of the temperature on ϕ_f and ϕ_p can also reflect changes of the excited state lifetime. One interesting example is provided by $Cr(CN)_6^{3-}$ (doublet–quartet energy gap of ~140 kJ/mol), which exhibits temperature- and solvent-dependent luminescence.[119–121] In a host crystal of

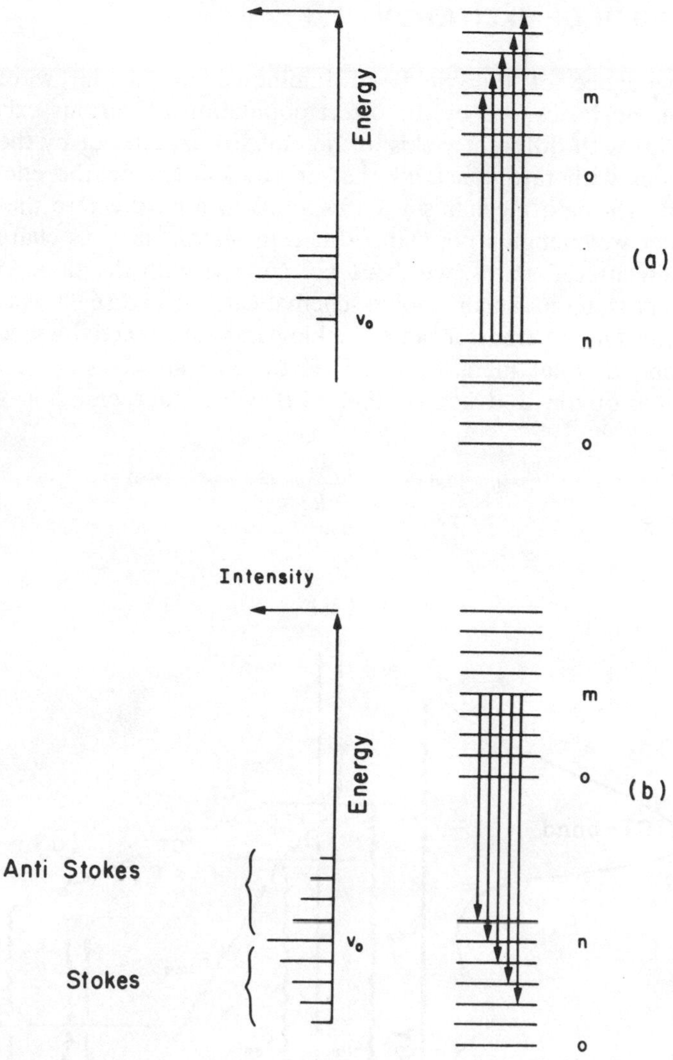

Figure 33. Vibronic transitions in (*a*) absorption and (*b*) emission (right side) leading to anti-Stokes lines in the emission spectrum (left side).

$K_3[Co(CN)_6]$, the phosphorescence lifetime changes from a limiting value of about 3.3×10^3 μs at $T < 50$ K to 0.05 μs at room temperature.[106,107,122] This change most likely reflects the dependence of k_p on temperature. A similar dependence is observed in other media. In solutions of methanol–water–ethylene glycol the phosphorescence lifetime changes from a low-temperature limiting value of 4.0×10^{-3} s to less than 10^{-6} s at room temperature.

4-2 DEPENDENCE OF PHOTOLUMINESCENCE YIELDS ON THE WAVELENGTH OF EXCITATION

The dependence of the yields of luminescence on the wavelength of excitation can be caused by the direct population of various excited states that convert with different yields to the emissive state(s) or by the existence of processes with rate constants that are dependent on the energy of the excitation. The first mechanism corresponds to a case where the excitation at different wavelengths populates different states, such as charge transfer and metal-centered states, which do not convert with the same yield to the luminescent state(s). This is shown schematically in Figure 34 where doublet and quartet charge transfer states lacking intrinsic reactivities coexist with quartet and doublet metal-centered states. The steady state excitation at wavelengths of the d–d bands (Fig. 34*a*) will induce emission with yields

$$\phi_{\mathrm{f}}(\mathrm{d–d}) = \frac{k_{\mathrm{f}}}{k_{\mathrm{f}} + k_{\mathrm{n}'} k_{\mathrm{isc}}} \tag{18}$$

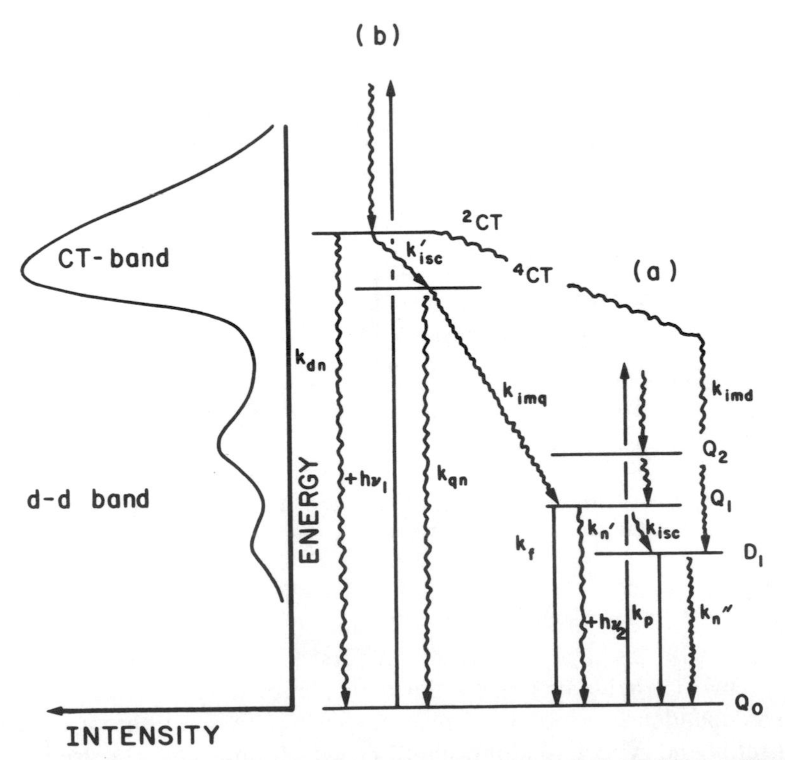

Figure 34. Photophysical processes associated with the population of various excited states (CT and dd) shown in the absorption spectrum (left side). Manifolds represent the set of (*a*) ligand field states and (*b*) charge transfer states.

$$\phi_p(\text{d–d}) = \left(\frac{k_{isc}}{k_f + k_n' + k_{isc}}\right)\left(\frac{k_p}{k_p + k_{n''}}\right) \tag{19}$$

The direct population of the charge transfer states (Fig. 34b) leads to the indirect population of the metal-centered states. Hence, fluorescence and phosphorescence will be observed with yields

$$\phi_f(\text{CT}) = \left(\frac{k_{isc}'}{k_d n + k_{isc}' + k_{imd}}\right)\left(\frac{k_{imq}}{k_{qn} + k_{imq}}\right)\phi_f(\text{dd}) \tag{20}$$

$$\phi_p(\text{CT}) = \left(\frac{k_{imd}}{k_{dn} + k_{isc}' + k_{imd}}\right)\left(\frac{k_f + k_{n'} + k_{isc}}{k_{isc}}\right) + \left(\frac{k_{isc}'}{k_{dn} + k_{isc}' + k_{imd}}\right)\left(\frac{k_{imq}}{k_{qn} + k_{imq}}\right)\phi_p(\text{dd}) \tag{21}$$

Equations 18 and 19 and eqs. 20 and 21 correspond to two limiting cases, namely for excitation at wavelengths where the spectrum corresponds to a pure d–d or CT band, respectively. For wavelengths where the overlap between the CT and d–d bands is significant, the quantum yields will depend on the wavelength of excitation.

The competition of the internal conversion of an upper state to the low-lying emissive state with fast wavelength-dependent processes is the second mechanism that explains the dependence of luminescence yields on excitation wavelength. The process must be fast to compete with the vibrational equilibration, that is, $k_r(\nu) \geq 10^{12}\ s^{-1}$, and we can assume that the value of the rate constant depends on the vibrational level that is achieved in the excitation. In the context, the yield of phosphorescence is

$$\phi_p = \left(\frac{k_0}{k_0 + k_r(\nu)}\right)\left(\frac{k_{isc}}{k_f + k_{n'} + k_{isc}}\right)\left(\frac{k_p}{k_{n''} + k_p}\right) \tag{22}$$

4-3 TIME-RESOLVED DETECTION OF THE EXCITED STATE (PROMPT AND P- OR E-DELAYED FLUORESCENCE)

The equations deduced in previous sections of this chapter were obtained by assuming a stationary state in the concentration of excited state(s). A common procedure, however, is the detection of the excited states by flash photolysis, that is, the determination of the absorption spectrum and the study of the kinetics of formation and decay of the excited state, in addition to the investigation of the photoluminescence by techniques such as flash fluorimetry or single-photon counting. The scheme in Figure 30 can be conveniently used to exemplify the differences between prompt and delayed fluorescence. It is possible to assume that as a result of a direct excitation to Q_1 or as a consequence of the rapid internal conversion of Q_2 to Q_1, the only species that will be observed immediately after the flash is Q_1. Hence, we can impose the initial conditions

$$[Q_2] = [D_1] = 0 , \qquad [Q_1] = [Q_2]_0 \quad \text{at } t = 0 \tag{23}$$

The complete rate equations for Q_1 and D_1 are

$$-\frac{\partial[Q_1]}{\partial t} = (k_{n'} + k_f + k_{isc})[Q_1] - k_{\overline{isc}}[D_1] \tag{24}$$

$$-\frac{\partial[D_1]}{\partial t} = (k_p + k_{n''} + k_{\overline{isc}})[D_1] - k_{isc}[Q_1] \tag{25}$$

The set of simultaneous differential equations is usually resolved by approximating a negligible contribution of $k_{\overline{isc}}[D_1]$ in eq. 24. Hence, the concentration of Q_1 changes in time according to

$$[Q_1] = [Q_1]_0 \exp(-t/\tau_f) \qquad \tau_f = 1/(k_{n'} + k_f + k_{isc}) \tag{26}$$

and the fluorescence intensity shows a time dependence

$$I_{pf} = k_f[Q_1]_0 \exp(-t/\tau_f) \tag{27}$$

where τ_f is the lifetime of Q_1 and also of the fluorescence. The change of D_1 concentration with time

$$[D_1] = k_{isc}[Q_1]_0\left(\frac{\tau_f\tau_p}{\tau_f - \tau_p}\right)\left(\exp\left(\frac{-t}{\tau_p}\right) - \exp\left(\frac{-t}{\tau_f}\right)\right)$$

$$\tau_p = \frac{1}{k_{n''} + k_p + k_{\overline{isc}}} \tag{28}$$

and the intensity of the phosphorescence

$$I_p = k_p k_{isc}[Q_1]_0\left(\frac{\tau_f\tau_p}{\tau_f - \tau_p}\right)\left(\exp\left(\frac{-t}{\tau_p}\right) - \exp\left(\frac{-t}{\tau_f}\right)\right) \tag{29}$$

show dual contributions corresponding to a growth with the fluorescence lifetime τ_f and a decay with the lifetime τ_p of D_1.

The back intersystems crossing gives some population to D_1 at long times, and some fluorescence will be observed with the same lifetime of the phosphorescence, namely with an intensity

$$I_{df} = k_f k_{isc} k_{\overline{isc}}\left(\frac{\tau_f\tau_p^2}{\tau_f - \tau_p}\right)\left(\exp\left(\frac{-t}{\tau_p}\right) - \exp\left(\frac{-t}{\tau_f}\right)\right) \tag{30}$$

Fluorescence with the same lifetime as that of the phosphorescence is called *delayed fluorescence* to differentiate it from normal fluorescence, which is called *prompt fluorescence*.[118] The delayed fluorescence caused by back intersystem crossing was initially investigated with eosin dyes and for this reason is known as E-type delayed fluorescence.[123] It is important to analyze the limitations of eq. 30 for the study of delayed fluorescence. There are cases where the contribution of $k_{\overline{\text{isc}}}[D_1]$ in eq. 24 cannot be ignored and eqs. 24 and 25 must be resolved as a set of simultaneous equations to find correct expressions for the intensities. For example, resolution of the rate equations by the Laplace transform gives

$$I_f = k_f[Q_1] = \frac{k_f[Q_1]_0}{n-m}\left(\left(\frac{n\tau_p+1}{\tau_p}\right)\exp(-nt) - \left(\frac{m\tau_p+1}{\tau_p}\right)\exp(-mt)\right) \tag{31}$$

$$I_p = k_p[D_1] = \frac{k_p k_{isc}[Q_1]_0}{n-m}(\exp(-mt) - \exp(-nt)) \tag{32}$$

where $(-m)$ and $(-n)$ are the two roots of $x^2 - (1/\tau_f + 1/\tau_p)x + (1/\tau_f\tau_p - k_{isc}k_{\overline{\text{isc}}})$. These solutions (eqs. 31 and 32) are valid if $(1/\tau_f + 1/\tau_p)^{2/4} > (1/\tau_f\tau_p - k_{isc}k_{\overline{\text{isc}}})$ and show that it may be difficult in certain cases to differentiate prompt fluorescence from delayed fluorescence (Fig. 35). A careful analysis of the fluorescence intensity as a function of time, however must show departurcs from the exponential decay expected for prompt fluorescence (eq. 27). There are a few Cr(III) complexes that show both

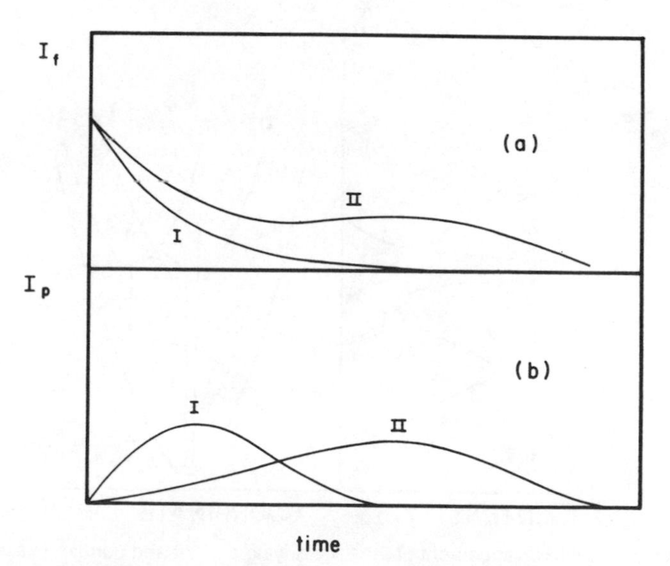

Figure 35. Profiles for the resolved emission (I_p, intensity of phosphorescence; I_f, intensity of prompt and delayed fluorescence). The two cases I and II correspond to differences in the rates of formation and decay of the emissive states.

fluorescence and phosphorescence, a condition that depends on the relative positions of the $^4T_2$ and ^2E states (Fig. 36). This photoluminescence is temperature dependent. For example, $Cr(urea)_6^{3+}$ fluoresces and phosphoresces at room temperature but only phosphoresces at low temperatures as one would expect if the origin of $^4T_2 \leftarrow {}^4A_2$ lies above the $^2E \leftarrow {}^4A_2$. From this assumption, it can be determined that, for complexes showing both fluorescence and phosphorescence, some amount of fluorescence could be E-type delayed fluorescence. It has been found that the fluorescence is partly prompt, with a short lifetime, and partly delayed, with the phosphorescence lifetime.

In large concentrations of a long-lived excited state it is possible to observe bimolecular reactions. This problem has been studied extensively with organic compounds where the formation of adducts can be summarized as[118]

$$A + h\nu \rightarrow A^* \xrightarrow{\text{A or A}^*} (AA)^* \qquad \text{(excimer)} \qquad (33)$$

$$A + h\nu \longrightarrow A^* \xrightarrow{\text{B}} (AB)^* \qquad \text{(exciplex)} \qquad (34)$$

$$A + h\nu \longrightarrow A^* \xrightarrow{\text{A}} A_2 \qquad \text{(photodimer)} \qquad (35)$$

The formation of excimers (eq. 33) is of interest here because of its

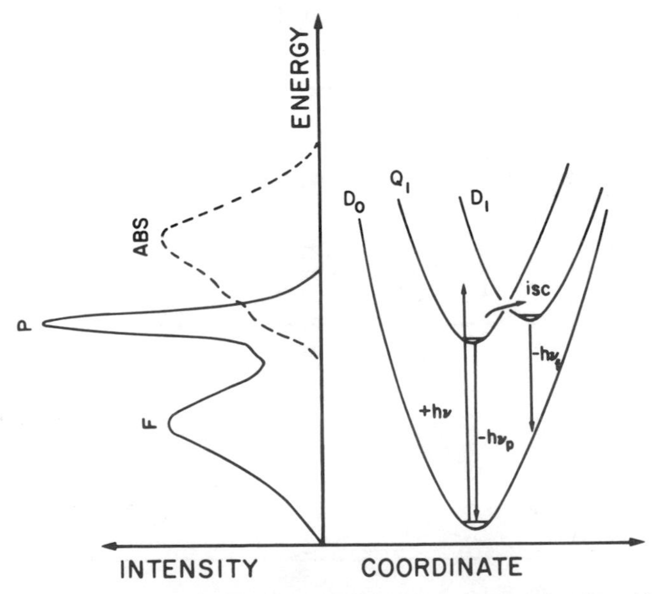

Figure 36. One possible arrangement of the lowest lying quartet and doublet excited state (D_1 and Q_1) potential surfaces in relationship to the ground state (D_0). This arrangement of surfaces is consistent with fluorescent (F) and phosphorescent (P) emissions in the spectrum of $Cr(urea)_6^{3+}$.

relationship with a form of delayed fluorescence known as P-type delayed fluorescence, that is, the delayed fluorescence exhibited by pyrene.[124,125] In addition, the formation of exciplexes (eq. 34) can be related to quenching processes that are described later in this chapter.

The P-type delayed fluorescence emitted from solutions of anthracene or phenanthrene can be attributed to the formation of a singlet excimer that dissociates in an excited singlet

$$^3A + {}^3A \longrightarrow {}^1A_2^* \longrightarrow {}^1A^* + {}^1A \qquad (36)$$

$$^1A^* \longrightarrow {}^1A + h\nu \qquad (37)$$

In this regard, the P-type delayed fluorescence will have the same spectrum as the prompt fluorescence and its intensity will change with time in a manner that corresponds to the second-order rate of formation and first-order rate of decay of the excimer.

The excimers do not necessarily decay into an excited singlet; they can dissociate into radical ions (eq. 38) or undergo radiative relaxation to the ground state (eq. 39) (Fig. 37)

$$^1A_2^* \longrightarrow A^{\cdot +} + A^{\cdot -} \qquad (38)$$

$$^1A_2^* \longrightarrow 2A + h\nu \qquad (39)$$

A given excimer can experience more than one form of relaxation. For example, the dissociation in radicals (eq. (38) will be favored in polar

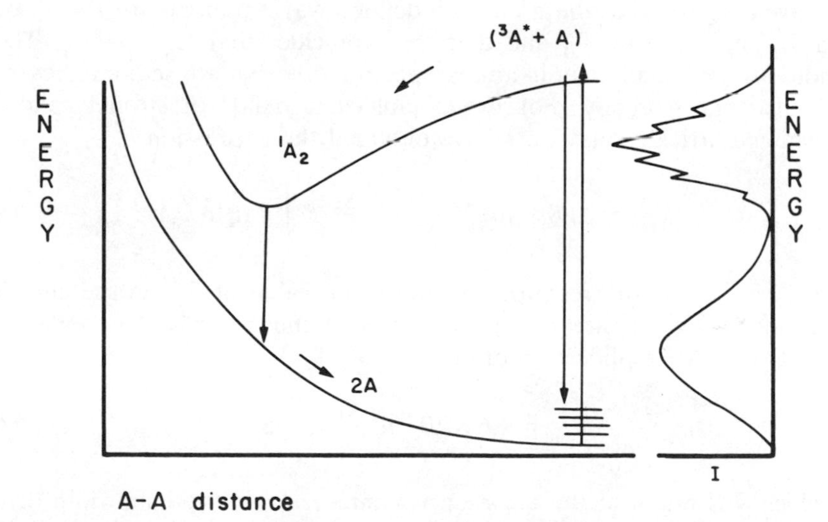

Figure 37. Potential surfaces describing the formation of an excimer $^1A_2$ from the triplet $^3A^*$. Vertical arrows show the excitation $^3A^* \leftarrow A$ and emissions from $^3A^*$ and $^1A_2$. The emission spectrum of the system is shown on the right.

solvents over the radiative relaxation (eq. 39) or the dissociation in an excited singlet (eq. (36). In coordination chemistry, the formation of excimers and exciplexes has been observed with phthalocyanine and porphyrins where the large π systems of the macrocycles allow the transfer of charge that stabilizes the dimers.[126–129] The generation of the long-lived $(^2E)CrL_3^{3+}$ (L = bipy, phen) also leads to self-quenching by a bimolecular reaction of the excited state.[130] The effect of such a bimolecular process on the phosphorescence is similar to that described above for P-delayed fluorescence.

4-4 RADIATIVE TRANSITIONS

The Einstein transition probability coefficients have already been associated with the rate constant for the emission of light (eq. 63 in Chapter 3). Moreover, the use of the oscillator strength (eqs. 65 and 66 in Chapter 3) allows the rate constant to relate to experimentally measurable extinction coefficients. It is customary to introduce in the expression of the oscillator strength a correction factor f that is related to the refractive index n of the medium, that is, $F = n^2$. In this context, the rate constant for the radiative relaxation is

$$k_{rad} = 2.88 \times 10^{-9} \frac{g_1}{g_2} \bar{\nu}^2 F \int \epsilon \, d\bar{\nu} \tag{40}$$

In the derivation of this equation, known as Lewis and Kasha's equation,[131] we have assumed that there are well-defined wave numbers for the absorption $\bar{\nu}_{12}$ and emission $\bar{\nu}_{21}$ and that they coincide, that is, $\bar{\nu}_{12} = \bar{\nu}_{21}$. These conditions are usually met in atomic spectroscopy, but are seldom obeyed in molecular spectroscopy. For broad molecular bands of strongly allowed transitions, Strickler and Berg have obtained the expression[132]

$$k_{rad} = 2.88 \times 10^{-9} \langle \bar{\nu}^{-3} \rangle^{-1} \frac{g_1}{g_2} F \int \epsilon \, d(\ln \bar{\nu}_a) \tag{41}$$

where $\langle \bar{\nu}^{-3} \rangle^{-1}$ is the reciprocal of the mean value of $\bar{\nu}_e^{-3}$ in the emission spectrum and g_1, g_2 are the degeneracies of the lower and upper states, respectively. A simplification of eq. 41

$$k_{rad} = 4.6 \times 10^{-9} \bar{\nu}^2 \frac{g_1}{g_2} \epsilon_{max} \Delta\bar{\nu} \tag{42}$$

in which $\bar{\nu}$ is taken at the emission maximum, $\Delta\bar{\nu}$ is the bandwidth of the emission, and ϵ_{max} is the extinction coefficient of the maximum absorption, is useful when we wish to estimate orders of magnitude for the radiative rate constant. Equations 41 and 42 cannot be applied to transitions where there is a noticeable difference between the equilibrium nuclear configurations of the ground and excited states, a fact that impedes their use with transition metal complexes. The failure of these equations with the radiative lifetimes of coordination complexes can be illustrated with $Ru(bipy)_3^{2+}$ where the values of the radiative rate constant obtained with these equations are larger than the experimental values. These equations work better when we estimate the phosphorescence lifetime of some Cr(III) complexes, but this agreement has not been proved to be general.[133] Other expressions for the radiative transition probability are obtained in Section 4-6.

4-5 RADIATIONLESS PROCESSES IN POLYATOMIC MOLECULES

For a molecule that is in an excited state, return to the ground state can be achieved by radiationless processes in addition to the emission of light.[134] In solution and in rigid media, the presence of the surrounding medium provides a heat bath that assures the irreversible electronic relaxation because of the subsequent degradation of the excess vibrational energy.[135–137] In isolated large molecules, the density of states is sufficiently large and the molecule can act as its own heat bath.[138–139]

The first observation required for the interpretation of the radiationless relaxation processes is that the system under consideration is in a compound state.[134] In this context, the states of the system can be partitioned into two or more subsets of zero-order states (Fig. 38). One subset, the sparse set, consists of a finite number of discrete energy levels corresponding to a small subset of the total number of degrees of freedom. The other set of zero-order states, the dense part, has a continuum spectrum and is as-

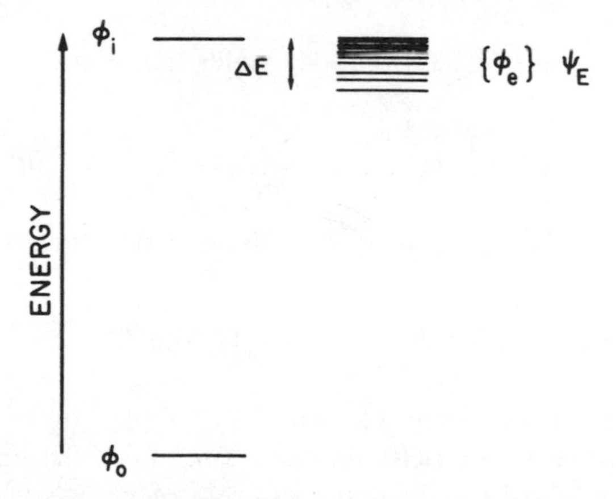

Figure 38. Compound state associated with the radiationless conversion from zero-order states ϕ_i to the sparse set Ψ_E consisting of a finite number of discrete energy levels or to the dense set $\{\phi_e\}$ associated with infinite or effectively infinite degrees of freedom.

sociated with infinite or effectively infinite degrees of freedom. Therefore a relaxation process takes place when a system in a compound state decays into the continuum. To develop an expression for the transition probability, there must be $\phi_i(X_1, X_2)$ $(i = 1, 2, \ldots)$ discrete states and $\Psi_E(X_1, X_2)$, $\Psi_{E'}(X_1, X_2), \ldots$ continuum states of the zero-order Hamiltonian. These are solutions of

$$\hat{\mathcal{H}}_T(X_1, X_2) = \hat{\mathcal{H}}_0(X_1, X_2) + \hat{V}_{12}(X_1, X_2)$$

with an interaction $\hat{V}_{12} = 0$. The exact states are derived from the interaction of the two sets and for simplicity we introduce only one zero-order state from each subset. In this approximation the wave function is

$$\Psi_E = a(E)\phi_1 + \int b_{E'}\Psi_{E'}\, dE'$$

It is possible to obtain a more explicit expression for the decay law using the time-dependent perturbation theory. In this analysis, the zero-order metastable state is $\phi_1(X_1, X_2)$ and by the superposition principle the state function at the time t is

$$\Psi(t) = a(t)\phi_1 e^{-iE_1 t/\hbar} + \int dE'\, C_{E'}(t)\Psi_{E'} e^{-iE't/\hbar}$$

The equations of motion for the expansion coefficients are

$$i\hbar\dot{a}(t) = \int C_{E'}(t)\langle \phi_1|\hat{\mathcal{H}}_T|\Psi_{E'}\rangle e^{i(E_1-E')t/\hbar}\, dE' \tag{43}$$

$$i\hbar\dot{C}_{E'}(t) = a(t)\langle \Psi_{E'}|\hat{\mathcal{H}}_T|\phi_1\rangle e^{-i(E_1-E')t/\hbar} \tag{44}$$

Therefore integration of $\dot{C}_{E'}(t)$ (eq. 44) yields the relationship

$$i\hbar C_{E'}(t) = \int_0^t a(t')\langle \Psi_{E'}|\hat{\mathcal{H}}_T|\phi_1\rangle e^{-i(E_1-E')t'/\hbar}\, dt'$$

which is introduced in eq. 43 to resolve the equation for $\dot{a}(t)$,

$$-h^2\dot{a}(t) = \int_0^t a(t')\, dt' \int |\langle \Psi_{E'}|\hat{\mathcal{H}}_T|\phi_1\rangle|^2 e^{-i(E_1-E')t'/\hbar}\, dE'$$

The replacement of the matrix element $\langle \Psi_{E'}|\hat{\mathcal{H}}_T|\phi_1\rangle$ by its average value and the introduction of the delta function

$$\delta(t - t') = \frac{1}{2\pi}\int_{-\infty}^{\infty} e^{iw(t-t')}\, dw$$

in the expression for $\dot{a}(t)$ leads to

$$-\hbar\dot{a}(t) = 2\pi|\langle\Psi_{E'}|\hat{\mathscr{H}}_{\mathrm{T}}|\phi_1\rangle|^2 a(t)$$

Solving this differential equation gives

$$a(t) = a(0)e^{-\kappa t}$$

where

$$\kappa = \frac{2\pi}{\hbar}|\langle\Psi_{E'}|\hat{\mathscr{H}}_{\mathrm{T}}|\phi_1\rangle|^2 = \frac{1}{\tau} \tag{45}$$

This expression for τ, that is, the reciprocal of the excited state lifetime, is known as Fermi's Golden Rule for the transition probability. Now, the evaluation of κ (eq. 45) requires the use of some scheme for the energy levels of the molecule. The nonradiative decay probability (eq. 45) of the sparse manifold (Fig. 38) can be expressed

$$\kappa = \frac{2\pi}{\hbar}\Big(\sum_i\sum_j \rho(si)|V_{si,lj}|^2\delta(E_i - E_j)\Big) \tag{46}$$

when we use functions based on the Born–Oppenheimer approximation. The elements of the electronic Hamiltonian in eq. 46 are $V_{si,lj} = \langle\phi_{si}|\hat{V}_{12}|\phi_{lj}\rangle$, and the probability that the system be in the zero-order state at time zero is $\rho(si)$. Provided that the relaxation and excitation rates exceed the radiationless decay probability, a state of thermal equilibrium prevails and $\rho(si) = (\exp[-E_{si}/kT])/\Sigma_i \exp[-E_{si}/kT])$.

Equation 46 can be recast in a form that is completely analogous to the expressions for the lineshape in the optical absorption in solids. Moreover, the nonradiative process can be treated as an optical excitation in the limit of zero excitation energy. Hence, the transition probability (eq. 46) can be expressed according to Freed and Jortner[138] in the form,

$$k = \frac{2\pi}{\hbar}F(0) \tag{47}$$

where k is the transition probability and $F(0)$ is a lineshape function[135] that can be defined by

$$F(E) = Z^{-1}\Big(\sum_i\sum_j |V_{si,lj}|^2 \exp(-E_{si}/kT)\delta(\Delta E + E_{lj} - E_{si} - E)\Big) \tag{48}$$

with $\Delta E = E_{s0} - E_{l0}$, and $Z = \Sigma_i \exp(-E_{si}/kT)$. It is also convenient to use the Fourier transform of $F(E)$

$$f(t) = \int_{-\infty}^{\infty} \exp\left(\frac{i(E - \Delta E)t}{\hbar}\right) F(E)\, dE$$

$$= Z^{-1}\left(\sum_i \sum_j V_{si,lj} \exp\left(\frac{iE_{lj}t}{\hbar}\right) V_{lj,si} \exp\left(\frac{iE_{si}(t - i\hbar/kT)}{\hbar}\right)\right) \tag{49}$$

In this context, the generalized lineshape function can be expressed

$$F(E) = (2\pi\hbar)^{-1} \int_{-\infty}^{\infty} f(t) \exp\left(-\frac{i(E - \Delta E)t}{\hbar}\right) dt$$

and the transition probability

$$\kappa = \frac{2\pi}{\hbar} F(0) = \hbar^{-2} \int_{-\infty}^{\infty} f(t) \exp\left(-\frac{i\Delta Et}{\hbar}\right) dt$$

For mathematical reasons, it is convenient to introduce the operator

$$V_{sl}(Q) = \int dq\, \phi_s(q, Q) V \phi_l(q, Q)$$

where q represents the electronic coordinates and Q the nuclear coordinates. Thus with the use of the Green functions

$$G_l(Q^{(s)}, \tau, Q^{(s)'}, 0) = \sum_i X_{si}(Q^{(s)}) X_{si}^*(Q^{(s)'}) \exp\left(-\frac{iE_{si}\tau}{\hbar}\right)$$

where $T = t - (i\hbar/kT)$, and

$$G_l(Q^{(l)}, t, Q^{(l)'}, 0) = \sum_j X_{lj}(Q^{(l)'}) X_{lj}^*(Q^{(l)}) \exp\left(-\frac{iE_{lj}t}{\hbar}\right)$$

the transform (eq. 49) can be recast

$$f(t) = Z^{-1} \int\int dQ\, dQ' \,[V_{sl}(Q) G_l][V_{sl}(Q) G_s]$$

The vibrational motions in both states can be expanded in terms of independent oscillators that need not have the same frequencies or be harmonic. For example, the Green function for the s state will be in this approximation

$$G_s\left(Q^{(s)}, \tau, Q^{(s)'}, 0) = \prod_p^N g_p(Q_p^{(s)}, \tau, Q_p^{(s)'}, 0\right) \tag{50}$$

To reduce the formal result into a simpler expression, it is common to introduce the following approximations:

1. A harmonic approximation is invoked for the potential energy. This approximation is similar to the one used in eq. 22 of Chapter 3.

2. The directions of the principal axis for the normal coordinates are identical in the two electronic states. It must be pointed out that this is a mathematical simplification that doesn't limit the validity of the treatment.

3. The operators V_{sl} are written in the form

$$V_{sl}(Q) = \sum_{k=1}^{p} C_{sl}^{k}[i\hbar/(M_k)]^{1/2}(\partial/\partial Q_k^{(l)}) \tag{51}$$

where M_k is the effective mass associated with the k mode, and C_{sl}^{k} is

$$C_{sl}^{k} = J_{sl}^{k} = \frac{\hbar}{M_k^{1/2}}\left\langle \phi_s(q, Q^{(s)}) \middle| \frac{i\partial}{\partial Q_k^{(l)}} \middle| \phi_l(q, Q^{(l)}) \right\rangle \tag{52}$$

for internal conversion

$$C_{sl}^{k} = \sum_{\nu \neq s,l}\left(\frac{\langle \phi_s | \hat{\mathcal{H}}_{\mathrm{so}} | \phi_\nu \rangle J_{\nu l}^{k}}{E_{\mathrm{so}} - E_{\nu 0}} + \frac{\langle \phi_\nu | \hat{\mathcal{H}}_{\mathrm{so}} | \phi_l \rangle J_{s\nu}^{k}}{E_{\nu 0} - E_{l0}} \right) \tag{53}$$

for intersystem crossing.

Thus, in eq. 53, $\hat{\mathcal{H}}_{\mathrm{so}}$ is the spin–orbit coupling operator and $E_{\mathrm{so}} - E_{\nu 0}$ or $E_{\nu 0} - E_{l0}$ are the energy gaps between the spin–orbit coupled states. The introduction of these expressions in the generating function (eq. 49) leads to

$$f(t) = z^{-1}\sum_{k}\sum_{k'}\left[\frac{\hbar^2}{(M_k M_{k'})^{1/2}}\right] C_{sl}^{k} C_{sl}^{*k'} \int\int \left(i\hbar \frac{\partial}{\partial Q_k^{(l)'}}\right) G_l(Q^{(l)'}, -t; Q^{(l)}, 0)$$
$$\times \left(i\hbar \frac{\partial}{\partial Q_{k'}^{(s)}}\right) G_s(Q^{(s)}, \tau; Q^{(s)'}, 0\; dQ\, dQ'$$

Factoring this function into contributions from independent oscillators (eq. 50) gives the formal representation of the nonradiative transition probability in the harmonic approximation

$$f(t) = \sum_{k=1}^{p} |C_{sl}^{k}|^2 \tilde{f}_k(t) \prod_{j \neq k} f_j(t) \tag{54}$$

where $\tilde{f}_k(t)$ corresponds to single-mode generating functions that involve the nuclear momentum operator for the promoting mode Q_k, that is, $i(\partial/\partial Q_k)$, and $f_k(t)$ corresponds to single-mode generating functions that arise from Franck–Condon vibrational overlap factors.

It is useful to consider some limiting cases of nonradiative relaxation that are determined by the coupling strength

$$G = \frac{1}{2} \sum_j \left(\frac{M_j W_j}{h} \right) (Q_j^{0(l)} - Q_j^{0(s)})^2 \left(2 \left[\exp\left(\frac{\hbar W_j}{kT} \right) - 1 \right]^{-1} + 1 \right)$$

(where W_j is the frequency of the j normal mode) and the molecular rearrangement energy in the excited state

$$E_M = \frac{1}{2} \sum_{j=1}^{N} M_j W_j (Q_j^{0(l)} - Q_j^{0(s)})^2 = \frac{1}{2} \sum_{j=1}^{N} \hbar W_j \Delta_j^2$$

The molecular rearrangement energy E_M corresponds to half the Stokes shift for the two electronic states under consideration (Fig. 39). In the strong

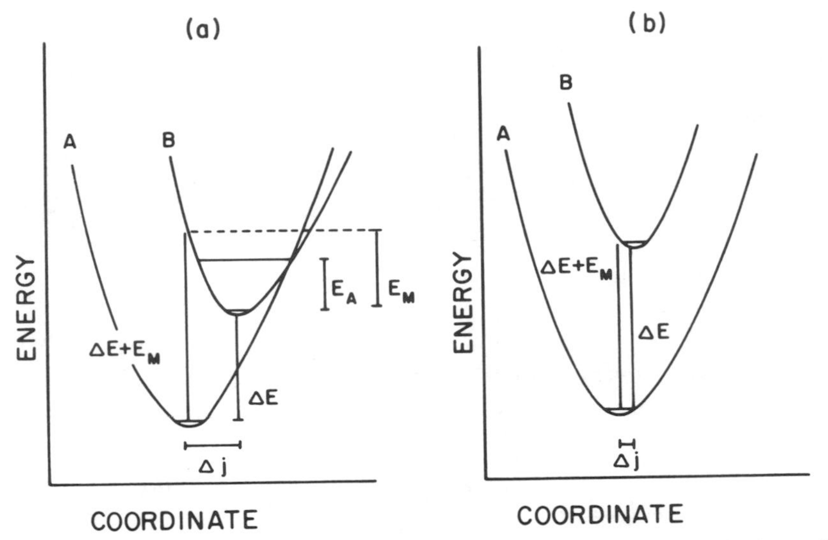

Figure 39. Potential surfaces illustrating the radiationless conversion of an excited state B into another A under the assumptions of (*a*) a strong coupling ($E_M \gg \Sigma_{j=1}^{N} W_j/N$, $\Delta j > (\hbar/M_j W_j)^{1/2}$) and (*b*) weak coupling ($E_M \sim \hbar W$).

coupling limit $G \gg 1$ and the displacements between the origins of the two electronic states exceeds the root mean square vibrational displacements $(\hbar/M_j W_j)^{1/2}$ of some normal modes. Therefore, the molecular rearrangement energy is larger than the mean vibrational energy

$$E_M \gg \hbar \tilde{W}\ ; \qquad \tilde{W} = \text{mean vibrational frequency} = \sum_{j=1}^{N} W_j/N$$

This case corresponds to a large displacement between potential surfaces as shown in Figure 39*a*. In the weak coupling limit $G \leq 1$ and $E_M \simeq \hbar \tilde{W}$, so the relative displacement of each normal mode is small (Fig. 39*b*).

The final expression for the nonradiative decay rate in the displaced potential surface model is

$$\kappa = \sum_k (|C_{se}^k|^2 W_k / 4\hbar) \exp(-G)$$
$$\times \left\{ (\coth y_k + 1) \int_{-\infty}^{\infty} dt \exp[-i(\Delta E/\hbar - W_k)t + G_+(t) + G_-(t)] \right.$$
$$\left. + (\coth y_k - 1) \int_{-\infty}^{\infty} dt \exp[-i(\Delta E/\hbar - W_k)t + G_+(t) + G_-(t)] \right\}$$

where

$$G_+(t) = \tfrac{1}{2} \sum_j \Delta_j^2 (\tilde{n}_j + 1) \exp(iW_j t) \tag{55}$$

$$G_-(t) = \tfrac{1}{2} \sum_j \Delta_j^2 \tilde{n}_j \exp(-iW_j t) \tag{56}$$

$$\tilde{n}_j = [\exp(\hbar W_j / kT) - 1]^{-1}$$

$$\Delta_j = (M_j W_j / \hbar)^{1/2} (Q_j^{0(l)} - Q_j^{0(s)})$$

Therefore, the transition probability in the weak coupling, low-temperature limit is reduced to a simple form

$$\kappa = \sum_k |C_{sl}^k|^2 W_k \exp(-G) \left[\tfrac{1}{2} \pi \hbar W_M Y_k \left(1 + \sum_m{}' \Omega_{m,k} \xi_m Y_k^{\xi m - 1} \right) \right]^{1/2}$$
$$\times \exp(-\nu_k \Delta E_k / \hbar W_M) \tag{57}$$

where the effective energy gap is

$$\Delta E_k = \Delta E - \hbar W_k$$

the ratio of the frequencies is

$$\xi m = \frac{W_m}{W_m} < 1$$

and M is an index that denotes a group of modes with maximum frequency W_M and $\Delta_M^2 \neq 0$. In eq. 57 the prime on the summation means that it is to be taken over all the groups of modes $m \neq M$. The dimensionless positive quantities

$$\Omega_{m,k} = d_m \Delta_m^2 W_m / d_M \Delta_M^2 W_M (d_M \Delta_M^2 h W_M / 2\Delta E_k)^{1-(W_m/W_M)}$$

contain the parameter d_i which gives the number of modes within the ith group, namely the degeneracy of each group. Finally, the last parameter is

$$\nu_k = \ln(\Delta E_k Y_k / a_M) - Y_k \left(1 + \sum_m{}' Q_{m,k} Y_k^{\xi_m - 1} / \xi_m \right)$$

with $a_i = \frac{1}{2} d_i \hbar W_i \Delta_i^2$. It is important to notice that in the cumbersome eq. 57, the matrix elements $C_{sl}k$ (eqs. 52 and 53) lead to selection rules that control the transition probability. Moreover, eq. 57 has the form of the so-called energy gap law, which establishes that the nonradiative transition probability decreases with increasing values of the energy gap ΔE_k.

In the strong coupling limit $G \gg 1$ and the functions $G_+(+)$ and $G_-(t)$ (eqs. 55 and 56) can be expanded in series. Retention of terms up to t^2 in these expansions lead to

$$G_+(+) + G_-(+) = G + \left(\frac{itE_M}{\hbar}\right) - \tfrac{1}{2} D^2 t^2$$

where

$$D^2 = \frac{1}{2} \sum_j W^2 j \Delta_j^2 (2\tilde{n}_j + 1)$$

Therefore the equation for the nonradiative transition probability in the strong coupling limit reduces to

$$\kappa = \sum_k \frac{|C_{sl}^k|^2 W_k (2\pi)^{1/2}}{(2E_M kT^*)^{1/2}} \coth\left(\frac{\frac{1}{2}\hbar\tilde{W}}{kT}\right) \exp\left(\frac{-E_A}{kT^*}\right) \tag{58}$$

where E_A is the point intersection of minimum energy for the two potential surfaces measured from the energy origin (Fig. 39*a*). The energy for the intersection of two surfaces satisfies the relationship

$$E_A = (\Delta E - E_M)^2 / 4E_M$$

The effective temperature T^* is defined by

$$T^* = \tfrac{1}{2}(\hbar\tilde{W}/kT) \coth(\tfrac{1}{2}\hbar\tilde{W}/kT)$$

Thus in the strong coupling limit the transition probability is determined by a sort of activation energy E_A and not by the energy gap law as in the case of the weak coupling limit.

The expressions for the rate constant of the excited state radiationless relaxation in the weak coupling (eq. 57) and strong coupling (eq. 58) approximations were also deduced by Englman and Jortner with a simpler scheme of energy levels.[139] In the strong coupling limit, for a molecule in an inert medium or for an isolated molecule with only one level excited the rate constant is

$$\kappa = \frac{C^2\sqrt{2\pi}}{\hbar(E_M kT^*)^{1/2}} \exp\left(\frac{-E_A}{kT^*}\right) \tag{59}$$

where the effective temperature is defined in the form

$$kT^* = \tfrac{1}{2}\hbar\tilde{W} \coth(\hbar\tilde{W}/2kT) \tag{60}$$

Equation 60 reaches a limit of $\frac{1}{2}\hbar\tilde{W}$ for very low temperatures, that is, for $\hbar\tilde{W}/2kT \gg 1$. hence, for low temperatures the expression for the rate constant in the strong coupling limit is

$$\kappa = \frac{C^2\sqrt{4\pi}}{\sqrt{E_m\hbar\tilde{W}}} \exp\left(\frac{-2E_A}{\hbar\tilde{W}}\right) \tag{61}$$

For the weak coupling limit the rate constant can be written

$$\kappa = \frac{C^2}{\hbar}\sqrt{\frac{2\pi}{\hbar W_M \Delta E}} \exp\left(-\frac{1}{2}\sum_j \Delta_j^2\right) \exp\left\{-\frac{\Delta E}{\hbar W_M}\left[\log\left(\frac{2\Delta E}{\sum_M \hbar W_M \Delta_M^2}\right) - 1\right]\right\} \tag{62}$$

It is possible to see in eq. 62 the form of the energy gap law. Furthermore, by introducing into eq. 62 the parameter

$$\nu = \log\left(\frac{\Delta E}{de_M} - 1\right) \tag{63}$$

where

$$de_M = \frac{1}{2}\sum_M \hbar W_m \Delta_M^2$$

is the contribution of the vibrational modes of maximum frequency, the expression of the rate constant is

$$\kappa = \frac{C^2}{\hbar}\sqrt{\frac{2\pi}{\hbar W_M \Delta E}} \exp\left(-\frac{\nu\Delta E}{\hbar W_M}\right) \tag{64}$$

The preexponential factor in eq. 64 exhibits a weak dependence in the energy gap ΔE and contains the coupling factor C^2. Such a coupling factor can be expressed for the internal conversion and intersystem crossing as $C_{CO} = \langle H_V \rangle$ and $C_{CR} = \langle H_V \rangle \cdot \langle H_{SO} \rangle / (E_{SO} - E_{\nu 0})$, respectively. The factors $\langle H_V \rangle$ and $\langle H_{SO} \rangle$ correspond to the vibronic and spin–orbit coupling matrix elements and $(E_{SO} - E_{\nu 0})$ is the energy gap between the two spin–orbit coupled states. In this context, for singlet–singlet and singlet–triplet radiationless transitions in organic compounds it has been estimated

$$\kappa_{CO} \simeq 10^{13\pm 2} \exp(-\nu\Delta E/\hbar W_M)$$

$$\kappa_{CR} \simeq 10^{3\pm 2} \exp(-\nu\Delta E/\hbar W_M)$$

for the weak coupling limit.

The success of the relationship developed here can be judged in terms of the bases that they provide for interpreting photophysical phenomena. One typical example is the anomalous fluorescence of azulene, namely fluorescence from the second excited singlet that violates Kasha's rule. Optical relaxation and line broadening experiments reveal that the rate of nonradiative relaxation of the first singlet is $\kappa(S_1 \rightarrow S_0) \sim 3 \times 10^{10}\ s^{-1}$. Measurements of quantum yields give $\kappa(S_2 \rightarrow S_1) \sim 6 \times 10^8\ s^{-1}$ for the internal conversion between singlets. With the use of the expressions obtained for the rate constants of the radiationless process, one can relate $\kappa(S_2 \rightarrow S_1)$ and $\kappa(S_1 \rightarrow S_0)$ by means of the relationship:

$$\frac{\kappa(S_1 \rightarrow S_0)}{\kappa(S_2 \rightarrow S_1)} = \exp\left[\frac{\Delta E}{\hbar W_M}(\nu_2 - \nu_1)\right]$$

where ΔE is the common gap between S_2 and S_1, and ν_i are the corresponding parameters (eq. 63). Since the energy gap $\Delta E \sim 14 \times 10^3\ cm^{-1}$, $\nu_2 \sim 1.6$, and $\nu_1 \sim 0.8$, the estimated ratio is

$$\frac{\kappa(S_1 \rightarrow S_2)}{\kappa(S_2 \rightarrow S_1)} \simeq 30$$

Such a result leads to a nearly unity quantum yield for the fluorescence from the second singlet, in good agreement with experimental observations.

The expressions deduced for the rate of radiationless relaxation have the vibrational frequencies of coupling modes. In consequence, intrinsic ligand vibrations and metal–ligand vibrations must be considered in the case of coordination complexes. For example, deuteration has a significant effect on lengthening the lifetime of the ^2E state of $Cr(NH_3)_6^{3+}$,[140] a fact that argues strongly in favor of the participation of NH_3 vibrational modes in radiationless relaxation of the excited state. In addition, metal–ligand vibrations must be important in radiationless transitions of halide complexes, for example, $CrCl_6^{3-}$. In these complexes, however, the mismatch between the values of the energy gap and the vibrational frequencies of coupling modes may lead to extremely slow rates of radiationless relaxation. Although the model described above for radiationless relaxation is appropriate for coordination complexes, some studies have revealed a number of complex features. For example, it has been demonstrated that the relaxation rates for the lowest energy doublet ^2E of Cr(III) complexes can be represented by $k_{re} = k_{re}^0 + k_{re}(T)$. The limiting low-temperature, excited state lifetime $(k_{re}^0)^{-1}$ is a molecular property that is independent of temperature and medium. The thermally activated decay is a function of temperature, solvent, and coordinated ligands. It has been suggested that the thermally activated path represents a channel or channels that involve a strongly vibronic coupled, but spin forbiddent crossing to the potential energy surface of a reaction intermediate (see Chapter 6).

4-6 ENERGY TRANSFER

P-delayed fluorescence is one of the mechanisms available for the transfer of electronic energy. Other such mechanisms are based on remote dipole–dipole or dipole–quadrupole electronic interactions and on a collisional mechanism that involves electronic exchange.[141–145] Mathematical treatments for evaluating rate constants have been reported by several authors (see below). In these mechanisms, it is possible to visualize the electronic energy transfer,

$$ {}^{*}\mathrm{S}(\text{donor}) + \mathrm{A}(\text{acceptor}) \rightarrow \mathrm{S} + {}^{*}\mathrm{A} \tag{65} $$

by means of potential surfaces (Fig. 40) that describe the process in terms of the donor–acceptor separation. The minima in these surfaces is sufficiently shallow to prevent the formation of the adducts described elsewhere in this chapter (eqs. 33–35) and the general shape is determined by the nature of the donor and acceptor.

For the process described in eq. 65, the probability of energy transfer can be formulated according to Fermi's Golden Rule (eq. 45).[143] If $\hat{\mathscr{H}}_1$ is the coupling Hamiltonian and p is a measure of the density of initial and final states that are coupled, the probability can be expressed

$$ \kappa_{({}^{*}\mathrm{S}+\mathrm{A}\rightarrow \mathrm{S}+{}^{*}\mathrm{A})} = \frac{2\pi}{\hbar}\, p\,|\langle \Psi_{{}^{*}\mathrm{S}} \Psi_{\mathrm{A}} | \hat{\mathscr{H}}_1 | \Psi_{\mathrm{S}} \Psi_{{}^{*}\mathrm{A}} \rangle|^2 \tag{66} $$

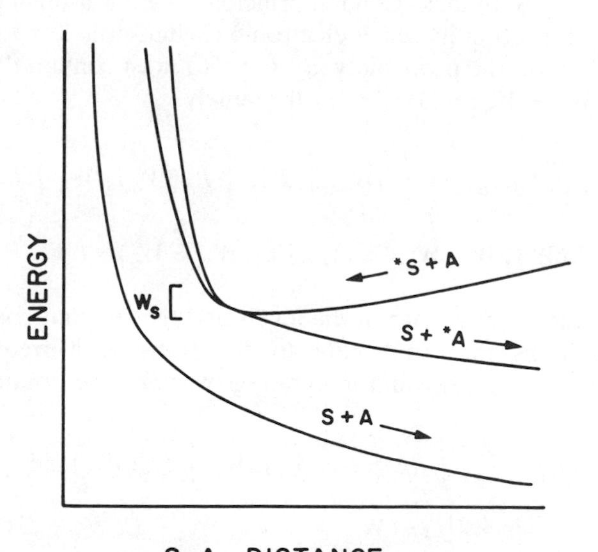

Figure 40. Potential surfaces describing an energy transfer process that does not involve the formation of stable adducts.

The wave functions for the excited state of S and ground state of A must be normalized

$$\langle \Psi_{*S} | \Psi_{*S} \rangle = \langle \Psi_A | \Psi_A \rangle = 1$$

Moreover, the final state functions are normalized in an energy scale

$$\frac{1}{\Delta W} \int_W^{W+\Delta W} \langle \Psi_S | \Psi_S \rangle \, dw = \frac{1}{\Delta W} \int_W^{W+\Psi W} \int_W^{W+\Delta W} \langle \Psi_A | \Psi_A \rangle \, dw = 1$$

The functions $p_{*S}(W_{*S})$ and $P_A(W_A)$ that give the probability of S being with a particular energy W_{*S} and A with W_A are also normalized in an energy scale

$$\int_0^\infty P_{*S}(W_{*S}) \, dw_{*S} = \int_0^\infty P_A(W_A) \, dW_A = 1$$

With these normalization conditions, the antisymmetrized wave functions

$$\sqrt{2}\Psi_I = \Psi_{*S}(r_1, W_{*S})\Psi_A(r_2, W_A) - \Psi_{*S}(r_2, W_{*S})\Psi_A(r_1, W_A)$$

$$\sqrt{2}\Psi_{II} = \Psi_S(r_1, W_S)\Psi_{*A}(r_2, W_{*A}) - \Psi_S(r_2, W_S)\Psi_{*A}(r_1, W_{*A})$$

can be used with expressions that account for the exchange and coulombic interactions.

According to the Franck–Condon principle, there is a small probability of an electronic transition in which electronic excitation is lost to the medium as heat. Therefore, the probability κ_{SA} (eq. 66) must contain the Dirac delta function $\delta[(W_{*S} - W_S) - (W_{*A} - W_A)]$, namely

$$\kappa_{SA} = \frac{2\pi}{\hbar} \sum_I \sum_F (g_{*S} g_A)^{-1} \int dW_{*A} \int dW_S \int P_A(W_A) \, dW_A \int P_{*S}(W_{*S}) \, dW_{*S}$$
$$\times |\langle H_1(W_{*S}, W_A; W_S, W_{*A}) \rangle_{IF}|^2 \delta[(W_{*S} - W_S) - (W_{*A} - W_A)]$$

where g_i are the degeneracies of the levels and the summations contemplate all the transitions that contribute to the transfer. Moreover by using $E = W_{*S} - W_S = W_{*A} - W_A$ and integrating over W_S, we obtain

$$\kappa_{SA} = \frac{2\pi}{\hbar} \int dE \int P_A(W_A) dW_A \int P_{*S}(W_{*S}) \, dW_{*S}$$
$$\times \{|\langle H_1(W_{*S}, W_A; W_S, W_{*S} - E, W_A + E) \rangle_{IF}|^2\}$$

For the matrix element $\langle H_1 \rangle$, the interaction must be expressed as the sum of all the Coulomb interactions of the outer and core electrons of A and S. In this context, when the coulomic sum is expanded in series about the vector R, the vector that gives the separation between S and A, we obtain

$$\mathscr{H}_1(R) = \frac{e^2}{KR^3}\left\{r_S \cdot R_A - \frac{3(r_S \cdot R)(r_A \cdot R)}{R^2}\right\}$$
$$+ \frac{3e^2}{2KR^4}\left\{\sum_{i=1}^{3} \frac{Ri}{R} r^2_{Ai} r_{Si}\left(-3 + 5\frac{R_i^2}{R^2}\right)\right.$$
$$+ 10\left(\frac{XYZ}{R^3}\right)(X_A Y_A Z_S + X_A Z_A Y_S + Y_A Z_A X_S)$$
$$\left. + \sum^{3}\sum_{i \neq j}\left[\left(\frac{R_j}{R}\right) - 5\frac{R_i^2 R_j}{R^3}\right] \wedge (-r^2_{Ai} r_{Sj} - 2r_{Ai} r_{Aj} r_{Si})\right\} + \cdots$$

The first set of curly brackets contains the same dipole–dipole interaction that gives rise to the van der Waal forces and the second set of curly brackets corresponds to the dipole–quadrupole interaction. Moreover $r_S = \Sigma_M r_{Sm}$ and $r_A = \Sigma_n r_{An}$ refer to all the electrons in S and A measured from their respective nucleus and e is the electronic charge. It should be noted that in the expansion of $\mathscr{H}_1(R)$ the higher orders can be omitted. The rates for the dipole–dipole, dipole–quadrupole, and exchange interactions are derived next.

Dipole–Dipole Interaction

The probability κ_{SA} can now be evaluated by using the expression for dipole–dipole interaction

$$K_{SA}^{dd} = \frac{2\pi}{\hbar}\sum_I \sum_F (g_{*S} g_A)^{-1}\left(\frac{e^4}{K^2 R^6}\right)\int dE \int P_A(W_A)\, dW_A$$
$$\times \int P_{*S}(W_{*S})\, dW_{*S} |\langle r_S\rangle\langle r_A\rangle - 3(\langle r_S\rangle R)(\langle r_A\rangle R)|^2$$

In this expression the dependence in R^{-6} is the same as that found in the van der Waal forces. The average of the square of the matrix element over all the possible orientations of R must now be introduced in κ_{SA}^{dd}. This average

$$\langle|\langle r_S\rangle\langle r_4\rangle - 3R^2\langle r_S\rangle\langle r_A\rangle|^2\rangle_{av} = \tfrac{2}{3}|\langle r_S\rangle|^2|\langle r_A\rangle|^2$$

gives a transfer probability

$$\kappa_{SA}^{dd} = \frac{4\pi e^4}{3\hbar K^2 R^6 g_{*A} g_A}\sum_I \sum_F \int dE\left\{\int P_{*S}(W_{*S})\, dW_{*S}|\langle r_S(W_{*S}, W_{*S} - E)\rangle|^2\right\}$$
$$\times\left\{\int P_A(W_A)\, dW_A|\langle r_A(W_A, W_A - E)\rangle|^2\right\} \quad (67)$$

The expressions in curly brackets in eq. 67 can be expressed in terms of the probability for a spontaneous radiative transition. It is necessary, however, to adapt the expression deduced in Section 3-6 for electronic transitions to the conditions encountered in our system. Therefore, we must introduce the probability function $P^*(W^*)\,dW^*$ that accounts for the lack of definition that vibrations of the surroundings bring into the energies of the initial and final states. In addition, the transition probability was obtained for a species in a vacuum. Since this probability depends on the square of the electric field, the corrected expression must contain a factor $(\epsilon_c/\epsilon)^2$, which gives the ratio of the electric field in the condensed medium ϵ_c to that in the vacuum ϵ. These corrections lead to

$$A(E) = \sum_I \sum_F \frac{4e^2E^3}{3\hbar^4c^3g_{*S}} \left(\frac{\epsilon_c}{\epsilon}\right)^2 \rho$$

$$\times \int |\langle r_{if}(W_{*S}, W_{*S} - E)\rangle|^2 \rho(W_{*S})\, dW_{*S} \tag{68}$$

where the sum is over all the transitions and ρ is the density of states in momentum space, that is, $\rho = \hbar\, dk$ where k is the photon propagation vector. Since the photon propagation vector is proportional to the refraction index that is, $k \propto n$, the density ρ in eq. 68 can be replaced by n^2.

The shape of the function $A(E)$ is given by the emission spectrum and is related to the decay constant of the level $1/\tau$ through the integral

$$1/\tau = \int A(E)\, dE$$

Let the normalized function $f(E)$ represent the observed shape of the emission band, for example, $\int fE(E)\, dE = 1$. Hence, the sum in eq. 68 can be expressed

$$\sum_I \sum_F \int |\langle r_{if}(W_{*S}, W_{*S} - E)\rangle|^2 \rho(S_{*S})\, dW_{*S} = \frac{3\hbar^4c^3g_{*S}}{4n^3e^2E^3} \left(\frac{\epsilon}{\epsilon_c}\right) \frac{1}{\tau}\, f(E) \tag{69}$$

An additional relationship can be obtained by using Einstein's coefficient for induced transitions

$$B = \frac{2\pi e^2}{3\hbar^2} |\langle r_{if}\rangle|^2$$

The expression for $B(E)$ must be modified to make it valid in the case of electronic transitions that give rise to broad absorption bands. In this context, the wave functions must be normalized and the transition probability can be expressed per unit energy range

$$B(E) = \sum_I \sum_F \frac{2\pi e^2}{3\hbar^2 g} \left(\frac{\epsilon_c}{\epsilon}\right)^2 \int p(W)\, dW |\langle r_{if}(W, W+E)\rangle|^2$$

It is convenient to introduce here the absorption cross section

$$\sigma(E) = \sum_I \sum_F \frac{4\pi^2 e^2 nE}{3\hbar cg} \left(\frac{\epsilon_c}{\epsilon}\right)^2 \int P(W)\, dW |\langle r_{if}(W, W+E)\rangle|^2 \quad (70)$$

and a normalized function $1 = \int F(E)\, dE$, which can be related to the cross section (eq. 70) by means of the integral

$$\sigma(E) = F(E) \int \sigma(E)\, dE$$

where $Q = \int \sigma(E)\, dE$ can be measured as the area under the absorption band. The summation in eq. 70 can now be recast as

$$\sum_I \sum_F \int p(W)\, dW |\langle r_{if}(W, W+E)\rangle|^2 = \frac{3\hbar^2 q}{2\pi e^2} \left(\frac{\epsilon}{\epsilon_c}\right)^2 QF(E) \quad (71)$$

Therefore, we can express the strength of the line in terms of the area under the absorption band

$$\sum_I \sum_F \int P_{*\mathrm{S}}(W_{*\mathrm{S}})\, dW_{*\mathrm{S}} |\langle r_{if}(W_{*\mathrm{S}}, W_{*\mathrm{S}} - E)\rangle|^2 = \frac{3g_{\mathrm{S}}\hbar c}{4\pi^2 e^2 nE} \left(\frac{\epsilon}{\epsilon_c}\right)^2 Q_{\mathrm{S}} f_{\mathrm{S}}(E) \quad (72)$$

Combining eq. 67 with eqs. 69, 71, and 72 gives

$$\kappa_{\mathrm{SA}}^{\mathrm{dd}} = \frac{3\hbar^4 c^4 Q_{\mathrm{A}}}{4\pi R^6 n^4 T_{\mathrm{S}}} \left(\frac{\epsilon}{K^{1/4}\epsilon_c}\right)^4 \int \frac{f_{\mathrm{S}}(E) F_{\mathrm{A}}(E)}{E^4}\, dE \quad (73)$$

where we have used the absorption data for A and emission data for S, and

$$\kappa_{\mathrm{SA}}^{\mathrm{dd}} = \frac{3\hbar c^2 Q_{\mathrm{S}} Q_{\mathrm{A}}}{4\pi^3 n^2 R^6} \left(\frac{g_{\mathrm{S}}}{g_{*\mathrm{S}}}\right) \left(\frac{\epsilon}{K^{1/2}\epsilon_c}\right)^4 \int \frac{f_{\mathrm{S}}(E) f_{\mathrm{A}}(E)}{E^2}\, dE \quad (74)$$

where we have also used absorption data for S. Notice that eq. 74 is similar to one derived by Forster with a more limited treatment.[141,142]

Kroger has pointed that in this mechanism, the energy transfer can be visualized on a configurational coordinate diagram (Fig. 40).[146] We conclude that energy transfer can occur only at the point where the curves cross, and that the point of crossing corresponds to specific vibrational states for both centers. Also, the energy transfer process has an activation energy W_{S} that gives an exponential dependence of the transition probability on the reciprocal of the temperature

$$\kappa \sim \exp\left(-\frac{W_S}{kT}\right)$$

Such an activation energy is the difference between the crossing point and the minimum of the curve representing the pair (S*, A).

Dipole–Quadrupole Interaction

The expression in the second set of brackets must be used in the determination of the transfer probability corresponding to allowed and forbidden transitions in *S and A, respectively. To evaluate the matrix element $|\langle H_1 \rangle|^2$ we must average over all possible orientations of R and omit exchange terms

$$|H_1|^2_{av} = \frac{9e^4}{4K^2R^8} |\langle r_S \rangle|^2 \left\{ \frac{71}{36} [|\langle X_A^2 \rangle|^2 + |\langle Y_A^2 \rangle|^2 + |\langle Z_A^2 \rangle|^2] \right.$$
$$+ \frac{4}{3} [|\langle X_A Y_A \rangle|^2 + |\langle X_A Z_A \rangle|^2 + |\langle Y_A Z_A \rangle|^2]$$
$$\left. - \frac{8}{21} [|\langle X_A^2 \rangle \langle Y_A^2 \rangle| + |\langle y^2 A \rangle \langle Z_A^2 \rangle| + |\langle X_A^2 \rangle \langle Z_A^2 \rangle|] \right\}$$

The usual expression for the quadrupole transitions probability is

$$|\langle N \rangle|^2 = |\langle X^2 \rangle|^2 + |\langle Y^2 \rangle|^2 + |\langle Z^2 \rangle|^2 + 2|\langle XY \rangle|^2 + 2|\langle YZ \rangle|^2 + 2|\langle XZ \rangle|^2$$

Therefore, the sum in $|\langle H_1 \rangle|^2_{av}$ can be recast in the form

$$|\langle H_1 \rangle|^2_{av} = \frac{9e^4 a}{4R^8K^2} |\langle r_S \rangle|^2 |\langle r_A \rangle|^2 \qquad (75)$$

where $a \simeq 1.266$. The combination of eqs. 66 and 75 leads to

$$\kappa_{SA}^{dq} = \frac{9e^4 \pi a}{2\hbar K^2 R^8 g_{*S} g_A} \sum_I \sum_F \int dE$$
$$\times \left\{ \int dW_{*S}\, p^* S(W_{*S}) |\langle r_S(W_{*S}, W_{*S} - E) \rangle|^2 \right\}$$
$$\times \left\{ \int dW_A\, P_A(W_A) |\langle N_A(W_A, W_A + E) \rangle|^2 \right\} \qquad (76)$$

The expressions in brackets can be related to emission or absorption data. Indeed, the expression associated with the sensitizer can be transformed into more tractable forms by using eq. 69 or 72. There is little hope, however, in using absorption data for quencher A since the absorption coefficients of a quadrupole transition are usually very small. It is necessary, in such a case, to use emission data, that is,

$$\sum_{I}\sum_{F}\int P_{A}(W_{A})\,dW_{A}|\langle N(W_{A}, W_{A}+E)\rangle|^{2} = \frac{10\hbar^{6}c^{5}g_{*A}}{E^{5}e^{2}T_{A}n^{3}}\left(\frac{\epsilon}{\epsilon_{c}}\right)^{2}F_{A}(E) \tag{77}$$

where the effects of medium and broadening have been included. Therefore, with eqs. 69 and 75–77, the transition probability can be expressed

$$\kappa_{SA}^{dq} = \frac{135\pi\hbar^{9}ac^{8}}{4n^{6}R^{8}\tau_{S}\tau_{A}}\left(\frac{g_{*A}g_{S}}{g_{A}g_{*S}}\right)\left(\frac{\epsilon}{K^{1/2}\epsilon_{c}}\right)\int\frac{f_{S}(E)F_{A}(E)}{E^{8}}\,dE$$

We can use eq. 72 when τ_S is not available

$$\kappa_{SA}^{dq} = \frac{135a\hbar^{6}c^{6}}{4\pi n^{4}R^{8}}\frac{Q_{S}}{\tau_{A}}\left(\frac{g_{*A}g_{S}}{g_{A}g_{*S}}\right)\left(\frac{\epsilon}{K^{1/2}\epsilon_{c}}\right)\int\frac{f_{S}(E)f_{A}(e)}{E^{6}}\,dE$$

It should be noted that the dipole–quadrupole transition probability shows a $1/R^8$ dependence in the donor–acceptor distance. It is also expressed as

$$\frac{\kappa_{SA}^{dq}}{\kappa_{SA}^{dd}} = \left(\frac{45a}{4\pi^{2}}\right)\left(\frac{n\lambda}{R}\right)^{2}\frac{\tau_{A}^{d}}{\tau_{A}^{q}}$$

and the quadrupole transition probability is of the order $(a/\lambda)^2$ times the probability of a dipolar radiative transition. Hence, the ratio of the dipole–quadrupole to the dipole–dipole energy transfer probability is

$$\kappa_{SA}^{dq}\kappa_{SA}^{dd} = \frac{a}{R^{2}}$$

It is possible to demonstrate that for close neighbors the difference in probabilitics amounts to an order of magnitude.

Exchange Mechanism

to introduce the exchange effects, the wave function can be explicitly written

$$\Psi(r, \sigma) = \Psi(r)X(\sigma)$$

where $X(\sigma)$ are the spin wave functions. Therefore

$$\begin{aligned}\langle H_{1}\rangle = \int {}^{*}\Psi_{*S}(r_{1})^{*}\Psi_{A}(r_{2})\hat{\mathscr{H}}_{1}\Psi_{S}(r_{1})\Psi_{*A}(r_{2})^{*}X_{*S}(\sigma_{S})^{*}X_{A}(\sigma_{A})X_{S}(\sigma_{S})\\ \times X_{*A}(\sigma_{A}) - \int {}^{*}\Psi_{*S}(r_{1})^{*}\Psi_{*A}(r_{2})\hat{\mathscr{H}}_{1}\Psi_{S}(r_{2})\Psi_{A}(r_{1})\\ \times {}^{*}X_{*S}(\sigma_{1})^{*}X_{A}(\sigma_{2})X_{S}(\sigma_{2})X_{*A}(\sigma_{1})\end{aligned} \tag{78}$$

The first term in $\langle H_1 \rangle$ corresponds to the coulombic interactions that were discussed above. The second integral describes the exchange interaction with $\hat{\mathscr{H}}_1 = e^2/Kr_{12}$. Further inspection of the integral shows that it corresponds to an electrostatic interaction between the charge clouds: $Q^1(r^1) = \Psi_{*S}(r_1)\Psi_A(r_1)$ and $Q(r_2) = \Psi_{*A}(r_2)\Psi_S(r_2)$.

Since $\hat{\mathscr{H}}_1$ does not operate on the spin functions, the conditions

$$X_{*S} = X_S \qquad \text{and} \qquad X_{*A} = X_A$$

determine that the spin must be preserved through the energy transfer process. The transfer probability can now be written

$$\kappa_{SA}^{E} = \frac{2\pi}{\hbar}\, \xi_{CR}^2 \int f_S(E)F_S(E)\, dE$$

where ξ_{CR}^2 is a quantity that cannot be directly related to optical experiments, namely

$$\xi_{CR}^2 = \sum_I \sum_F \frac{e^4}{g_{*S}g_A K^2} \left| \int Q^1(r_1)\left(\frac{1}{r_{12}}\right) Q(r_2)\, dr_{12} \right|_{IF}^2$$

The functions Q and p are normalized over space and not energy. Moreover, for a small separation between S and A the functions Q must be sizable in the same region of space for the factor ξ_{CR}^2 to have the required weight. This factor, which contains the dependencies on concentration and distance, varies according to

$$b\, \frac{e^4}{K^2 R_0^2} \exp\left(-\frac{2R}{L}\right)$$

where L is an effective Bohr radius for S and A, and b is a dimensionless quantity smaller than unity. For coordination complexes, it has been demonstrated that the description of the exchange mechanism is formally similar to the description used for outer-sphere electron transfer reactions.[144,145] In this context, the reaction standard free energy (chemical potential) and reorganizational energies determine the reaction rate.

The mechanisms discussed above have been tested with organic donors and acceptors possessing dipole–multipole electronic transitions. With coordination complexes, the transitions to charge transfer states are dipole allowed, and can be used in the transfer of energy through dipole–multipole interactions. The metal-centered or d–d transitions in coordination complexes are Laporte forbidden and fell out of the classes considered above for energy transfer. Thus, it is necessary to return to eq. 66 and express the matrix element $\langle H_1 \rangle$ in terms of these vibronic allowed metal-centered transitions. Endicott et al. have reformulated the problem in terms of a radiationless electronic transition within a collision complex.[147] In this treatment, the expressions previously obtained for the rate constant of radiationless transitions (eqs. 57 and 58) can be reduced to

$$k_{\mathrm{q}} = \frac{2\pi}{\hbar} K_0 |\langle V \rangle|^2 N\rho$$

where $|\langle V \rangle|$ is the electronic coupling element, N is the Franck–Condon factor, and K_0 is the equilibrium constant for the formation of the collision complex. Despite the similarities between radiationless relaxation and the energy transfer reactions, it is not possible to discuss the energy transfer mechanism in the strong and weak coupling limits of radiationless transitions and we must consider intermediate cases. Endicott has described four cases:

(*a*) *Strong Coupling*. this case characterized by very different bond distances or bond angles for the products and reactants (Fig. 41*a*). The condition gives an activation energy

$$E_{\mathrm{a}} = \frac{\lambda}{4}\left(1 + \frac{\Delta E}{\lambda}\right)^2, \qquad \lambda = \frac{1}{2}\,\bar{k}(\Delta Q_0)^2$$

where $\bar{k}$ is the mean value of the force constant associated with the vibration mode ΔQ_0.

(*b*) *Nested Surfaces* (*Weak Coupling*). The energy transfer process takes place with a very small change in nuclear coordinates (Fig. 41*b*). In this context, the transition between surfaces depends on electron and nuclear tunneling. The rate, therefore, is temperature independent and decreases with increasing energy gap ΔE This limit seems to apply to the relaxation rate of the ^{2}Eg of Cr(III) complexes at low temperature.

(*c*) *Large Displacement and Large Energy Gap*. This case has been discussed in relationship with electron transfer reactions. The strong coupling limit predicts a decrease of the electron transfer rate constant with increasing ΔE for very exothermic reactions. For a large displacement of the potential surfaces of the reactants and products one can expect crossing of the surfaces, a condition that gives place to activation energy as shown in Figure 41*c*.

(*d*) *Small Displacement and Small Energy Gap*. Very small reactant–product energy gaps are frequently found in energy transfer reactions. For example, energy transfer from (^{2}E)Cr(III) donors to ($^4A_{2g}$)Cr(III) acceptors are expected to have $\lambda \sim 0$ and ΔE equal to or approaching zero. Another type of energy transfer mechanism operates when a charge transfer state of polypyridylruthenium(II) complexes functions as a donor and ($^4A_{2g}$)Cr(III) complexes function as acceptors.[148] Although these processes meet the requirements for dipole–multipole coupling, no evidence has been

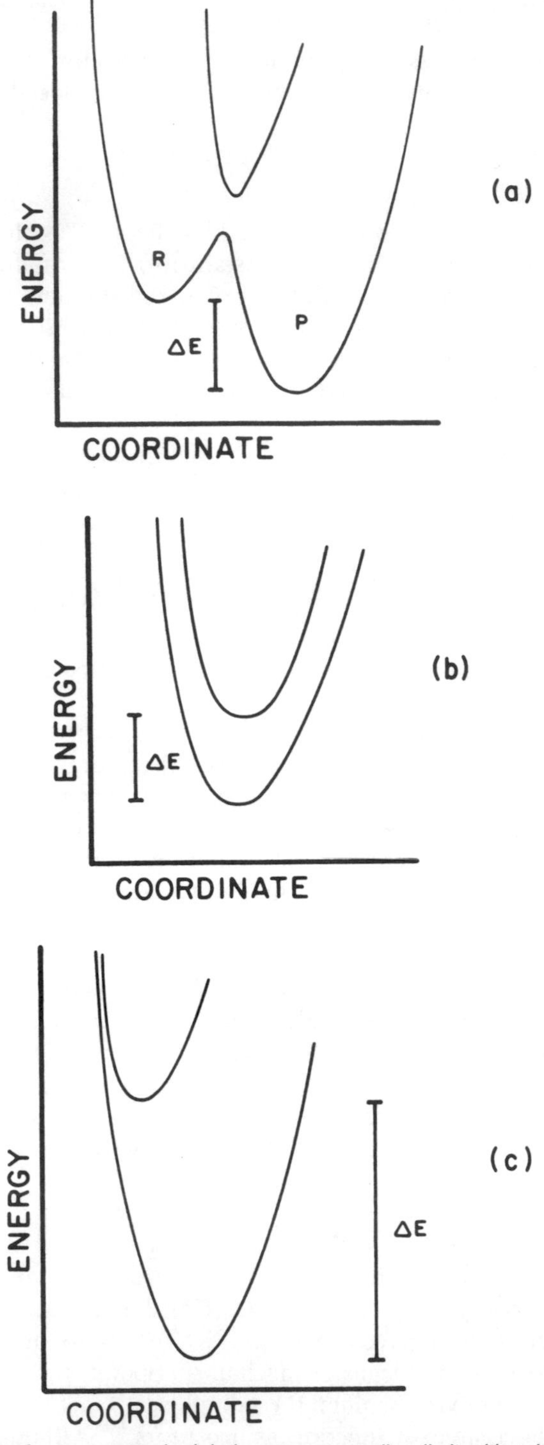

Figure 41. Energy transfer processes in (*a*) the strong coupling limit with a large reorganization energy, (*b*) the weak coupling limit (nested surfaces) with small nuclear reorganization, and (*c*) large nuclear displacements and energy gaps.

found for such a mechanism. All the rates are smaller than the expected diffusion limit and exhibit a shallow dependence on the donor–acceptor energy gap for Cr(III) complexes where only doublet states are accessible. However, the nuclear reorganizational energies in these systems are small and the energy transfer processes must correspond to transitions between approximately nested reactant and product potential energy surfaces. The dependence of the quenching rate constant on ΔE corresponds to class (a) when there is not significant spectral overlap between the emission spectrum of the Ru(II) donor and the absorption spectrum of the Cr(III) acceptor. However, one can expect that in these reactions ^2E and $^2T_1$ work as acceptors in Cr(III) complexes. In this context, the combined contributions of the two states to the observed rate, and the various high-frequency vibrational modes, that is C–H stretching, cause a shallow dependence of the quenching rate constant on ΔE.

For the quenching of excited Cr(III) polypyridyl complexes by various Co(III) compounds,[149] the quenching rates can be much slower, by three orders of magnitude, than those expected for diffusion-controlled processes and independent of the energy gap. It is possible to ascribe such properties to a rate controlled by electronic rather than Franck–Condon factors. The distance dependence of the rates is compatible with expectations for an exchange mechanism while correlation with the energy of CT transitions suggests the possibility of donor and acceptor wave functions polarized by CT-induced dipolc moments.

5

PHOTOREDOX REACTIONS

5-1 MULLIKEN'S CONCEPT OF CHARGE TRANSFER COMPLEXES

Iodine forms solutions in organic solvents that can be brown to violet in color. The variation of the spectra, Table 5, is associated with the formation of molecular complexes between I_2 and aromatic molecules. Indeed, the 300.0 nm absorption in aromatic solvents is not found in inert solvents or the aromatic molecules; it can be assigned to an electronic transition of the complex. Moreover, the visible band at ~ 500 nm can be assigned as an electronic transition in the aromatic part of the complex which become allowed as a result of the complexation.

Mulliken[150,151] has considered two limiting cases corresponding to (a) weak complexes, where the main stabilizing interaction can be various types of polarizations, and (b) strong complexes.

Weak Complexes

In this treatment of the molecular complex, the donor A (e.g., I_2) and the acceptor B (e.g., an aromatic molecule) are in totally symmetrical singlet ground states. The wave function of the ground state for the molecular compound AB is

$$\Psi_{GS} = a\Psi_1 + b\Psi_2$$

where Ψ_1 is a "no bond" wave function that can be written

$$\Psi_1 = \hat{a}(\Psi_A \Psi_B) + \cdot \cdot \cdot$$

TABLE 5 Charge Transfer Transitions in I_2–Hydrocarbon Complexes[a]

Hydrocarbon	$\lambda_{max}(\epsilon)^b$	$E_{HOMO}{}^c$
Cyclohexane	527 (1.07×10^3)	
Benzene	500 (1.04×10^3)	
	297	1.0
Naphthalene	500 (1.0×10^3)	
	360	0.618
Anthracene	~500	
	430	0.414
Pyrene	~500	
	420	0.445

[a]Values from R. Bhattacharyya, and S. Basu, *Trans. Faraday Soc.* **1958,** *54,* 1286.
[b]Wavelength of the maximum (nm) and the corresponding extinction coefficient ϵ (M^{-1} cm^{-1}).
[c]Energy of the highest occupied molecular orbital.

The operator $\hat{a}$ denotes that the product $\Psi_A \Psi_B$ of the wave functions of A and B must be antisymmetric in all the electrons. Moreover, Ψ_2 is a "dative wave function" corresponding to the transfer of one electron

$$\Psi_2 = \Psi(A^- - B^+) + \text{other terms}$$

The energy of the ground state E_{GS} can be estimated by using second-order perturbation theory

$$E_{GS} = W_{00} - \frac{(W_{01} - SW_{00})^2}{(W_{11} - W_{00})} \tag{1}$$

with $H_{ij} = \langle \Psi_i | \mathscr{H} | \Psi_j \rangle$, $\hat{H}$ being the exact Hamiltonian for the entire set of electrons and nuclei, and $S = \langle \Psi_1 | \Psi_2 \rangle$. Also, second-order perturbation theory gives a ratio of the a and b coefficients for Ψ_{GS}

$$\frac{b}{a} = -\frac{(W_{01} - SW_{00})}{(W_{11} - W_{00})}$$

which can be combined with the normalization condition

$$1 = \langle \Psi_{GS} | \Psi_{GS} \rangle = a^2 + b^2 + 2abS$$

It is possible to write an excited state wave function

$$\Psi^* = a^* \Psi_2 - b^* \Psi_1$$

where $a^* \sim a$ and $b^* \sim b$. The energy of the excited state is

$$E^* = W_{11} + \frac{(W_{01} - SW_{11})^2}{(W_{11} - W_{00})} \tag{2}$$

and the ratio of the coefficients

$$\frac{b^*}{a^*} = \frac{(W_{01} - SW_{11})}{(W_{11} - W_{00})}$$

Therefore, for the weak complex, a^2 is larger than b^2 and the electronic transition between these states causes a shift of the electronic density from nearly around B to around A (Fig. 42).

Strong Complexes

In this case eqs. 1 and 2 are no longer more than qualitatively correct. A better description of the adduct's electronic structure can then be obtained with a molecular orbital treatment. Consider now that the molecular orbital between the donor A and acceptor B can be expressed as a linear combination of atomic orbitals,

$$\Psi = C_a \Psi_A + C_b \Psi_B$$

This secular equation is

$$\begin{bmatrix} H_{AA} - E & H_{AB} - SE \\ H_{AB} - SE & H_{BB} - E \end{bmatrix} = 0$$

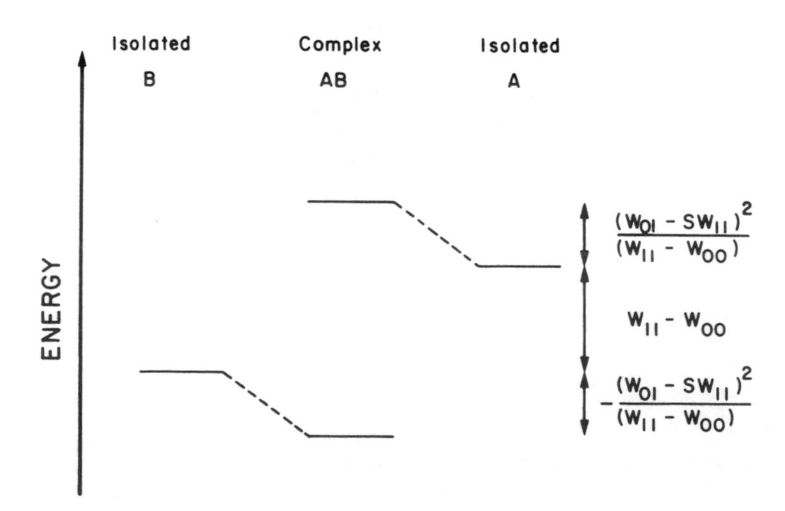

Figure 42. Electronic levels for a charge transfer complex (weak overlap limit).

If the nuclear distance is sufficiently large the overlap integral S may be taken equal to zero. Furthermore, the eigenvalues for the particular case $A_{AA} > H_{BB}$ are

$$E = \tfrac{1}{2}\{(H_{AA} + H_{BB}) \pm \sqrt{(H_{AA} - H_{BB})^2 + 4H_{AB}^2}\} \qquad (3)$$

and for $H_{AA} = H_{BB}$

$$E = H_{AA} \mp H_{AB} \qquad (4)$$

The solution $H_{AA} - H_{AB}$ gives $C_A = -C_B$ and $H_{AA} + H_{AB}$ gives $^*C_A = {}^*C_B$. It must be noticed that the electronic transition does not give rise to a radial distribution of the electronic density, that is, from B (in the ground state) to A (in the excited state). Therefore, such an electronic transition cannot be properly described as a charge transfer transition. For $H_{AA} > H_{BB}$, the positive sign in eq. 3 leads to $^*C_a^2 > {}^*C_b^2$ and the negative sign to $C_a^2 < C_b^2$. The proper linear combinations are

$$\Psi_{GS} = C_A \Psi_A + C_B \Psi_B$$

$$\Psi^* = {}^*C_A \Psi_A - {}^*C_B \Psi_B$$

and the electronic transition causes again a drift of the electronic density from B to A.

Although Mulliken's treatment of the charge transfer complexes considers only complexes between molecules containing nonmetallic elements, the existence of charge transfer transitions in coordination complexes is considered a possibility. A model for charge transfer transitions can therefore be developed by using an approximate molecular orbital treatment (see Appendix V) in the same form as that used above in the strong complex limit. Figure 43 shows several configurations corresponding to the ground state (*a*) and ligand to metal charge transfer states (*b* and *c*) of a d^3 metal ion complex with an octahedral symmetry. The energy for the ground state $\epsilon(^2T_{2g})$ is

$$\epsilon(^2T_{2g}) = 2\epsilon(a_{1u}) + \epsilon(t_{2g}) + J(a_{1u}, a_{1u}) + 2J(a_{1u}, t_{2g}) \qquad (5)$$

where $\epsilon(i)$ is the energy of the ith orbital, and $J(i, j)$ is the coulombic integral that expresses the interelectronic repulsion energy of a pair of electrons, that is, the $1/r_{ij}$ terms of the Hamiltonian. The charge transfer excited states in Figure 43 are generated by promoting one electron from a molecular orbital a_{1u} (largely localized around the ligand) to orbitals t_{2g} or e_g (largely localized around the metal). The excitation to t_{2g} gives a configuration $(a_{1u})^1(t_{2g})^2$, and the direct product of the irreducible repre-

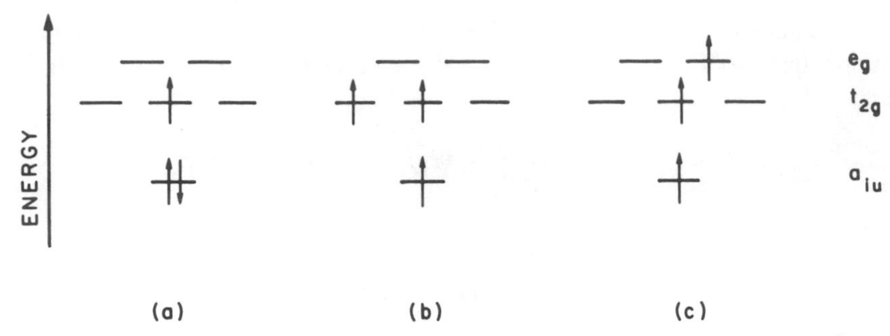

Figure 43. Charge transfer transitions from a ground state configuration: (*a*) $(a_{1u})^2(t_{2g})$ to excited state configurations; (*b*) $(a_{1u})(t_{2g})^2$; and (*c*) $(a_{1u})(t_{2g})(eg)$.

sentations, $a_{1u} \times t_{2g} \times t_{2g}$, leads to the states associated with such a configuration, namely the states A_{1u}, E_u, and T_{2u} which are symmetric in the permutation of two given electrons, and the state T_{1u} which is antisymmetric. Therefore, the electronic states $^2A_{1u}$, ^2E_u, $^2T_{1u}$, $^4T_{1u}$, and $^2T_{2u}$ are expected to have different energies and give rise to a number of charge transfer bands. An additional number of charge transfer bands can be associated with the population of the states generated with the configuration $(a_{1u})^1(t_{2g})^1(e_g)^1$. For example, the energy of the $^4T_{1u}$ state is

$$\epsilon(^4T_{1u}) = \epsilon(a_{1u}) + 2\epsilon(t_{2g}) + J(t_{2g}, t_{2g}) + 2J(a_{1u}, t_{2g}) - K(t_{2g}, t_{2g}) - 2K(a_{1u}, t_{2g}) \tag{6}$$

where $K(i, j)$ is the exchange integral. The promotion of the electron that generates the $^4T_{1u}$ state, a spin forbidden transition, is expected to take place with an energy

$$\epsilon(^4T_{1u}) - \epsilon(^2T_{2g}) = \epsilon(t_{2g}) - \epsilon(a_{1u}) + J(t_{2g}, t_{2g}) - J(a_{1u}, a_{1u}) - K(t_{2g}, t_{2g}) - 2K(a_{1u}, t_{2g}) \tag{7}$$

In practice, only one charge transfer band of a given type is observed in complexes with light ligands; a fact that has been attributed to the similar energies of the charge transfer transitions corresponding to a given electronic configuration. This suggests that multicenter electrostatic integrals, especially the exchange integral $K(i, j)$ have small values due to a poor (weak) overlap between orbitals. In this context, it is possible to introduce simplifications for the calculations of the energies of charge transfer transitions.

5-2 JØRGENSEN'S SCALE OF OPTICAL ELECTRONEGATIVITIES

Insofar as the energy of the charge transfer transition ${}^4T_{1u} \leftarrow {}^2T_{2g}$ (eq. 7) receives a large contribution from the difference $\epsilon(t_{2g}) - \epsilon(a_{1u})$ and minor contributions from the Coulomb and exchange integrals, it is possible to assume that the difference in orbital energy is proportional to the electronegativity difference between the donor and acceptor orbitals. Therefore the energy for the maximum of the charge transfer transition can also be related to the electronegativities. The constants for the orbitals of the metal and the ligand that when combined give the energy of the charge transfer transition are called optical electronegativities (Table 6).[152–154] For the optical electronegativity of a metal ion to have a general validity in complexes of various symmetries, the values are obtained for the spherically perturbed shell of the metal ion. For example, in octahedral symmetry the t_{2g} orbitals of a first-row transition metal ion remain close in energy to the spherically perturbed shell if the π-bonding between ligands and metal is negligible (Fig. 44 and Appendix V). Therefore, the difference between the energies of the t_{2g} and a_{1u} orbitals is

$$\epsilon(t_{2g}) - \epsilon(a_{1u}) \simeq A(\chi_L - \chi_M) \tag{8}$$

where χ_L and χ_M are the corresponding optical electronegativities of the ligand L and the metal M. The constant $A = 35.8$ kJ relates the scale of the optical electronegativities to Pauling's scale of electronegativities; that is, the optical electronegativity of F^- is arbitrarily chosen to have the same value, $S_{F^-} \sim 3.9$, as the electronegativity scale. In addition, eq. 8 is applied to charge transfer transitions that involve ligand orbitals with σ or π character.[155] For ligands with σ and π orbitals available for the charge transfer transitions, distinct values of the optical electronegativity must be used for charge transfer transitions from each particular orbital. Inspection of the values in Table 6 reveal that the optical electronegativities for π orbitals are about 10% smaller than the optical electronegativities for σ orbitals.

An additional term, Δ, must be added to eq. 8 in charge transfer transitions to d orbitals that do not belong to the spherically perturbed shell. Moreover, the change of the interelectronic repulsion energy, δ(SPE), caused by the electronic transition must be introduced as another term (eq. 9) in the evaluation of the energy $\Delta\epsilon_{CT}$ of the charge transfer transition

$$\Delta\epsilon_{CT} = 35.8(\chi_L - \chi_M) + \Delta + \delta(\text{SPE}) + aE + b\xi \tag{9}$$

For example, in the octahedral complex that we have so far considered (Fig.

TABLE 6 Optical Electronegativity Data[a]

Species	χ	Orbital or Symmetry Assignment[b]
Ti(IV)	2.05	
	1.08	Tetrahedral
V(III)	1.9	
	2.1	Tetrahedral
V(IV)	2.6	
Cr(III)	1.8–1.9	
Cr(IV)	2.65	
Mn(III)	2.0	
Mn(IV)	2.7–3.0	
Fe(II)	1.8	Tetrahedral
Fe(III)	2.1	Low spin
	2.5	High spin
Co(II)	1.8–1.9	Tetrahedral
Co(III)	2.3	Low spin
Ni(II)	2.0–2.1	Tetrahedral
	2.2	Low spin–square planar
Ni(III)	3.05	
Ni(IV)	3.4	
Cu(I)	~1.6	For $L \rightarrow (s, p)_{Cu}$ Transitions in planar or octahedral complexes
	~1.8	For $d_{Cu} \rightarrow L$ transitions in planar or octahedral complexes
Cu(II)	2.0–2.2	Planar or distorted octahedral
	2.2–2.5	Tetrahedral
Cu(III)	2.74–2.8	Planar or distorted octahedral
	3.5	
Zn(II)	1.2	Tetrahedral
Hg(II)	1.5	
Bi(III)	1.7	
Nb(V)	1.85	
Ta(V)	1.7	
Zr(IV)	1.6	
Mo(III)	1.7	
Mo(IV)	1.95	
Mo(V)	2.0	
Mo(VI)	2.1	
W(V)	1.95	
W(VI)	2.0	
Re(IV)	2.1	
Re(VI)	2.0–2.1	
Tc(IV)	2.25	
Ru(III)	2.0–2.1	Low spin
Ru(IV)	2.45	
Rh(III)	2.3–2.4	Low spin
Rh(IV)	2.65	
IR(III)	2.25	
Ir(IV)	2.3–2.4	Low spin
Os(III)	1.95	
Os(IV)	2.2	

TABLE 6 (*Continued*)

Species	χ	Orbital or Symmetry Assignment[b]
Os(VI)	2.6	
Pt(II)	2.3	Low spin–square planar
Pt(IV)	2.6–2.7	Low spin
Pt(V)	3.0	
Pt(VI)	3.2	Low spin
Pd(II)	2.3–2.4	Low spin–square planar
Au(III)	2.9	Low spin–square planar
$U(VI)O_2$	1.8	
U(VI)	2.4	
Np(VI)	2.6	
Pu(VI)	2.85	
F^-	3.9	π orbital
	4.4	σ orbital
Cl^-	2.89	π orbital
	3.12	σ orbital
Br^-	2.76	π orbital
	3.99	σ orbital
I^-	2.60	π orbital
	2.87	σ orbital
CN^-	2.8	π orbital
SCN^-	2.6	π orbital
NCS^-	2.9	π orbital
$SeCN^-$	2.85	π orbital
$NCSe^-$	2.8	π orbital
NCO^-	3.0	π orbital
N_3^-	2.8	π orbital
NH_3	3.28	σ orbital
RNH_2	3.20	σ orbital
H_2O	3.5	π orbital
ROH	3.1	π orbital
S^{2-}	2.5	π orbital
R_2S	2.9	π orbital
R_3P	2.6	σ orbital
R_3A_5	2.5	σ orbital
SO_4^{2-}	3.2	π orbital
RCO_2^- (e.g., acetate, malonate, oxalate EDTA)	3.1	Average
N (e.g., amine ligands as $Me_6[14]aneN_4$, ethylene diamines)	2.98	σ orbital
C═N (isolated imino groups, e.g., $Me_6[14]$-4,11-dieneN_4)	3.02	π orbital
C═N (isolated imino groups, e.g., $Me_6[14]$-4,11-dieneN_4)	0.4–0.3	π^* orbital
PyNO (e.g., 4-Methylpyridine *N*-oxide, 2-Picoline *N*-oxide)	0.8	π^* orbital
4-Nitropyridine *N*-oxide	0.9	π^* orbital

[a]Sources: A. B. P. Lever, *Inorganic Electronic Spectroscopy*, 2nd, ed., Amsterdam: Elsevier, **1984**, pp. 221, 222; G. Ferraudi and S. Muralidharan, *Coord. Chem. Rev.* **1981,** *36,* 45, and references therein.
[b]Octahedral symmetry unless specially noted.

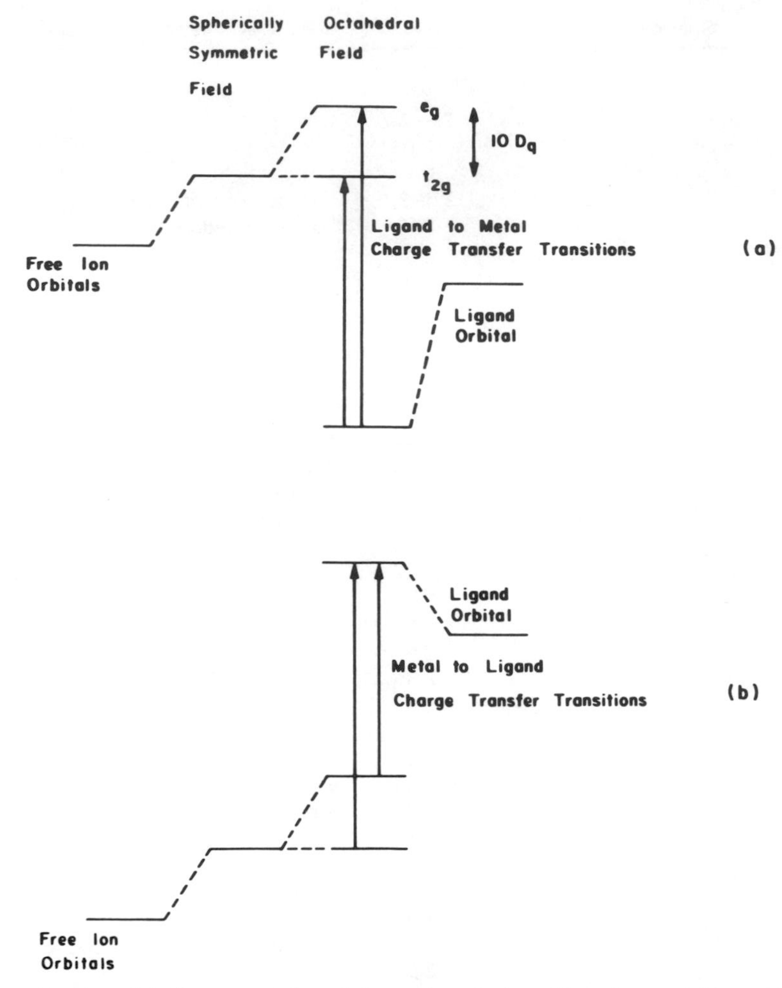

Figure 44. Ligand field representation of charge transfer transitions: free metal ion orbitals are sequentially perturbed by ligand fields of spherical and octahedral symmetry. The vertical arrows represent charge transfer transitions ((*a*) ligand to metal and (*b*) metal to ligand) between the perturbed levels of the metal ion and orbitals with a strong ligand character.

44) $\Delta = 0$ for the charge transfer transition to the t_{2g} orbitals and $\Delta = 10D_q$ for a similar transition to the e_g orbitals.

The spin pairing energy (SPE) for a given configuration can be represented by

$$(\text{SPE}) \simeq [(S(S+1))_{\text{av}} - S(S+1)]D_{\text{SP}} \tag{10}$$

where $(S(S+1))_{\text{av}}$ is the average value of $S(S+1)$ and S is the spin quantum number. For a configuration l^n, with orbital quantum number l and a number of electrons n, the average value of $S(S+1)$ is

$$(S(S+1))_{av} = \frac{n(n+2)}{4} - \frac{n(n-1)(l-1)}{4l+1} \tag{11}$$

The factor D_{SP} in eq. 10 is the spin pairing energy parameter which is approximately equal to $7B$, where B is the Racah parameter of crystal field theory.

The (SPE) is independent of the complex symmetry and the ligands. Indeed, the change in the interelectronic repulsion within the ligand orbitals for a given ligand can be assumed to remain constant. However, in case where a small change is caused by the electronic transition, such a variation can be assimilated in the value of the optical electronegativity. Values for the spin pairing energies of d^n are reported in Table 7.

Since the expression for the SPE (eq. 10) is valid only for the baricenter of a given multiplicity, the aE term in eq. 9 is important when the ground state and excited states have the same multiplicity. The constant E is the Racah parameter B for d electrons and B^3 for f electrons. Moreover, $b\xi$ in eq. 9 is important only in cases where the spin–orbit coupling is extremely strong, for example, in complexes of rare-earth ions.

It is possible to use the optical electronegativities in Table 6 and the SPE in Table 7 to predict the charge transfer transition maxima. For example, for $Co(NH_3)_3X^{2+}$ and $Cr(NH_3)_5X^{2+}$, $X = Cl^-$, Br^-, and I^-, we can expect charge transfer transitions from σ and π orbitals of the ligands to e_g orbitals of the Co(III) or e_g and t_{2g} orbitals of the Cr(III). For Co(III) complexes, the term Δ for the transition to t_{2g} orbitals has a value of about 28 kJ and a

TABLE 7 Spin Pairing Energy of d^n Configurations (in D_{SP} units)[a]

n	Energy (S)
1 or 9	0 (1/2)
2 or 8	4/3 (0)
	− 2/3 (1)
3 or 7	1 (1/2)
	− 2 (3/2)
4 or 6	2 (0)
	0 (1)
	− 4 (2)
5	4/3 (1/2)
	− 5/3 (3/2)
	−20/3 (5/2)

[a]The spin pairing energy parameter D_{SP} of eq. 10 depends on the orbital quantum number, $D_{SP} = (15/4)F_2$ for p electrons and $D_{SP} = (7/6)\,[(5/2)B + C] \simeq 7B$, where F_2 is one of the integrals defined by Condon and Shortley and B and C are Racah parameters (Appendix IV). A definition of D_{SP} is given by C. K. Jorgensen, *Orbitals in Atoms and Molecules*, London: Academic, **1962.**

SPE proportional to $B \sim 0.75$ kJ. These values can be introduced into eqs. 9–11 to obtain an expression for the position of the charge transfer maxima in the spectra of the cobalt(III) complexes

$$\Delta\epsilon_{CT} \sim (35.8\chi_X - 59.5) \tag{12}$$

Since the optical electronegativity of Cr(III) is smaller than the optical electronegativity of Co(III) ($\chi_{Cr(III)} \sim 1.3$ versus $\chi_{Co(III)} \sim 2.3$) and the value of the other parameters ($\Delta \sim 25.6$ kJ and $B \sim 0.2$ kJ) cannot compensate for the difference in optical electronegativities, the charge transfer transitions in Cr(III) complexes are placed at higher energies. The difference in energies is therefore expressed in eq. 13,

$$\Delta\epsilon_{CT} \approx (35.8\chi_X - 31.0) \tag{13}$$

for Cr(III) complexes.

Although the use of eq. 9 has been demonstrated on compounds that exhibit ligand to metal charge transfer transitions, it is also possible to use eq. 9 for metal to ligand charge transfer transitions. The existence of unoccupied orbitals of the ligand at energies larger than the energies of the metal (Fig. 44) is required for the observation of metal to ligand charge transfer transitions. Such a condition is fulfilled when the optical electronegativities obey the inequality

$$\chi_L < \chi_M - \Delta/35.8 \tag{14}$$

For example, with low-spin cobalt(III) complexes, where $\Delta \sim 30$ kJ and $\chi_{Co} \sim 2.3$, the condition for the observation of metal to ligand charge transfer transitions is $\chi_L < 1.5$.

5-3 CHARGE TRANSFER TRANSITIONS IN ION PAIRS

Mulliken's concept of charge transfer transitions is based in a weak overlap between orbitals of the donor and the acceptor. Such a condition is fulfilled when the donor and the acceptor are held together mainly by electrostatic interactions such as those present in ion pairs. According to Bjerrum's theory of ion association, the concentration of ions with a charge Z_2e around a central ion with a charge Z, e is given by a Boltzmann distribution

$$\frac{dn_2}{dv} = n_0 \exp\left(-\frac{Z_1Z_2e^2}{\mathcal{K}rkT}\right) \tag{15}$$

where r is the distance to the central ion, $\mathcal{K}$ is the dielectric constant of the medium, and n_0 is the bulk concentration of the ions surrounding the central ion. Equation 15 can be used to evaluate the number of ions in shells of thickness dr at a distance r from the center. The concentration of counterions in these shells decreases with r and is at a minimum at a distance

$$q = \frac{Z_1 Z_2 e^2}{2\mathscr{K}kt}$$

To obtain the degree of association $(1-\alpha)$ we must find the fraction of counterions inside a shell that has for boundaries two spheres of radius $r = q$ and $r = a$, where a is the contact distance of the ions. This is accomplished by integrating eq. 15

$$(1-\alpha) = 4\pi n_0 \int_a^q r^2 \exp\left(-\frac{Z_1 Z_2 e^2}{\mathscr{K}rkT}\right) dr \tag{16}$$

Tabulated values of the integral or fast computers are usually used to evaluate the integral. For an association equilibrium

$$\begin{matrix} M^+ & + & X^- & \rightleftharpoons & M^+, X^- \\ \alpha & & \alpha & & (1-\alpha) \end{matrix}$$

the equilibrium constant K_a can be related to the degree of association by means of

$$K_a = \frac{[M^+, X^-]}{[M^+][X^-]} = \frac{(1-\alpha)N\nu_0}{\alpha^2 n_0 1000 \nu^2} \tag{17}$$

where ν_0 and ν are the activity coefficients of the ion pair and the free ions, respectively. An approximate expression for the equilibrium constant is

$$K_a \sim K_a^0 \exp\left(-\frac{Z_1 Z_2 e^2}{\mathscr{K}kTa}\right) \tag{18}$$

Equation 18 shows that the equilibrium constant depends on the solvent, the temperature, and the contact distance of the ions. For example, the constant for the equilibria[157–158]

$$Co(NH_3)_6^{3+} + X^- \rightleftharpoons Co(NH_3)_6^{3+}, X^- \tag{19}$$

has the values $K_a = 74 (X = Cl^-)$, $K_a = 46\ (X = Br^-)$, and $K_a = 17\ (X = I^-)$. A much larger value of the constant is obtained with divalent anions, for example, $K_a \sim 2.2 \times 10^3$ for the association of SO_4^{2-} with $Co(NH_3)_6^{3+}$.[159,160]

The association of ions in ion pairs can introduce modifications in the spectra of the isolated ions; such an association gives rise to electronic transitions that are characteristic of the ion pair.[157,161,162] In general, these transitions involve a radial distribution of the electronic density between the ions in the pair and can be described as charge transfer transitions. In eq. 19 the position of the charge transfer transition shifts to lower energies in direct relationship to the decrease of the reduction potential of the couple $X^{\cdot}/X^-$. Hence, if the oxidation of the anion demands too much energy the new transitions can be covered by the intense transitions in the spectra of the isolated ions, as in the case of the $Co(NH_3)_6^{3+}$, F^- pair.

The photochemistry exhibited by ion pairs is also in agreement with the assignment of the new absorption bands as charge transfer transitions of the ion pairs. Indeed, irradiations at wavelengths of these bands usually induce photoredox processes of the type[164]

$$Co(NH_3)_6^{3+}, I^- \underset{\leftsquigarrow}{\xrightarrow{h\nu}} CTTI \longrightarrow \{Co(NH_3)_6^{2+}, I^{\cdot}\} \quad (20)$$

where CTTI designates the charge transfer state and the bracketed species as a radical–ion pair. The energetics of such photoprocesses is discussed in the next section.

5-4 REACTIVITY OF CTTM AND CTTL EXCITED STATES

The transition to CTTM or CTTL excited states produces a radial redistribution of the electronic density. In this context, the CTTM excited state can be regarded as a radical coordinated to the reduced metal center in the limit of a net one-electron transference from the ligand to the metal. Moreover, the increased antibonding character in the excited state is believed to assist in the homolytic dissociation of the complex

$$M^{III}L_5X^{(3-n)} \underset{\leftsquigarrow}{\xrightarrow{h\nu}} CTTM \longrightarrow [M^{II}L_5^{(3-n)}, X] \quad (21)$$

where the bracketed species represents the geminate radical–ion pair.

The rather simplistic description of the CTTM state ignores several features that must be incorporated in the mechanism of the photoredox reactions. In the mechanism proposed by Adamson et al. for the photoredox decomposition of cobalt(III) complexes the nature are yields of the photochemical products are determined by the reactivity of geminate and secondary radical–ion pairs[164–166]

$$Co^{III}L_5X^{(3-n)} \longrightarrow [Co^{II}L_5^{(3-n)}, X] + \Delta H \quad (22)$$

$$[Co^{II}L_5^{(3-n)}, X] \longrightarrow Co^{III}L_5X^{(3-n)} (\Delta H \text{ small}) \quad (23)$$

$$[Co^{II}L_5^{(3-n)}, X] \longrightarrow [Co^{II}L_5(S)^{(3-n)}, X] (\Delta H \text{ large}) \quad (24)$$

$$[Co^{II}L_5(S)^{(3-n)}, X] \longrightarrow Co^{2+} + 5L + X \quad (25)$$

$$[Co^{II}L_5(S)^{(3-n)}, X] \longrightarrow Co^{III}L_5(S)^{(4-n)} + X^- \tag{26}$$

The quantity ΔH (eq. 22) represents the amount of light energy absorbed in excess of that necessary for the electron transfer. For a large excess of energy, X may diffuse far enough from the metallo fragment to allow a solvent molecule S to penetrate in the solvent cage (eq. 24). If ΔH is small, the restricted diffusion of the products out of the solvent cage favors recombination (eq. 23). In other words, it is assumed that the excess energy ΔH appears as kinetic energy of the products. This hypothesis is valid in the photodissociation of diatomic molecules, but is much less valid for polyatomic molecules where the excess energy of polyatomic fragments can be dissipated by the activation of appropriate vibrational modes. The formation of radical–ion pairs has been verified, however, in flash photolysis where these species exhibited lifetimes of several hundred picoseconds.[167] In this regard, the Adamson mechanism can be regarded, with suitable changes, as a limiting mechanism. Indeed, a modified mechanism must account for reactions such as the photoinduced linkage isomerization

$$Co^{III}(NH_3)_5NO_2^{2+} \xrightarrow{h\nu} Co^{III}(NH_3)_5ONO^{2+} \tag{27}$$

and the dependence of the product yields on excitation energy and medium viscosity.[168–170] It seems unlikely that any single radical–ion pair model can accommodate all the experimental observations. For example, one alternative to Adamson's mechanism is the change of eq. 20 by eq. 28, namely the recombination of the secondary radical–ion pair leading to the linkage isomerization

$$[Co^{III}L_5(S)^{(3-n)}, X] \longrightarrow Co^{III}L_5X^{(3-n)} + S \tag{28}$$

Besides, the elimination of eq. 26 leads to a simple model for a limiting photodissociative behavior that is exhibited by some complexes. An additional point that must be considered in a more complex mechanism is the dependence of the rate constants for nonradiative conversion, for example, intersystem crossing and internal conversions, on medium conditions and molecular parameters. For example, Meyer et al. have found that the rate constant for the relaxation of the charge transfer state of Os(III) polypyridine complexes (Table 8) depends on the ground to excited state energy gap as predicted by Jortner's mechanism (Chapter 4).[171] This also suggests that there are several mechanisms by which the rates of formation and relaxation of charge transfer states must be dependent on medium conditions. Indeed, the point to notice is that in the majority of the complexes the charge transfer transitions produce a large change of dipolar momentum that can be equated with the dependence of the charge transfer spectra in experimental parameters such as the dielectric constant (Chapter 3). A more dramatic example of the participation of the medium on the dynamics

TABLE 8 Relationship between Energy Gap and the Rate of Radiationless Relaxation[a]

Complex[b]	E_{max}[c] (kJ)	$10^{-6}k_{nr}$[c] (s^{-1})	$10^{-5}k_r$[c] (s^{-1})
$Os(bipy)_3^{3+}$	15.97	16.6	77.1
$Os(bipy)_2(CH_3CN)_2^{3+}$	16.83	6.29	0.839
trans-$Os(bipy)_2(PPh_3)_2^{3+}$	17.83	2.41	2.17
$Os(bipy)_2$(*cis*-dpp-ene)$_2^{3+}$	18.91	1.86	1.40
$Os(bipy)_2(Me_2SO)_2^{3+}$	20.72	0.41	2.55

[a]Data from Ref. 171.

[b]All the complexes listed in the table were PF_6^- salts and all the measurements were carried out in deaerated CH_3CN solutions at 23°C.

[c]E_{max} are the corrected energies for the emission maxima, k_r and k_{nr} are the rate constants for the radiative and radiationless relaxation, respectively.

of the excited state can be illustrated with the long-lived CTTL excited state of $Ru(bipy)_3^{2+}$. According to the information obtained by resonance Raman spectroscopy and by investigating the polarization of the emission, the state exhibits a symmetry lower than D_{3d}, that is, a symmetry lower than that of the ground state or any precursor state where the electronic density is equally distributed between the three ligands.[172,173] Moreover a quantitative comparison of the resonance Raman spectra of the excited state and the anion radical bipy$^{\cdot -}$shows similarities which indicate that, in the excited state, the electron is localized in one bipyridine ligand and not delocalized over the three ligands. However, the investigation of the excited state in a nanosecond time domain, in glassy solutions, has demonstrated that the localization of the electron in a particular ligand seems to take place during the relaxation of the solvent.[174] In this context, the localization of the electron can be described as a solvent around the excited state in a process that increases its stability (Fig. 45). Further studies have demonstrated that the behavior is rather general with bipyridine complexes of d^6 metal ions.[175]

It is now possible to relate the energy of photoredox processes (eqs. 20 and 22) to several empirical parameters by using a model developed by Endicott et al.[176–178] For complexes where the photoredox quantum yield rises rapidly with increasing photonic energy, the curves ϕ_{redox} versus excitation energy (Fig. 46) may be extrapolated to $\phi_{redox} \simeq 0$; alternatively, the energy corresponding to $\phi_{redox} \simeq 0$ may be approximated by the largest excitation energy corresponding to a ϕ_{redox} that is 1% of the maximum quantum yield. The energies obtained by either procedure are usually designated the *photoredox threshold energy* E_{th}. Since the thermodynamic parameters are more easily defined for ion pairs, the treatment of CTTI transitions is a convenient point of departure. In this context, the oxidation reaction (eq. 29) can be combined with the association equilibria (eqs. 30 and 31) to obtain the enthalpy corresponding to the charge transfer transition within the ion pair (eq. 32)

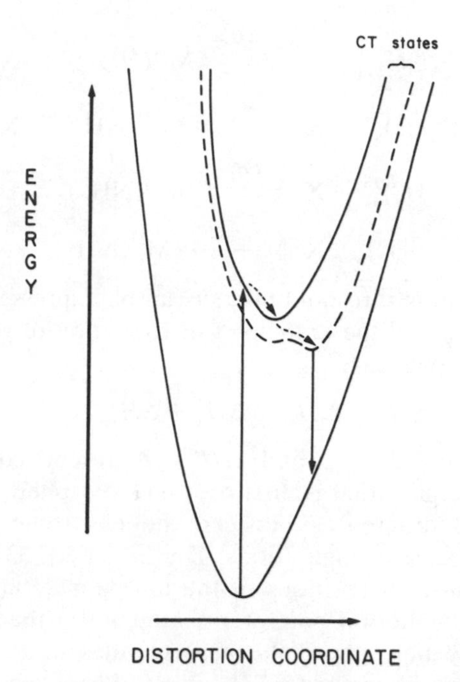

Figure 45. Potential surfaces for the ground and charge transfer states of $Ru(bipy)_3^{2+}$. The dashed curve has a minimum corresponding to the localization of the charge in one of the ligands as a consequence of solvent repolarization. The distortion coordinate gives a measure of such electron localization.

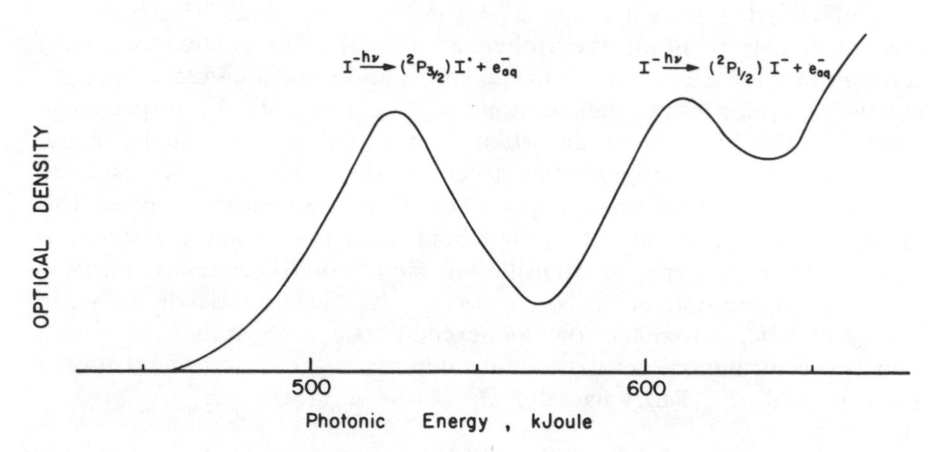

Figure 46. Charge transfer spectrum of I^- in aqueous solutions. Assignments for each absorption are indicated in the figure.

$$M^{III}(NH_3)_6 + X^- \overset{\Delta H_R^0}{\rightleftharpoons} M^{II}(NH_3)_6^{2+} + X^{\cdot} \tag{29}$$

$$M^{III}(NH_3)_6^{3+} + X^- \overset{\Delta H_f^0}{\rightleftharpoons} \{M^{III}(NH_3)_6^{3+}, X^-\} \tag{30}$$

$$M^{II}(NH_3)_6^{2+} + X^- \overset{\Delta H_f^{0'}}{\rightleftharpoons} \{M^{II}(NH_3)_6^{2+}, X^-\} \tag{31}$$

$$\{M^{III}(NH_3)_6^{3+}, X^-\} \xrightarrow{E_{th}} \{M^{II}(NH_3)_6^{2+}, X^{\cdot}\} \tag{32}$$

Indeed, the photoredox threshold energy can be expressed in terms of the redox enthalpy ΔH_R^0 and the enthalpies of formation of the ion pair, ΔH^0, and the radical–ion pair, $\Delta H_f^{0'}$,

$$E_{th} = \Delta H_R^0 + \Delta H_f^{0'} - \Delta H_f^0 + \Delta H_{spin}^0 + \Delta H_{FC}^0 \tag{33}$$

The additional terms ΔH_{spin}^0 and ΔH_{FC}^0 represent contributions from Franck–Condon energies, that is, first coordination sphere nuclear reorganization energy, and differences between the electronic repulsion in the ground and excited state configurations. The use of eq. 33 can be illustrated with the charge transfer transitions of the ion pairs $\{Ru(NH_3)_6^{3+}, I^-\}$ and $\{Co(NH_3)_6^{3+}, I^-\}$. In the ion pair of ruthenium(III) the CTTI absorption maximum $\lambda_{max} \sim 330$ nm, where the absorptivities of the free anion and cation are negligible. To evaluate ΔH_R^0 in eq. 33 we can use the standard reduction potentials of $Ru(NH_3)_6^{3+}/Ru(NH_3)_6^{2+}$, $\epsilon^0 = 0.1$ V, and $I^{\cdot}/I^-$, $\epsilon^0 = 1.42$ V, and the respective standard entropies, $\Delta S^0 = 39$ and 9.1 e.u. Moreover, the formation equilibrium constants for the ion pair and the radical–ion pair are 10 and 0.1 M^{-1}, respectively. With these parameters, the estimated value $E_{th} \sim 190$ kJ/mol is in good agreement with the experimental value $E_{th} \sim 214$ kJ/mol, as we would expect if the contributions from $\Delta H_{spin}^0 + \Delta H_{FC}^0$ in eq. 33 are negligible.

Although the thermodynamic parameters for the $\{Co(NH_3)_6^{3+}, I^-\}$ and $\{Ru(NH_3)_6^{3+}, I^-\}$ pairs are similar, E_{th} is larger (by about 95 kJ/mol) for the cobalt pair than for the ruthenium pair. In this context, we must consider that the excited states of the cobalt and ruthenium pairs leave the metals with electronic configurations $(t_{2g})^6(e_g)$ and $(t_{2g})^6$, respectively. Insofar as the Jahn–Teller distortion is expected in a $(t_{2g})^6(e_g)$ configuration, the excited state potential surface of the cobalt must be displaced with respect to the corresponding surface of the ruthenium complex. This displacement is expected to bring about Franck–Condon contributions ΔH_{FC}^0 to the charge transfer transition in the cobalt(III) complex. Furthermore, the ground state of $Co(NH_3)_6^{2+}$ is $(t_{2g})^5(e_g)^2$, which has a difference in energy of ΔH_{spin}^0 from the low spin-excited state configuration $(t_{2g})^6(e_g)$. The two contributions must therefore determine the differences between E_{th}(Co) and E_{th}(Ru), namely, $E_{th}(Co) - E_{th}(Ru) = \Delta H_{spin}^0 + \Delta H_{FC}^0 = 90$ kJ/mol.

The approach, based on eq. 33, for the evaluation of the photoredox threshold energies of ion pairs can be extended to the charge transfer transition discussed above (eq. 22). We must assume that at the threshold of the photoredox dissociation the primary products are geminate radical pairs $\{M^{II}L_5^{2+}, X^{\cdot}\}$. The threshold energy for the charge transfer process (eq. 22) can be evaluated by breaking down the process into several steps

$$M^{III}L_5X^{2+} \xrightarrow{\Delta H^{\ddagger}} \{M^{III}L_5^{3+}, X^-\} \tag{34}$$

$$\{M^{III}L_5^{3+}, X^-\} \xrightarrow{\Delta H^0_{IP}} \{M^{II}L_5^{2+}, X^{\cdot}\} \tag{35}$$

Equation 34 represents the formation of an ion pair by the removal of one of the ligands from the coordination sphere of the complex. Although these ion pairs are not known as stable species, they participate as intermediates in the thermal hydrolysis of the complexes

$$M^{III}L_5X^{2+} \xrightleftharpoons{\Delta H^{\ddagger}} \{M^{III}L_5^{3+}, X^-\} \xrightleftharpoons[X^-]{H_2O}$$
$$\{M^{III}L_5^{3+}, OH_2\} \longrightarrow M^{III}L_5OH_2^{3+} \tag{36}$$

The energy of formation of the ion pair (eq. 34) can be approximated by the activation energy $\Delta H^{\ddagger}$ of the thermal hydrolysis (eq. 36). The energy associated with the electron transfer (eq. 35) contains a contribution ΔH^0_M from the couple $M^{III}L_5^{3+}/M^{II}L_5^{2+}$ and a contribution ΔH^0_X from the couple $X^{\cdot}/X^-$. A Franck–Condon energy ΔH^0_{FC} must also be incorporated to account for the fact that the charge transfer transition produces the radical–ion pair $\{M^{II}L_5^{2+}, X^{\cdot}\}$ with the ground state nuclear configuration while the enthalpies ΔH^0_M and ΔH^0_X correspond to the formation of thermalized products. In this regard, eq. 37 relates the various thermodynamics parameters that contribute to the value of the photoredox threshold energy

$$E_{th} = (\Delta H^0_M - \Delta H^0_X) + \Delta H^0_{FC} + \Delta H^{\ddagger} - E_B \tag{37}$$

The possibility of some metallo–radical bonding in the excited state is considered by introducing a residual bonding energy E_B in eq. 37. Experimental data allow us to evaluate the sum $\Delta H^0_X + \Delta H^{\ddagger}$, while the other terms, $\Delta H^0_M + \Delta H^0_{FC} - E_B$, must be evaluated by various routes. In the evaluation of ΔH^0_M, it is possible to assume that the standard entropy changes for the couples $M^{III}L_5^{3+}/M^{II}L_5^{2+}$ and $M^{III}L_6^{3+}/M^{III}L_6^{2+}$ are similar. In this approximation the difference between the redox potentials of the couples can be correlated with differences in crystal field stabilization energies, that is, in a ligand field model, or in a molecular orbital calculation. For example, for low-spin cobalt(III) complexes with a symmetry D_{4h}, for example *trans*-$Co^{III}LX_2^+$ where L is a tetradentated cycle tetraaza ligand, the potentials of the $Co^{III}LX_2^+/Co^{II}LX_2$ couples increase with decreasing crystal field strengths of the axial ligands. Hence, the potentials of the $Co^{III}L_5^{3+}/Co^{II}L_5^{2+}$ couples must be more positive than the potentials of the $Co^{III}L_6^{3+}/Co^{II}L_6^{2+}$ couples. It is easy to demonstrate with the ligand field theory that the potential of the $Co^{III}(NH_3)_5^{3+}/Co^{II}(NH_3)_5^{2+}$ couple is 1 V more positive than the potential of the $Co^{III}(NH_3)_6^{3+}/Co^{II}(NH_3)_6^{2+}$, where the corresponding cobalt(II) products are assumed to have a low-spin configuration. In addition, the differences in the energy between the low- and high-spin configurations of $Co(NH_3)_6^{2+}$, that is, $\Delta H^0_{spin} \sim 48$ kJ/mol, suggests that the potential of $Co^{II}(NH_3)_5^{3+}$/(low spin) $Co^{II}(NH_3)_5^{2+}$ is 0.5 V more positive than the potential of the high-spin couple, $Co^{III}(NH_3)_5^{3+}$/(high spin) $Co^{II}(NH_3)_5^{2+}$. In a semiquantitative manner, at least, the contribution from $(\Delta H^0_M - \Delta H^0_X)$ in eq. 37 can be related to various parameters. With the help of Born cycles (Appendix VI) it is possible to show how these energies can be expressed in terms of the electroaffinity of the atom X, the ionization potentials of the metal, solvation energies, and ligand field stabilization energies.

An approach for the evaluation of Franck–Condon contributions ΔH^0_{FC} is to consider the energy associated with the changes in internuclear distances. In reconstructing the initial Franck–Condon excited state from the radical pair the energetic contributions fall into two categories: λ_i representing the energy changes caused by the compression (to ground state bond lengths) of the metal–ligand distances in the reduced metallo fragment, and λ_o corresponding to energy changes related to the reorientation of solvent molecules until the solvent environment of the ground state is regenerated. For the evaluation of λ_o and λ_i, one can follow a procedure that is commonly used with condensed phase electron transfer reactions.[179–183] The reorganizational energy can be represented as $\lambda/4$, where $\lambda = \lambda_o + \lambda_i$. The solvent reorganization energy

$$\lambda_o = e^2\left(\frac{1}{2a_1} + \frac{1}{2a_2} - \frac{1}{r}\right)\left(\frac{1}{\mathcal{K}_{op}} - \frac{1}{\mathcal{K}_s}\right) \tag{38}$$

depends on the covalent radii a_1 and a_2 of the reduced metallo fragment and radical $X^{\cdot}$, respectively, and the distance r between these species in the particular Franck–Condon state considered.[183] In this regard, the distance r must be very close in value to the corresponding distance in the ground state, that is, it can be approximated by the sum of the covalent radius of the oxidized metallo fragment and the ionic radius of the ligand X^-. The other parameters in eq. 38 are the optical dielectric constant $\mathcal{K}_{op}$, which is the square of the medium refractive index, that is, $\mathcal{K}_{op} = n^2$, and the static dielectric constant of the medium $\mathcal{K}_s$. For the inner sphere reorganization energy λ_i it is possible to use an expression (eq. 39) which is based on the force constants f_i and f_i^P of the vibrational coordinates that experience a significant change Δr_i^0 in each of the species that participate in the outer sphere electron transfer reaction[180,182]

$$\lambda_i = \sum_j \frac{f_j f_j^p}{f_j + f_j^p} (\Delta r_j^0)^2 \tag{39}$$

In our photochemical problem, eq. 39 must be applied to the metallo fragment in changing the nuclear coordinates from those of the reduced metallo fragment to the coordinates of the ground state. For the acidopentaammine complexes of Ru(II) and Co(III) the contributions from $\lambda/4$ amount to 30 and 50% of the respective E_{th} values.

5-5 MODELS FOR CHARGE TRANSFER TO SOLVENT (CTTS) TRANSITIONS

In the gas phase, the ionization potential of I^-, $I_{I^-} \sim 290$ kJ/mol, is smaller than the energy associated with the $5p^5 6s^1 \leftarrow 5p^6$ electronic transition.[184,185] In other words, there are no stable binding states above the ground state for iodide or the rest of the halides. However, the absorption spectrum of aqueous iodide exhibits two peaks, 225.9 and 200.3 nm, at much higher energies than the ionization potential (Fig. 46). The energy difference between peaks is in agreement with the formation of $(^2P_{1/2})I^{\cdot}$ and $(^2P_{3/2})I^{\cdot}$ in electronic transitions that must also liberate electrons

$$I^-_{aq} \xrightarrow{h\nu} [(^2P_{1/2})I, e^-]_{aq} \tag{40}$$

$$I^-_{aq} \xrightarrow{h\nu} [(^2P_{3/2})I, e^-]_{aq} \tag{41}$$

The difference in the energy between these transitions and the ionization potential, about 210 kJ/mol, can be interpreted as the energy required (in addition to the ionization potential) for the creation of the electron in a stationary state. These transitions are called charge transfer to solvent (CTTS)[186] and can be observed in monoatomic and polyatomic anions, such as Cl^-, OH^-, $H_2PO_4^-$, and coordination complexes, such as $Fe(CN)_6^{4-}$, $CuCl_2^-$.[187–189] Internal transitions, $\pi = \pi^*$ and $n\text{–}\pi^*$, coexist with CTTS transitions in polyatomic molecules, a point that is illustrated in Table 9.

The excitation of photolytes at wavelengths of CTTS absorption bands procedures excited states (eq. 42) that undergo radiationless relaxation to the ground state and dissociation into oxidized fragments and solvated electrons

$$X^-_{aq} \xrightarrow{h\nu} \text{CTTS} \tag{42}$$

$$\text{CTTS} \longrightarrow X^-_{aq} \tag{43}$$

$$\text{CTTS} \longrightarrow [X^{\cdot}, e^-]_{aq} \tag{44}$$

TABLE 9 Charge Transfer to Solvent Spectra of Anions[a]

Species	$\lambda_{max}(\epsilon_{max})$[b]	Assignment[c]
I^-	228 (1.29×10^4)	$(^2P_{3/2})I$ + 2s-type electron
	196 (1.32×10^4)	$(^2P_{1/2})I$ + 2s-type electron
	183 (4×10^3)	$(^2P_{3/2})I$ + 3s-type electron
	171 (2×10^4)	
Br^-	198 (1.3×10^4)	$(^2P_{3/2})Br$ + 2s-type electron
	186 (1.35×10^4)	$(^2P_{1/2})Br$ + 2s-type electron
	175 (5×10^3)	$(^2P_{3/2})Br$ + 3s-type electron
Cl^-	174 (1.69×10^4)	$(^2P_{3/2}, {}^2P_{1/2})Cl$ + 2s-type electron
CN^-	<182	Unknown
NCO^-	<182	Unknown
NCS^-	222 (3.5×10^3)	
$NCSe^-$	235 (3.0×10^3)	
$NCTe^-$	270 (3.0×10^3)	
N_3^-	233 (4×10^2)	Internal
	203 (4×10^3)	
	189 ($>10^4$)	Internal
OH^-	187 (3.85×10^3)	
SH^-	230 (7.0×10^3)	
	210 (1.0×10^3)	*d*
	185 (5.0×10^3)	*d*
	163	*d*
S^{2-}	360 (2×10^2)	
Se^{2-}	417	Unknown
SO_4^{2-}	175 (3.0×10^2)	
HSO_4^-	167 ($>10^2$)	*d*
$Fe(CN)_6^{4-}$	256	
Cu^+_{aq}	~ 200	*d*
$CuCl_2^-$	271 (4.3×10^3)	
$CuBr_2^-$	276 (8.5×10^3)	

[a]In aqueous solutions at 25°C.
[b]Wavelengths in nm and extinction coefficients in M^{-1} mc^{-1}.
[c]CTTS unless explicitly stated.
[d]Tentatively assigned as CTTS.

The electron in the CTTS excited state seems to be closely confined to a limited volume around the anion. Several models have been proposed for the treatment of the CTTS spectra. In the Franck–Platzman model[190] the electron is placed in a spherically symmetric orbital that is located in the center of the anion's cavity. A combination of eqs. 45–48 in a Born cycle is used for the final expression of the energy for the absorption maximum

$$X^-_{aq} \xrightarrow{A} X^-_g \tag{45}$$

$$X^-_g \xrightarrow{B} X^{\cdot}_g + e^-_g \tag{46}$$

$$X^{\cdot}_g + e^- \xrightarrow{C} [X^{\cdot}, e^-]_{aq} \tag{47}$$

$$X^-_{aq} \xrightarrow{E_{max}} [X^{\cdot}, e^-]_{aq} \tag{48}$$

The energy A (eq. 45) is the opposite to the solvation energy of the anion, ΔH^0_{S}. To evaluate this energy, the water molecules in the first solvation sphere are oriented so that they are coplanar with the anion, with OH--- X^- linear and with the second hydrogen available for hydrogen bonding with bulk water. Therefore, ΔH^0_{S} includes the contribution associated with the rearrangement of the solvent from the structure in the immediate vicinity of the anion to the structure of the bulk solvent. The contribution C is the solvation energy of the halide atom, ΔH^0_{S}, which is taken as one-half of the solvation energy of the halogen molecule, that is, 19.2 kJ/mol for $Cl^{\cdot}$, $Br^{\cdot}$, and $I^{\cdot}$ in aqueous solutions. There is also a contribution ΔH^0_{FC} from restoring the polarization of the solvent around $X^{\cdot}$ to the same condition that existed around the parent anion X^-, a condition imposed by the Franck–Condon principle. Such energy is of the order of 77 kJ/mol for the halides. A last contribution associated with C (eq. 47) is the energy $\Delta H^0_{\mathrm{e}^-}$ liberated when the electron is transferred from the gas phase to the centrosymmetric excited state. Process B (eq. 46) represents the ionization of the gaseous ion and the energy change associated with such a process is the ionization potential I_{X^-}. Hence, the value of E_{max} for a univalent anion in water is expressed as a sum of various contributions

$$E_{\mathrm{max}} = \Delta H^0_{\mathrm{S}} - \Delta H^{0'}_{\mathrm{S}} + I_{\mathrm{X}^-} - \Delta H^0_{\mathrm{e}^-} + \Delta H^0_{\mathrm{FC}} \tag{49}$$

A mathematical model of the solvated electron is required to evaluate the energy $\Delta H^0_{\mathrm{e}^-}$. In the Polaron treatment,[191–194] the electron is localized to a space region by the self-polarization of the solvent. This treatment, when applied to the CTTS state, yields a series of stationary levels with an effective atomic number

$$Z_{\mathrm{eff}} = \frac{1}{\mathcal{K}_{\mathrm{op}}} - \frac{1}{\mathcal{K}_{\mathrm{s}}} \tag{50}$$

that is related to the optical $\mathcal{K}_{\mathrm{op}}$ and static $\mathcal{K}_{\mathrm{s}}$ dielectric constants. The energies of the levels with quantum number n are given by

$$E_n = \frac{2\pi^2 m e^4 Z^2_{\mathrm{eff}}}{h^2 n^2} \tag{51}$$

A state with $n = 2$ has $E_2 \sim 96$ kJ/mol and a mean radius $\langle R_2 \rangle \sim 580$ pm, which gives an acceptable penetration of the electron in the water. An additional contribution of ~ 48 kJ/mol is caused by the electron-induced polarization of the solvent. This brings the value of ΔH^0_{e} to ~ 146 kJ/mol and the value of E_{max} to 513 kJ/mol for the halides in water. Despite the success of the model in the evaluation of E_{max} (the experimental value is 517 kJ/mol) it does not account for the dependence of E_{max} on temperature.

The diffuse model is similar in certain aspects to the Franck–Platzman model.[195] However, in the diffuse model, the energy A in eq. 45 cannot be associated with the solvation energy of the anion since the structure of the cavity is kept constant in the generation of the excited state. In this context, the energy A represents the energy of the interaction of the anion with the cavity and contains contribution from the atomic and dipolar polarizations, E_s, and electric polarization, S_s. The energy E_s is therefore associated with the polarization of the solvent molecules around the ion under a coulombic field with the Z_{eff} of eq. 50, and at a distance R_0 from the center of the cavity

$$E_s = \frac{e^2}{2R_o}\left(\frac{1}{\mathscr{K}_{op}} - \frac{1}{\mathscr{K}_s}\right) \tag{52}$$

Moreover, the energy S_s is associated with the polarization of the solvent in the excited state

$$S_s = \frac{e^2}{2R_o}\left[1 - \frac{1}{\mathscr{K}_{op}}\right] \tag{53}$$

The energy C in eq. 47 can now be associated with the replacement of $X^{\cdot}$ into the preformed cavity, namely with the enthalpy $\Delta H_S^{0'}$ of the radical solvation.

Combination of the energies (eqs. 51, 52, 53) in an expression similar to eq. 49 gives

$$E_{max} = I_{X^-} - \Delta H_S^{0'} + \frac{e^2}{R_o}\left[\frac{1}{2} - \frac{1}{2\mathscr{K}_{op}} - \frac{1}{\mathscr{K}_s}\right] - \frac{e^2}{2R_o}\left[1 - \frac{1}{\mathscr{K}_{op}}\right] - \frac{\pi^2 me^4}{2h^2}\left[\frac{1}{\mathscr{K}_{op}} - \frac{1}{\mathscr{K}_s}\right]^2 \tag{54}$$

which in aqueous solutions at 25°C can be reduced, in kilojoule units, to

$$E_{max} = I_{X^-} + \Delta H_S^{0'} + \frac{4.1 \times 10^2 e^2}{R_o} - 3.5 \times 10^3 \tag{55}$$

The values of R_o were determined by using eq. 55, experimental values of E_{max}, and estimated values of $\Delta H_S^{0'}$. These calculations show that R_o is related to the ionic radius r_i by means of the product

$$R_o = 1.25\, r_i \tag{56}$$

It has been suggested that such a constancy of the ratio might reflect a common mode of packing water molecules around the anion.

For polyvalent ions, the CTTS transition is

$$(X^{-2})_{aq} \xrightarrow{h\nu} (X^{-(Z-1)}, e^-)_{aq} \tag{57}$$

and the interaction energies of the anion X^{-z} and the corresponding radical $X^{-(z-1)}$ must consider the multiple charges of the species.[196] In this context, the energy change in replacing $X^{-(z-1)}$ inside the cavity is

$$\Delta H_S^{0'} = \left[\frac{(Z-1)Ze^2}{R_0}\right]\left[\frac{1}{\mathcal{K}_{op}} - \frac{1}{\mathcal{K}_S}\right] + \left[\frac{Z-1)^2e^2}{2R_o}\right]\left[1 - \frac{1}{\mathcal{K}_{op}}\right] \tag{58}$$

and is reduced by $11Z^3/R_o^5$ as a consequence of dielectric saturation near the polyvalent anion. The expression for a divalent anion in aqueous solutions at 25°C is

$$E_{max} = I_x + \frac{465}{r_i} - \frac{29}{r_i^5} - 35 \tag{59}$$

where I_x represents the vertical ionization potential of the divalent anion.

The confined model regards the electron as placed in a spherical well of variable radius R_0, which remains constant through a cycle similar to that described in eqs. 45–48.[197–199] In this approximation the enthalpy of solvation of the anion is counterbalanced by the enthalpy of interaction of the electron with the cavity. Such a model gives an energy for the maximum absorption

$$E_{max} = I_x + \frac{h^2}{8mR_0^2} \tag{60}$$

where the second term represents the kinetic energy of an electron when it is confined to a potential well whose walls are assumed to be infinitely steep.

Another model for CTTS transitions is based on Mulliken's charge transfer theory.[200–202] The acceptor orbital of the solvent is assumed to be a simple spherical shell with an electron affinity ϵ_s and the anion is regarded as an electron donor in the charge transfer complex formed by the anion and the solvated electron–acceptor shell. Thus the energy of the maximum is

$$E_{max} = I_X - \epsilon_s + \frac{\sigma^2}{(I_X - \epsilon_s)R_o^2} \tag{61}$$

where σ is the overlap term that depends on the overlap integral and polarization terms.

The models presented above have many drawbacks that have not yet been resolved. These drawbacks range from the failure of the Franck–Platzman model to explain the dependence of E_{max} on T to the absence of an explicit description of solute–solvent interactions in Mulliken's model.

Table 10 shows the spectral properties of a number of anions where CTTS transitions are the only feature or they coexist with internal transitions n–π^* and π–π^*. N_3^- is one case where a CTTS transition, λ_{max} = 203 nm with $\epsilon_{max} = 4.0 \times 10^3\ M^{-1}\ cm^{-1}$, coexists with internal transitions,

TABLE 10 Optical Spectra and Photodissociation Modes of Polyatomic Anions[a]

Anion	Spectra		Photochemical Primary Process
	$\lambda_{max}(\epsilon_{max})$[b]	Assignment[c]	
ClO^-	291 (3.6×10^2)		$ClO^- \xrightarrow{h\nu} Cl^- + (^3P)O$
			$\xrightarrow{h\nu} Cl^{\cdot} + O^-$
BrO^-	333 (3.3×10^2)		$BrO^- \xrightarrow{h\nu} Br^- + (^3P)O$
			$\xrightarrow{h\nu} Br^{\cdot} + O^-$
BrO_2^-			$BrO_2^- \xrightarrow{h\nu} BrO_2^- + (^3P)O$
			$\xrightarrow{h\nu} BrO + O^-$
BrO_3^-	<200 ($>10^3$)		$Bro_3^- \xrightarrow{h\nu} BrO_2^- + (^3P)O$
			$\xrightarrow{h\nu} BrO_2 + O^-$
IO^-	365 (31)		$IO^- \xrightarrow{h\nu} I^{\cdot} + O^-$
	255 (4×10^2)		
	<238 ($>10^3$)		
IO_3^-	<200 ($>10^3$)		$IO_3^- \xrightarrow{h\nu} IO_2 + O^-$
ClO_3^-	<200($>10^3$)		$ClO_3^- \xrightarrow{h\nu} ClO_2 + O^-$
$S_2O_3^{2-}$	244 (2×10^2)		$S_2O_3^{2-} \xrightarrow{h\nu} S_2O_3^- + O^-$
	215 (4×10^3)		
	~ 192 ($>4 \times 10^3$)		
NO_2^-	355 (22.5)	$n_0-\pi^*$	$NO_2^- \xrightarrow{h\nu} NO + O^-$
	286 (9.4)	$n_0-\pi^*$	
	210 (5.4×10^3)	$\pi-\pi^*$	
	~ 192	CTTS (?)	
NO_3^-	303 (7)	$n-\pi^*$	$NO_3^- \xrightarrow{h\nu} NO_2 + O^-$
	202 (9.9×10^3)	$\pi-\pi^*$	$\xrightarrow{h\nu} NO + O_2$
			$\xrightarrow{h\nu} ONOO^-$

[a]Data in aqueous solution at 25°C extracted from A. Treinin, *Isr. J. Chem.* **1970,** *8,* 103.
[b]Wavelengths in nm and extinction coefficients in M^{-1} cm^{-1}.
[c]Assigned as internal unless explicitly stated.

$\lambda_{max} = 232$ nm with $\epsilon_{max} = 4.0 \times 10^2\ M^{-1}\ \text{cm}^{-1}$ and $\lambda_{max} = 189$ nm with $\epsilon_{max} > 10^4\ M^{-1}\ \text{cm}^{-1}$. Another example where CTTS and internal transitions coexist is SCN^-. The spectrum of this pseudohalide exhibits an intense absorption band, $\lambda_{max} = 222$ nm with $\epsilon_{max} = 3.5 \times 10^3\ M^{-1}\ \text{cm}^{-1}$, which has been labeled a CTTS transition. Moreover, the wavelength-dependent photochemistry of SCN^- can be described in terms of the two primary photoprocesses[203]

$$SCN^- \xrightarrow{h\nu} SCN^{\cdot} + e_{aq}^- \quad (62)$$

$$SCN^- \xrightarrow{h\nu} S^{\cdot} + CN^- \quad (63)$$

The halate anions, XO_3^- with X = Cl, Br, or I, do not exhibit CTTS transitions in their spectra and give (in UV photolysis) solvated electrons with negligible yields. The main photoprocess is the dissociation in oxide ions as shown in eq. 64 for bromate ions[204]

$$BrO_3^- \xrightarrow{h\nu} BrO_2 + O^- \quad (64)$$

The oxide ions rapidly form $^{\cdot}OH$ radicals by reacting with water unless previously scavenged by reactions with other species

$$O^- + H_2O \longrightarrow OH^- + {}^{\cdot}OH \tag{65}$$

The formation of O^- can be demonstrated in flash photolysis by the characteristic reactions with Br^-, CO_3^{2-}, or SCN^-, which give products with very distinctive spectra

$$^{\cdot}OH + 2Br^- \longrightarrow OH^- + Br_2^- \tag{66}$$

$$^{\cdot}OH + 2SCN^- \longrightarrow OH^- + (SCN)_2^- \tag{67}$$

$$^{\cdot}OH + CO_3^{2-} \longrightarrow OH^- + CO_3^{\cdot -} \tag{68}$$

Some anions, namely, MnO_4^- and NO_3^-, are photoreduced through the elimination of molecular oxygen (eq. 69).[205–207] For example, the irradiation of nitrate within the high-energy band (~ 195 nm) gives oxygen elimination and also photoisomerization to pernitrite[207]

$$NO_3^- \xrightarrow{h\nu} NO^- + O_2 \tag{69}$$

$$NO_3^- \xrightarrow{h\nu} ONOO^- \tag{70}$$

The number of anions that liberate $(^3P)O$ upon excitation (eq. 71) is larger, however, than those photogenerating molecular oxygen. In this context the formation of $(^3P)O$ can be distinguished from the formation of molecular oxygen in primary processes by using scavengers that are specific to $(^3P)O$, for example, allyl alcohol, and will suppress the liberation of oxygen

$$XO^- \xrightarrow{h\nu} X^- + (^3P)O \tag{71}$$

$$(^3P)O + (^3P)O \longrightarrow O_2 \tag{72}$$

$$(^3P)O + CH_2{=}CHCH_2OH \longrightarrow OCH_2\dot{C}HCH_2OH \tag{73}$$

It is possible to see from the examples presented above that internal π–π^* and n–π^* transitions in inorganic ions induce photodissociation. Organic anions exhibit a different behavior and $\pi\pi^*$ excited states can dissociate in solvated electrons and radicals. This is the case of phenolate ions, which in UV photolysis undergo photoionization

$$C_6H_5O^- + h\nu \longrightarrow C_6H_5O^{\cdot} + e^-_{aq} \tag{74}$$

5-6 SPECIFIC MEDIUM EFFECTS ON CTTM PHOTOCHEMISTRY

It is now possible to consider the effect of the solvent on the charge transfer transitions.[169,170,176,208] For example, the position of the $X^- \rightarrow M$ charge transfer transitions in the spectra of $M(NH_3)_5X^{n+}$ ($M = Co^{III}$ or Ru^{II} and $n = 2$ or 1, respectively) are solvent dependent. This dependence can be associated with the change in electronic density and dipolar momentum in passing from the ground to the excited state. Two limiting cases have been considered: (1) a repolarization of the solvent to accommodate the change in dipole momentum and charge density, and (2) oxidation of the solvent in a charge transfer process.[176,208] The first case has already been discussed in general terms in Chapter 3. The second case is easily observed in complexes where the charge transfer transitions are not associated with large changes, for example, in $Co(NH_3)_6^{3+}$ and $Ru(NH_3)_6^{3+}$. In this case, the spectral shifts can be related to a displacement of electronic charge from the solvent to the complex in a manner that resembles the generation of a CTTS excited state. For $Ru(NH_3)_6^{3+}$ the change of the solvent from water to a 50% glycerol–water mixture induces a red shift and an increase in the intensity of the 274-nm LMCT band that cannot be related to Ru(III)$\leftarrow$glycerol CTTM transitions. It is far more likely that the red shift and the intensity enhancement are manifestations of the role of the solvent environment in the photophysics of the charge transfer transitions. The Mulliken charge transfer theory that was applied to CTTS transitions (eq. 61) can be modified and introduced as an additional term in eq. 9 to account for the effect of the solvent. This approach suggests that the solvent contribution could take the form

$$a(I_{\text{solv}} - E_{\text{complex}}) + \frac{\sigma^2}{(I_{\text{solv}} - E_{\text{complex}})} \left(\frac{1}{r_{\text{i}}^2}\right) \tag{75}$$

where I_{solv} is the ionization potential of the solvent, E_{complex} is the electron affinity of the coordination complex, and σ is the overlap term, which depends on the overlap integral and polarization terms. The constant a has been introduced to diminish the contribution of the first term since the first contribution to the energy of the CTTM transition must come from the ligands in the first coordination sphere.

The observation and analysis of solvent perturbations of the charge transfer transitions suggest that primary photooxidation of the solvent could be possible for high-energy excitations. This point has been verified in the photochemistry of several acidopentaammine cobalt(III) complexes. In general, the effect of the solvent is not the same for excitation at different wavelengths and it is convenient to consider several spectral regions (Fig. 47). For Co(III) complexes of the pentaammine series there are photoredox and photoaquation reactions induced by excitation at wavelengths of the

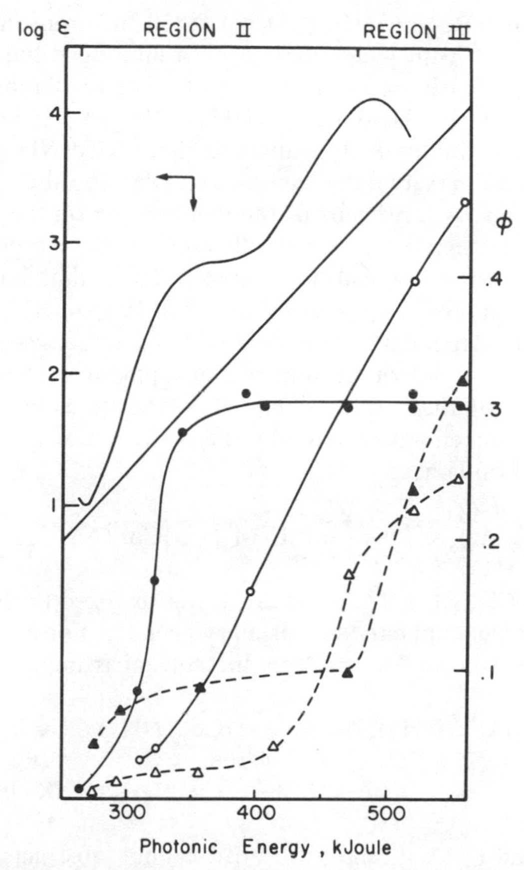

Figure 47. Typical charge transfer spectrum of $Co(NH_3)_5X^{2+}$ complexes (top) and dependence of the photoredox quantum yield on excitation energy and medium conditions (bottom): $Co(NH_3)_5Br^{2+}$ in aqueous acidic solutions (●) and in aqueous acidic solutions with 50% glycerol (▲); $Co(NH_3)_5NCS^{2+}$ in aqueous acidic solutions (○) and in aqueous acidic solutions with 50% glycerol (△).

d–d bands. In the region known as Region I, $Co(NH_3)_5NCS^{2+}$ exhibits very small, wavelength-dependent product yields. The yield of the photoaquation product $Co(NH_3)_5OH_2^{3+}$ diminishes in media containing glycerol, a component of the mixed solvent that induces a large increase of the bulk viscosity. Large yields of ammonia aquation have been determined for photolyses of $Co(NH_3)_5N_3^{2+}$ in Region I, a photobehavior that stands in contrast to the behavior of other acidopentaammine complexes. The effect of glycerol in the media is to reduce the yield of the photoaquation product $Co(NH_3)_4(OH_2)N_3^{2+}$ in a manner similar to that observed with $Co(NH_3)_5Br^{2+}$ and $Co(NH_3)_5(ONO)^{2+}$. Many of these features can be related to the population of d–d states with the characteristic photophysical and photochemical properties discussed in Chapter 6. Irradiations of

$Co(NH_3)_5N_3^{2+}$ in Region II ($254 \le \lambda \le 450$ nm) produce Co^{2+}_{aq} and $Co(NH_3)_4(OH_2)N_3^{2+}$ with yields that reach a limiting value for the 425- to 330-nm region. Increasing concentrations of glycerol diminish the yield of Co^{2+}_{aq} and increase the yield of $Co(NH_3)_4(OH_2)N_3^{2+}$. The behavior of $Co(NH_3)_5NCS^{2+}$ is qualitatively similar to that of $Co(NH_3)_5N_3^{2+}$. In more general terms, these observations can be extended to other members of the pentaammine series. A large part of the dependence on the solvent, for the photochemistry in Region II, may be dictated, in the simplest of analyses, by reactions of various radical–ion pairs, a point that has already been discussed in this chapter. The photobehavior in Region III ($\lambda_{excit} < 254$ nm) is strongly solvent dependent. The yield of Co^{2+}_{aq} increases rapidly with the concentration of glycerol or acetonitrile in aqueous mixtures without the formation of any of the products related to the oxidation of the ligands. Such behavior has been associated with the oxidation of the solvent S in the second coordination sphere

$$[Co^{III}(NH_3)_5X^{2+}]S \underset{\leftsquigarrow}{\xrightarrow{h\nu}} CTTM \longrightarrow [Co^{II}(NH_3)_5X^+, S^{\cdot +}] \tag{76}$$

The ligand, $X = Cl^-$, Br^-, N_3^-, or NCS^-, is not oxidized to the point that the corresponding radical appears as a primary product. For example, a process where an excited radical $^*X^{\cdot}$ oxidizes the solvent is implausible

$$[Co^{III}(NH_3)_5X^{2+}]S \xrightarrow{h\nu} [Co^{II}(NH_3)_5, {}^*X^{\cdot}]S$$
$$\longrightarrow [Co^{II}(NH_3)_5^{2+}, X^-]S^{\cdot +} \tag{77}$$

It is known that $(^3P_{3/2})$I and $(^3P_{3/2})$Br cannot abstract hydrogen from alcohols. Only $(^3P_{1/2})$Br, $(^3P_{3/2})$Cl and $(^3P_{1/2})$Cl react with alcohols. These observations suggest that the indirect photooxidation of the solvent (eq. 77) should make negligible contributions to the photochemistry of the cobalt-(III) complexes. The photooxidation of the solvent is more likely a consequence of the solvent repolarization that is, case 2 discussed above. The energetics of such oxidations were investigated in connection with the photochemistry of $Co(NH_3)_5NCS^{2+}$ by Endicott et al.[208] More specifically, these authors have considered the energy of the photoprocess

$$\{Co(NH_3)_5NCS^{2+}, H_2O\} \xrightarrow{\Delta H^0} \{Co(NH_3)_5NCS^+, H_2O^+\} \tag{78}$$

in which a solvating water molecule is oxidized. Since the proton affinity of H_2O is $\Delta H \sim 708$ kJ/mol

$$H_2O + H^+ \longrightarrow H_3O^+ \tag{79}$$

the enthalpy of formation of liquid water is $\Delta H = 284.7$ kJ/mol, and the OH bond energy in H_3O^+ is nearly $\Delta H \sim 460$ kJ/mol, one can estimate an enthalpy $\Delta H^0 \sim 240$ kJ/mol for the formation of H_2O^+. These data can be used in thermochemical equations to evaluate the standard potential of the H_2O^+/H_2O couple, $\epsilon^0 \sim 5.2$ V, and the enthalpy of eq. 80, $\Delta H \sim$ 510 kJ/mol. In addition, we can use arguments similar to those discussed in connection with the energetics of charge transfer transitions in ion pairs of cobalt(III) complexes to evaluate the onset for the $H_2O \rightarrow Co(III)$ charge transfer transition. This onset is expected to be at about 570 kJ, an energy that makes the direct population of $^1\{Co(NH_3)_5NCS^+, H_2O^+\}$ very unlikely. Hence, the dominant absorbance in Region III cannot arise from a charge transfer transition that involves solvent in the second coordination sphere. We can argue, however, that CTTM transitions involving first coordination sphere ligands may produce weakly bound excited states while CTTM excited states involving second coordination sphere species are not likely to be significantly bound. Therefore, conversion of the CT state, populated by the absorption of radiation in region III, to the dissociative $H_2O \rightarrow Co(III)$ CT state must form $\{Co(NH_3)_5NCS^+, H_2O^+\}$ species as primary products.

An alternative mechanism for $Co(NH_3)_5NCS^{2+}$ in the deep ultraviolet region might be proposed to involve the primary photooxidation of an ammine ligand

$$Co(NH_3)_5NCS^{2+} \xrightarrow{h\nu} \{Co(NH_3)_4NCS^+, NH_3^+\} \tag{80}$$

However, the observed photoreaction stoichiometries and the intensitivity of the photochemistry in Region III to changes in first coordination sphere make eq. 80 an improbable mechanism.

6

LIGAND FIELD PHOTOCHEMISTRY

6.1 GENERAL FEATURES OF LIGAND FIELD PHOTOCHEMISTRY

In photoreactive transition-metal coordination complexes, the population of metal-centered excited states induces the reorganization of the coordination sphere, that is photosolvation, photoisomerization, photoracemization, and photoanation reactions. The dark reactions of the d^3 and d^6 metal ions, such as Cr(III), Co(III), and Rh(III), are slower than the corresponding photochemical processes facilitating the photochemical studies and interesting experimental information has been obtained with their acido-ammine (or amine) complexes (Tables 11 and 12). The photochemical aquation processes of Cr(III) are different from the thermal aquation reactions with regard to the type of ligand labilized and the reaction stereochemistry. For example, $Cr(NH_3)_5X^{2+}$ ions (Table 11) exhibit two photoprocesses: the photoaquation of ammonia (eq. 1) and the photoaquation of the acido ligand (eq. 2)

$$Cr(NH_3)_5X^{2+} \xrightarrow{h\nu} cis\text{-}Cr(NH_3)_4(OH_2)X^{2+} + NH_3 \quad (1)$$

$$Cr(NH_3)_5X^{2+} \xrightarrow{h\nu} Cr(NH_3)_5OH_2^{3+} + X^- \quad (2)$$

Although the aquation of ammonia (eq. 1), has the largest yield of the two photoprocesses (Table 11), the aquation of X^- is the only thermal reaction observed with these compounds. The difference between thermal and photochemical reactions can also be illustrated with acido-tetraammine complexes, such as *trans*-$Cr(NH_3)_4(NCS)Cl^+$.[272] The thermal aquation hydrolyzes chloride with retention of the configuration

TABLE 11 Excited State Properties of d^3 Metal Complexes:

Compounds with O_h Microsymmetry

Compound	ϕ^a	$E_{max}(\tau)^b$		Refs.
		^2E_g	$^4T_{2g}$	
$Cr(urea)_6^{3+}$	0.1	$16.9(10^{-1})$	17.3	20, 111, 209
$Cr(OH_2)_6^{3+}$	0.01–0.03	$17.9(1.8 \times 10^{-3})$	19.5	106, 107, 210
$Cr(C_2O_4)_3^{3-}$	0.09	17.3	17.9	211, 212
$Cr(NCS)_6^{3-}$	0.26	$15.4(10^{-2})$	18.3	104, 213
$Cr(NH_3)_6^{3+}$	0.45	18.1(2.2)	23.5	105–107, 214, 217
$Cr(en)_3^{3+}$	0.37	17.9(1.2)	23.2	102–104, 106, 107, 214, 215, 216
$Cr(CN)_6^{3-}$	0.11	$14.7(1.2 \times 10^{-1})$	28.6	120–122, 218–220
$Cr(phen)_3^{3+}$	0.04	$16.3\ (3.3 \times 10^2)$	25.0	102, 213, 221, 222
$Cr(bipy)_3^{3+}$	0.18	$16.3(7.3 \times 10^1)$	24.7	217, 221, 223, 224
$Cr(acac)_3$	0.01	15.3(1.8)	18.4	106, 107, 225, 226
$Cr(sep)^{3+}$	<0.03	$17.9(4.5)^b$	23.0	227

Compounds with Symmetries Lower than O_h

Compound	ϕ^a		$E_{max}(\tau)^{b,c}$			Refs.
	X	L	^2E	$^2T_1$	^4E	
$Cr(NH_3)_5I^{2+}$			17.5	17.8	19.1	230, 231
$Cr(NH_3)_5Br^{2+}$	0.010	0.36	$17.6(5.9 \times 10^1)$	17.9	19.7	232–234
$Cr(NH_3)_5Cl^{2+}$	0.007	0.37	$17.7(4.1 \times 10^1)$	17.9	20.3	232, 234–236
$Cr(NH_3)_5NCS^{2+}$	0.025	0.47	$17.3(9.7 \times 10^1)$		21.5	212, 232, 233
$Cr(NH_3)_5CN^{2+}$	$<5 \times 10^{-4}$	0.33	17.5		23.5	237, 238
$Cr(NH_3)_5F^{2+}$	0.004	0.43	17.3		20.7	214, 239
trans-$Cr(NH_3)_4Cl_2^+$	0.44	0.003	$17.3(<5 \times 10^{-3})$	17.5	17.3	214, 241
trans-$Cr(en)_2Cl_2^+$	0.32	$<10^{-3}$	17.3		17.8	241–243
trans-$Cr(en)_2(NCS)F^+$	0.27	0.07	$16.5(3.2 \times 10^{-1})$		20.1	210, 212
trans-$Cr(en)_2F_2^+$	<0.08	0.35	15.6(1.9)		19.5	210, 244–248
trans-$Cr(en)_2(NCS)_2^+$	0.23	0.13	16.3(6)	16.7	20.3	210, 212, 249
trans-$Cr(en)_2(CN)_2^+$	<0.05	0.58			26.2	229–231, 250
trans-$Cr(NH_3)_4(CN)_2^+$	<0.005	0.1	(3.3×10)		22.6	229–231, 249

[a]Photoaquation yields determined in aqueous solutions at room temperature.
[b]Emission maximum E_{max} in kJ and lifetimes in μs. Determinations at room temperature.
[c]Lifetimes determined in DMF–$CHCl_3$ glasses at 77 K.

$$trans\text{-}Cr(NH_3)_4(NCS)Cl^+ \rightarrow trans\text{-}Cr(NH_3)_4(NCS)OH_2^{2+} + Cl^- \quad (3)$$

The photoaquation of Cl^- and SCN^- takes place, however, with similar yields and configuration inversion[272]

$$trans\text{-}Cr(NH_3)_4(NCS)Cl^+ \xrightarrow{h\nu} \begin{cases} cis\text{-}Cr(NH_3)_4(OH_2)Cl^{2+} + NCS^- & (4) \\ cis\text{-}Cr(NH_3)_4(OH_2)NCS^{2+} + Cl^- & (5) \end{cases}$$

TABLE 12 Excited State Properties and Photosubstitution in d^6 Metal Complexes

Compounds with O_h Microsymmetry

Compound		ϕ[a] $^1T_{1g}\leftarrow$	ϕ[a] $^1T_{2g}\leftarrow$	$\tau(^3LF)$[b]	τ_{isc}	Refs.
$Co(NH_3)_6^{3+}$		3.1×10^{-4}	5.4×10^{-3}			252, 253
$Co(en)_3^{3+}$		$<10^{-5}$	1.8×10^{-4}			253
$Co(CN)_6^{3-}$		~0.3		2–600[c]		254
$Fe(CN)_6^{3-}$		0.4[d]				255
$Rh(NH_3)_6^{3+}$		0.075[e]		2.1×10^{-2}		256, 257
$Rh(phen)_3^{3+}$				1.9×10^{-2}		258, 259
		Compounds with Symmetries Lower Than O_h				
$Co(NH_3)_5F^{2+}$	NH_3	1.9×10^{-3}				252
	F	5.5×10^{-4}				252
$Co(NH_3)_5CN^{2+}$	NH_3	1.6×10^{-3}	4.6×10^{-3}			260
	CN	1.2×10^{-3}	1.7×10^{-3}			260
$Co(NH_3)_5Cl^{2+}$	NH_3	5.0×10^{-3}				166, 252
	Cl	1.7×10^{-3}	1.1×10^{-2}			166, 252
$Co(NH_3)_5Br^{2+}$	Br	1.4×10^{-3}	6.7×10^{-2}[f]			166, 261
$Co(NH_3)_5NCS^{2+}$	NCS	5.4×10^{-4}	1.5×10^{-2}[f]			166, 261
$Co(NH_3)_5N_3^{2+}$	NH_3	2×10^{-1}[f]				262
	N_3	$<10^{-4}$	$<10^{-3}$			262
$Co(CN)_5Cl^{3-}$	Cl	7×10^{-2}				263
$Co(CN)_5Br^{3-}$	Br	2×10^{-1}				264
$Co(CN)_5I^{3-}$	I	1.7×10^{-1}				265
$Co(CN)_5N_3^{3-}$	N_3	4×10^{-2}				262
$Rh(NH_3)_5Cl^{2+}$	NH_3	4×10^{-2}		1.4×10^{-2}	2.2×10^{-4}	257, 258, 266, 267,
	Cl	1.5×10^{-1}				257, 258 266, 267
$Rh(NH_3)_5Br^{2+}$	NH_3	1.8×10^{-1}		1.3×10^{-2}	2.2×10^{-4}	257, 258, 268
	Br	2×10^{-2}				257, 258, 268
$Rh(NH_3)_5I^{2+}$	NH_3	7.9×10^{-1}		1.3×10^{-3}	4.0×10^{-5}	257, 258, 266
	I	$<10^{-2}$				257, 258, 266
cis-$Rh(NH_3)_4Cl_2^+$	Cl	1.4×10^{-1}		1.3×10^{-3}	7.0×10^{-5}	257, 258, 269
	NH_3	$<2 \times 10^{-3}$				257, 258, 269
	Isom	2.5×10^{-2}				
trans-$Rh(NH_3)_4Cl_2^+$	Cl	1.4×10^{-1}		1.8×10^{-3}	9.0×10^{-5}	257, 258, 269
	NH_3	$<2 \times 10^{-3}$				257, 258, 269
	Isom	2.5×10^{-2}				257, 258, 269
cis-$Rh(NH_3)_4Br_2^+$	Br	2.4×10^{-1}		1.0×10^{-3}	6.0×10^{-5}	257, 258
	NH_3	6.0×10^{-2}				257, 258
	Isom	$<2.4 \times 10^{-1}$				257, 258
trans-$Rh(NH_3)_4Br_2^+$	Br	1.0×10^{-1}		1.5×10^{-3}	9.0×10^{-5}	257, 258
	NH_3	2.0×10^{-3}				257, 258
	Isom	0.00[g]				257, 258

[a]Photoaquation yields determined in aqueous solutions and at room temperature.
[b]Lifetimes in μs.
[c]Variable with temperature in microcrystalline samples.
[d]λ_{exc} ~ 365 nm.
[e]λ_{exc} ~ 313 nm.
[f]Overlap between LF and CT absorptions.
[g]Not observed.

Photolabilization reactions of d^6 metal ion complexes, such as Co(III), Rh(III), Ir(III), Fe(II), and Ru(II), have been studied to a large extent and, as a result, there is more information available about these d^6 complexes than for any but the d^3 transition metal complexes. The photoreactivity of the d^6 complexes shows considerable variation from metal to metal. For example, cobalt(III) ammines or amines are photoinert ($\phi \sim 10^{-3}$) compared with similar complexes of heavier metal ions (Table 12). Ammonia and chloride aquation have been detected in the ligand field photochemistry of $Co(NH_3)_5Cl^{2+}$. The yields of the photoprocesses are largely dependent on the excitation wavelength

$$Co(NH_3)_5Cl^{2+} \xrightarrow{h\nu} Co(NH_3)_4(OH_2Cl^{2+} + NH_3 \qquad (6)$$

$$Co(NH_3)_5Cl^{2+} \xrightarrow{h\nu} Co(NH_3)_5OH \quad + Cl^- \qquad (7)$$

The photochemical behavior of the cobalt(III) ammines stands in contrast with the behavior of pentacyano complexes, $Co(CN)_5X^{3-}$ with $X = CN^-, I^-, Br^-, Cl^-, N_3^-$. Indeed, the photosubstitution reaction

$$Co(CN)_5X^{3-} \xrightarrow{h\nu} Co(CN)_5OH_2^{2-} + X^- \qquad (8)$$

exhibits a large yield (Table 12), which is independent of the excitation wavelength for excitations within the ligand field bands.

A number of systematic features in the ligand field photochemistry of Cr(III) complexes led to the formulation of empirical rules known also as Adamson's rules.[273] The two rules are: (1) Consider the six ligands to lie in pairs at the ends of three mutually perpendicular axes. The axis with the weakest average ligand field will be the one labilized, and the total quantum yield will be about that for an octahedral complex of the same average field. (2) For a labilized axis with two different ligands, the ligand of greater field strength is preferentially aquated.

The application of the rule 1 to $Cr(NH_3)_5Cl^+$ (eq. 1) shows that aquation of ammonia from the $H_3N–Cr–NCS$ axis, that is, the axis with the weakest average field, must be the dominant photoprocess. Moreover, *trans*-diacidotetraammines and *trans*-diacidotetraammines, that is, CrN_4XY^+ with $N_4 = (en)_2$ or $(NH_3)_4$ and $X, Y = I^-, Br^-, Cl^-, NCS^-, N_3^-$, undergo labilization along the X–Cr–Y axis according to rule 1. For *trans*-$Cr(NH_3)_4(NCS)Cl^+$ the photoaquation of NCS^- has a larger yield than the photoaquation of Cl^- (eqs. 4 and 5) in good agreement with both rules. That the rules have less than general validity within the family of Cr(III) compounds is illustrated by the photochemical behavior of *trans*-$Cr(en)_2F_2^+$, where the photoaquation of ethylenediamine is the main photoprocess (Table 11). The success and failure of the empirical rules in the prediction of the Co(III) ligand field photochemistry can be established from the photochemical behavior of $Co(NH_3)_5X^{2+}$ and $Co(CN)_5X^{3-}$, $X = Cl^-, Br^-, CN^-$.

For X = Cl^- and CN^-, the photoaquation of ammonia is almost three times larger than the photoaquation of X^- and the labilization of Br^- is the only process observed within the pentaammine series (Table 12). The photochemistry of the cyano complexes, $Co(CN)_5X^{3-}$, departs from the behavior predicted by the rules in that the photoaquation of X^- is the main photoprocess (Table 12).

The examples presented above demonstrate the wide scope of photoreactivities that must be accommodated within the mechanism(s) of photosubstitution reactions. The various mechanistic propositions will be discussed in the following pages.

6-2 PHOTOREACTIVE EXCITED STATES OF Cr(III) COMPLEXES

Early studies on the photosubstitution processes of octahedral complexes of Cr(III) assigned the photochemical reactivity to the lowest lying ^2E_g (Fig. 48).[274] It was suggested that such state, placed below the quartet states, has lifetimes sufficiently long and therefore has more chance for initiating dissociative or associative processes. Although, it was also argued that the formation of intermediates

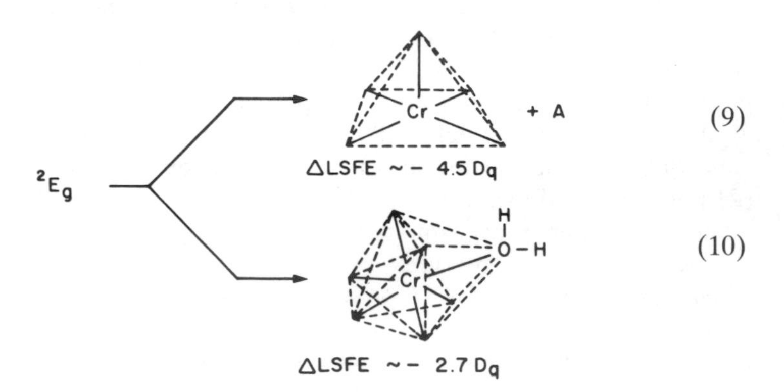

by dissociative (square pyramid) or associative (pentagonal bipyramid) mechanisms from the ^2E_g involve negative contributions, ΔLFSE, to the activation energy, there is no compelling evidence supporting any relationship between ΔLFSE and the photoreactivity. There is some evidence about the involvement of the doublet states (^2E_g, $^2T_{1g}$ in O_h) in the photosubstitution reactions of some Cr(III) complexes. Doublet quenching in $Cr(NH_3)_6^{3+}$ results in ~70% quenching of product formation. In this context, it is necessary to highlight the properties of the doublet states. In an

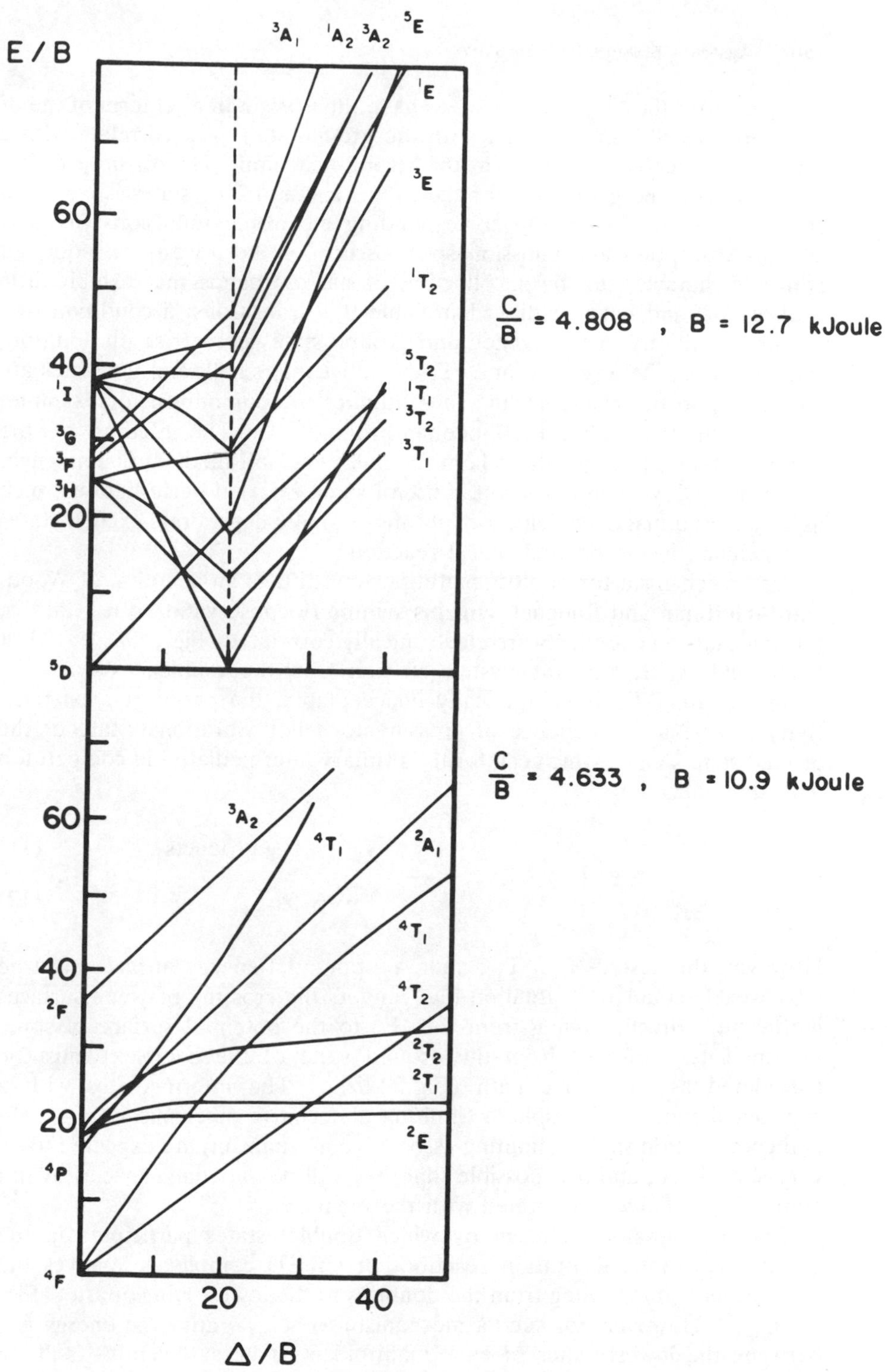

Figure 48. Tanabe–Sugano energy level diagrams for d^3 (bottom) and d^6 (top) metal ions.

O_h symmetry, the $^2T_{1g}$ and ^2E_g states have their origin in a ^2G term of the d^3 free ion (Fig. 48) and together with the ground state $^4A_{2g}$ correlate with a $(t_{2g})^3$ electronic configuration in the strong field limit. Hence, little difference between the geometries of the $^2T_{1g}$, ^2E_g, and $^4A_{2g}$ states is expected from the standpoint of the corresponding electronic repulsions in these states. Absorption and emission spectroscopy of the lowest lying doublet states is characterized by a collection of narrow bands that exhibit little Stokes shift and have readily identifiable 0–0 transitions; a condition that requires similarity of the excited and ground state geometries. In addition, the energy gap between ^2E_g and $^2T_{1g}$ is sufficiently small, and the states are considered to be in equilibrium in a number of compounds, for example, $Cr(bipyridine)_3^{3+}$ and $Cr(1,10\text{-phenanthroline})_3^{3+}$.[275] The lifetime of the doublets exhibit a particularly large temperature and medium dependence. In this regard, we can envision the use of small external perturbations, such as an external magnetic field, for splitting orbitally degenerate excited states that mediate in the photochemical reaction.[276]

One mechanism for photosubstitutions constitutes an adaption of Woodward–Hoffman and Longuet–Higgins symmetry conservation rules, that is, the reactants and products are electronically correlated (Fig. 49*a*).[277,278] For compounds where back intersystem crossing from the doublets $^2T_{1g}$ and ^2E_g to the quartet $^4T_{2g}$ is energetically unacceptable, the photochemical reactivity can be a consequence of crossing to a hot vibrational state of the ground state $^4A_{2g,v'}$, which can form a primary intermediate I in competition with thermalization

$$(^2E_g, {}^2T_{1g}) \rightarrow {}^4A_{2g,v'} \begin{cases} \rightsquigarrow I \rightarrow \text{Products} & (11) \\ \rightsquigarrow {}^4A_{2g,v} & (12) \end{cases}$$

However, the nested ^2E_g, $^2T_{1g}$, and $^4A_{2g}$ potential energy surfaces must be only weakly coupled, a situation that renders the crossing between surfaces inefficient. Direct crossing from the ^2E_g to the potential surface of some intermediate (different from the ground state of the complex) must be considered as alternative path (Fig. 49*b*).[261] The intermediates with a contracted coordination sphere (limiting dissociative mechanism) or expanded coordination sphere (limiting associative mechanism) are expected to be very short-lived, and it is possible that they will be undefined species by not being vibrationally equilibrated with the medium.

One additional mechanism by which doublet states participate in the quenchable photosubstitution reactions of Cr(III) complexes involves the back intersystem crossing from the doublets to the lowest lying quartet (Fig. 49*c*).[212,223] However, for such a mechanism to be operative the energy gap between the lowest states of each manifold must be sufficiently small. In many Cr(III) complexes the energy gap is too large, preventing the back thermal population of the quartet. This can be the case of $Cr(NH_3)_6^{3+}$ where

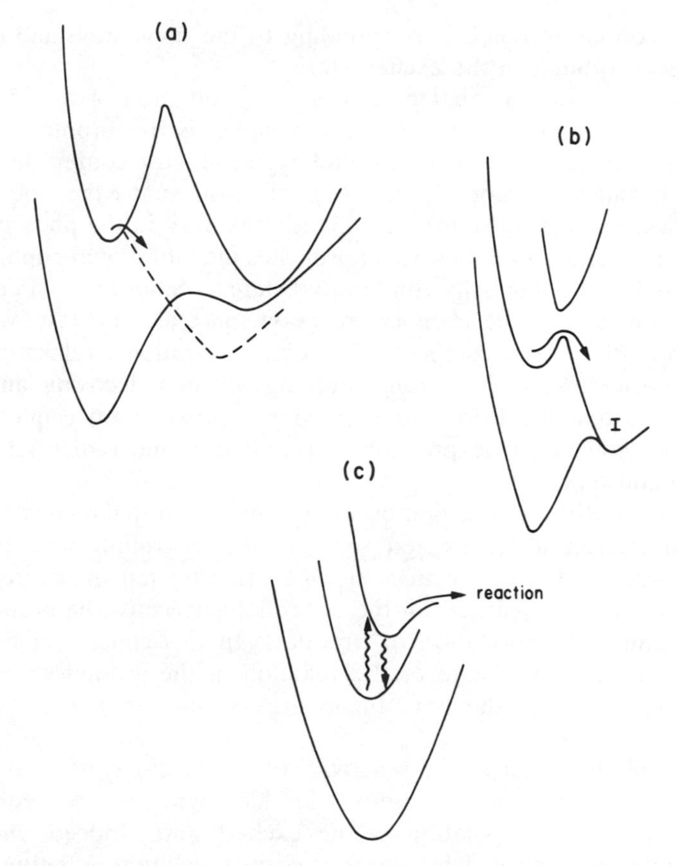

Figure 49. Potential curves describing photosubstitution mechanisms.

the energy gap between ${}^4T_{2g}$ and ${}^2E_g$, ${}^2T_{1g}$ is expected to be ~48 kJ/mol. For some complexes, however, the gap is small and the population of the lowest excited quartet via back intersystem crossing can make significant contributions to the overall photochemistry. $Cr(C_2O_4)_3^{3-}$ is one example where the separation between the quartet and doublet excited states is small. Moreover, a minute trigonal distortion from O_h symmetry, that is, less than 2.4 kJ/mol, must lead to a small splitting of ${}^4T_{2g}$ into ${}^4E$ and ${}^4A$, and a consequent decrease in the energy gap between the doublet and quartet manifolds.

A fraction of the Cr(III) photosubstitutions are not associated with the doublet states and cannot be quenched by quenching the doublets. This "prompt photochemistry" has been ascribed to the quartet state. In this regard, the mechanisms that have been proposed involve (a) vibrational modes, activated by absorption, mediating in the excited state reaction;[279] (b) internal conversion to a vibrationally excited species on the potential surface of the ground state;[261] and (c) photoreactivity associated with a selected weakening of bonds corresponding to the population and depopulation of given orbitals in the excited state.[230,231,280–286]

The absorption of light that promotes the population of the $^4T_{2g}$ in an octahedral or pseudooctahedral Cr(III) complex is a vibronic transition. That is, vibrational modes with t_{1u} and t_{2u} symmetry confer the correct character, through the vibronic coupling mechanism, to the upper state. These modes, in mechanism (a), are directly involved in the photochemical transformations as the excited state approaches the vibrational population of normal modes in the thermally equilibrated state.[279] Since thermal equilibration is extremely fast, the chemical reaction must also be fast, with rate constants $k > 10^{11}\ s^{-1}$, to compete with such a vibrational relaxation. This condition presupposes such a strong coupling between electronic and vibrational motions that the Born–Oppenheimer approximation cannot be applied in the deduction of expressions for radiation and radiationless relaxation rate constants.

In mechanism (b), the reaction occurs following internal conversion from the excited quartet to an excited species on the ground state potential energy surface.[261] Hence, the nuclear motions activated in the relaxation process and the point reached on the potential surface of the ground state would determine the formation of products. In this context, arrival at a point near the transition state of the reaction in the ground state can be regarded as entry onto the potential energy surface of a reaction intermediate.

It is possible to correlate the reactivity of the lowest lying quartet state (Fig. 49*c*) with weaker bonds, those that are lengthened as a consequence of the particular orbital population in the excited state. Indeed, the vibrationally equilibrated lowest lying quartet, corresponding to a limiting strong field configuration $(t_{2g})^2(e_g)^1$, and the ground state have different symmetries. The excited state exhibits an equatorial expansion and an axial contraction of the metal ligand bonds, which can be associated with distortion coordinates belonging to e_g (equatorial expansion) and a_{1g} (axial contraction) symmetry species (Appendix III). In this context, the relaxation of the Franck–Condon excited state along a "natural coordinate" of symmetry e_g must precede the reaction of the equilibrated quartet. Theoretical models of the mechanism, based on the MO and AOM theories, have used parameters appropriate to the ground state geometry.[230,231,280–286] In the Vanquickenborne–Coulemans model, bond energies are defined according to[230]

$$I(\text{M–L}) = -\sum_i n_i E_i \tag{13}$$

where the summation runs over the orbitals i (bonding and antibonding), n_i is the occupation number, and E_i is the energy of the ith orbital (Table 13). It is possible to associate the photolabilization of a given ligand to the change in bond energy $I(\text{M–L}) - I^*(\text{M–L})$ in passing from the ground state

TABLE 13 Bond Energies *I(M–L) in Excited Configurations t→e of d[3] and d[6] Complexes[a]**

		e			
		σ or π donor ($\pi \geq 0$)[b]		π acceptor ($\pi < 0$)[b]	
	t	z^2	x^2-y^2	z^2	x^2-y^2
		d[3] Complexes			
$I^*(M-L_{ax})$	*xy*	$\sigma_{ax} + 2\pi_{ax}$	$2\sigma_{ax} + 2\pi_{ax}$	$\sigma_{ax} - 2\pi_{ax}$	$2\sigma_{ax} - 2\pi_{ax}$
	xz, yz	σ_{ax}	$2\sigma_{ax} + 3\pi_{ax}$	$\sigma_{ax} - \pi_{ax}$	$2\pi_{ax} - \pi_{ax}$
$I^*(M-L_{eq})$	*xy*	$7/4\sigma_{eq} + 3\pi_{eq}$	$7/4\sigma_{eq} + 3\pi_{eq}$	$7/4\sigma_{eq} - \pi_{eq}$	$5/4\sigma_{eq} - \pi_{eq}$
	xz, yz	$7/4\sigma_{eq} + 5/2\pi_{eq}$	$7/4\sigma_{eq} + 5/2\pi_{eq}$	$7/4\sigma_{eq} - 3/2\pi_{eq}$	$5/4\sigma_{eq} - 3/2\pi_{eq}$
		d[6] Complexes			
$I^*(M-L_{ax})$	*xy*	σ_{ax}	$2\sigma_{ax}$	$\sigma_{ax} - 4\pi_{ax}$	$2\sigma_{ax} - 4\pi_{ax}$
	xz, yz	$\sigma_{ax} + \pi_{ax}$	$2\sigma_{ax} + \pi_{ax}$	$\sigma_{ax} - 3\pi_{ax}$	$2\sigma_{ax} - 3\pi_{ax}$
$I^*(M-L_{eq})$	*xy*	$7/4\sigma_{eq} + \pi_{eq}$	$5/4\sigma_{eq} + \pi_{eq}$	$7/4\sigma_{eq} - 3\pi_{eq}$	$5/4\sigma_{eq} - 3\pi_{eq}$
	xz, yz	$7/4\sigma_{eq} + 1/2\pi_{eq}$	$5/4\sigma_{eq} + 1/2\pi_{eq}$	$7/4\sigma_{eq} - 7/2\pi_{eq}$	$5/4\sigma_{eq} - 7/2\pi_{eq}$

[a]Data from Ref. 283 and references therein.
[b]The t→e excitation corresponds to the electronic transitions (*xy* or *xz, yz*)→(z^2 or x^2-y^2)

to the excited state. In this sense, the bond of the complex that experiences the largest positive increase in bond energy is regarded as the one that will dissociate. The photodissociation can also be related to the bond energies I^*(M–L) in the excited state. The labilized ligand is, hence, the one ligated by the bond with the most positive bond energy in the excited state. A good agreement has been found between experimental results and predictions about the labilized ligand based on I^*(M–L). The use of ground state parameters in the manner defined in the AOM model (Appendix IV and V; Fig. 50), allows us to obtain simple expressions for the ground and excited state bond energies. In this regard, we can specify parameters σ_2 and π_L for each ligand L. These parameters for ligands in opposite positions, for example, σ_x and σ_{-x}, appear paired, $\sigma_x + \sigma_{-x}$, in the perturbation matrix elements of a metal ion in an octahedral environment. It is convenient to replace such parameters by the corresponding average values, $2\bar{\sigma}_x = \sigma_x + \sigma_{-x}$. In an approximate D_{4h} symmetry the average values along two axes are nearly equal, $\bar{\sigma}_x \sim \bar{\sigma}_y = \sigma_{eq}$ and $\bar{\sigma}_z = \sigma_{ax}$. The one-electron energies of the five d orbitals are

$$E(d_z2) = 2\bar{\sigma}_{ax} + \bar{\sigma}_{eq} \tag{14}$$

$$E(d_x^2 - y^2) = 3\bar{\sigma}_{eq} \tag{15}$$

$$E(d_{xz}) = 2\bar{\pi}_{ax} + 2\bar{\pi}_{eq} \tag{16}$$

$$E(d_{yz}) = 2\bar{x}_{ax} + 2\bar{\pi}_{eq} \tag{17}$$

$$E(d_{xy}) = 4\pi_{\text{eq}} \tag{18}$$

In the AOM approximation, the ligand orbitals are considered to be stabilized to the same extent that the metal orbitals are destabilized. Table 13 gives the excited state bond energies for d^3 (Cr(III)) and d^6 (Co(III)) complexes. Such energies are obtained as a sum of the one-electron energies in eqs. 14–18. For the determination of the dominant configuration in the excited state, however, we must introduce interelectronic repulsion effects. The excited states ${}^4T_{2g}$ and ${}^4T_{1g}$ corresponding to the t^2e configuration in O_h symmetry are separated by an energy $12B$ and their strong field wave functions are

$${}^4B_{2g} = |(xz)(yz)(x^2 - y^2)| \tag{19}$$

$${}^4E_g(T_{2g}) = |(xz)(xy)\{\sqrt{3/2}(z^2) - \tfrac{1}{2}(x^2 - y^2)\}| \tag{20}$$

$$= |(yz)(xy)\{\sqrt{3/2}(z^2) - \tfrac{1}{2}(x^2 - y^2)\}| \tag{21}$$

$${}^4A_{2g} = |(xz)(yz)(z^2)| \tag{22}$$

$${}^4E_g(T_{1g}) = |(xz)(xy)\{-\tfrac{1}{2}(z^2) - \overline{3/2}(x^2 - y^2)\}| \tag{23}$$

$$= |(yz)(xy)\{-1/2(z^2) - \sqrt{3/2}(x^2 - y^2)\}| \tag{24}$$

The tetragonal field splittings are

$$E({}^4B_{2g}) - E({}^4E_g; {}^4T_{2g}) = 2(\bar{\pi}_{\text{ax}} - \bar{\pi}_{\text{eq}}) - \tfrac{3}{2}(\bar{\sigma}_{\text{ax}} - \bar{\sigma}_{\text{eq}})$$
$$= \tfrac{1}{2}(10\overline{Dq}_{\text{eq}} - 10\overline{Dq}_{\text{ax}}) \tag{25}$$

$$E({}^4A_{2g}) - E({}^4E_g; {}^4T_{1g}) = 2(\bar{\pi}_{\text{ax}} - \bar{\pi}_{\text{eq}}) + \tfrac{3}{2}(\bar{\sigma}_{\text{ax}} - \bar{\sigma}_{\text{eq}}) \tag{26}$$

where $10\overline{Dq} = 3\bar{\sigma} - 4\bar{\pi}$. The interaction between the ${}^4E_g$ state is

$$H_{ij} = \langle \Psi({}^4E_g; {}^4T_{2g}) | \hat{\mathscr{H}}^1 | \Psi({}^4E_g; {}^4T_{1g}) \rangle = \sqrt{3}(\bar{\sigma}_{\text{eq}} - \bar{\sigma}_{\text{ax}})/2 \tag{27}$$

where $\hat{\mathscr{H}}^1$ represents the Hamiltonian for the tetragonal perturbation. The mixing coefficient c of the perturbed wave function for the excited state

$$\Psi = N(\Psi({}^4E_g; {}^4T_{2g}) + c\Psi({}^4E_g; {}^4T_{1g}) \tag{28}$$

is given by

$$c = \frac{\sqrt{3}(\bar{\sigma}_{\text{eq}} - \bar{\sigma}_{\text{ax}})}{24B}$$

according to perturbation theory. The expression for c allows us to obtain the weight of each orbital d in the excited state wave function. One finds, for example, that the contribution x of the dz^2 orbital is

$$x = \frac{1}{N^2}\left(\frac{\sqrt{3}-c}{2}\right)^2 = \frac{1}{1+c^2}\left(\frac{\sqrt{3}-c}{2}\right)^2$$

Equation 25 shows that if $10Dq_{ax} < 1 - Dq_{eq}$, ^4E_g is the photoactive state. If $10Dq_{ax} > 10Dq_{eq}$, $^4B_{2g}$ is the photoactive state. For each excited state and each ligand one can estimate $I^*(M\text{–}L)$. In the case of *trans*-$Cr(NH_3)_4(NSC)Cl^+$, $\overline{Dq}_{ax}$ and $\overline{Dq}_{eq}$ are defined according to

$$Dq_{ax} = \tfrac{1}{2}[Dq(\mathrm{SCN}) + Dq(\mathrm{Cl})] \tag{29}$$

$$\overline{Dq}_{eq} = Dq(\mathrm{NH_3}) \tag{30}$$

The values for the spectrochemical parameters σ and π (Table 14) and $B \sim 8.34$ kJ for Cr(III) define an excited state with 84.6% from $(xz, yz)^1(xy)^1(z^2)^1$ and 15.4% from $(xz, yz)^1(xy)^1(x^2-y^2)^1$. Hence, the bond energies (expressed in kJ/mol) are

$$\begin{aligned} I^*(\mathrm{Cr\text{–}NCS}) &= 0.846[\sigma(\mathrm{NCS}) + 3\pi(\mathrm{NCS})] + 0.154[2\sigma(\mathrm{NCS}) + 3\pi(\mathrm{NCS})] \\ &= 101.8 \end{aligned} \tag{31}$$

$$\begin{aligned} I^*(\mathrm{Cr\text{–}Cl}) &= 0.846[\sigma(\mathrm{Cl}) + 3\pi(\mathrm{Cl})] + 0.154[2\sigma(\mathrm{Cl}) + 3\pi(\mathrm{Cl})] \\ &= 108.9 \end{aligned} \tag{32}$$

$$\begin{aligned} I^*(\mathrm{Cr\text{–}NH_3}) &= [0.846(7/4) + 0.154(5/4)]\sigma(\mathrm{NH_3}) \\ &= 143.1 \end{aligned} \tag{33}$$

TABLE 14 Spectrochemical Parameters for a Number of Cr^{III}–Ligand Interactions[a]

Ligand	$\Delta\sigma_L$ [b]	π_L [b]	$10Dq$ [b]
I^-	−3.098	0.775	13.29
Br^-	−2.503	1.013	14.12
Cl^-	−1.907	1.073	15.67
H_2O	−1.478	0.596	18.86
F^-	0.536	2.026	19.19
NCS^-	−0.918	0.453	21.12
OH^-	1.120	1.656	22.42
NH_3, en	0	0	25.68
CN^-	1.549	−0.346	31.71

[a] Data from Ref. 283 and references therein.
[b] Energies given in KJ/mol and the σ parameters are expressed with respect to NH_3, that is, $\Delta\sigma_L = \sigma_L - \sigma_{NH_3}$, while $\pi_{NH_3} \sim 0$. See also Table 13.

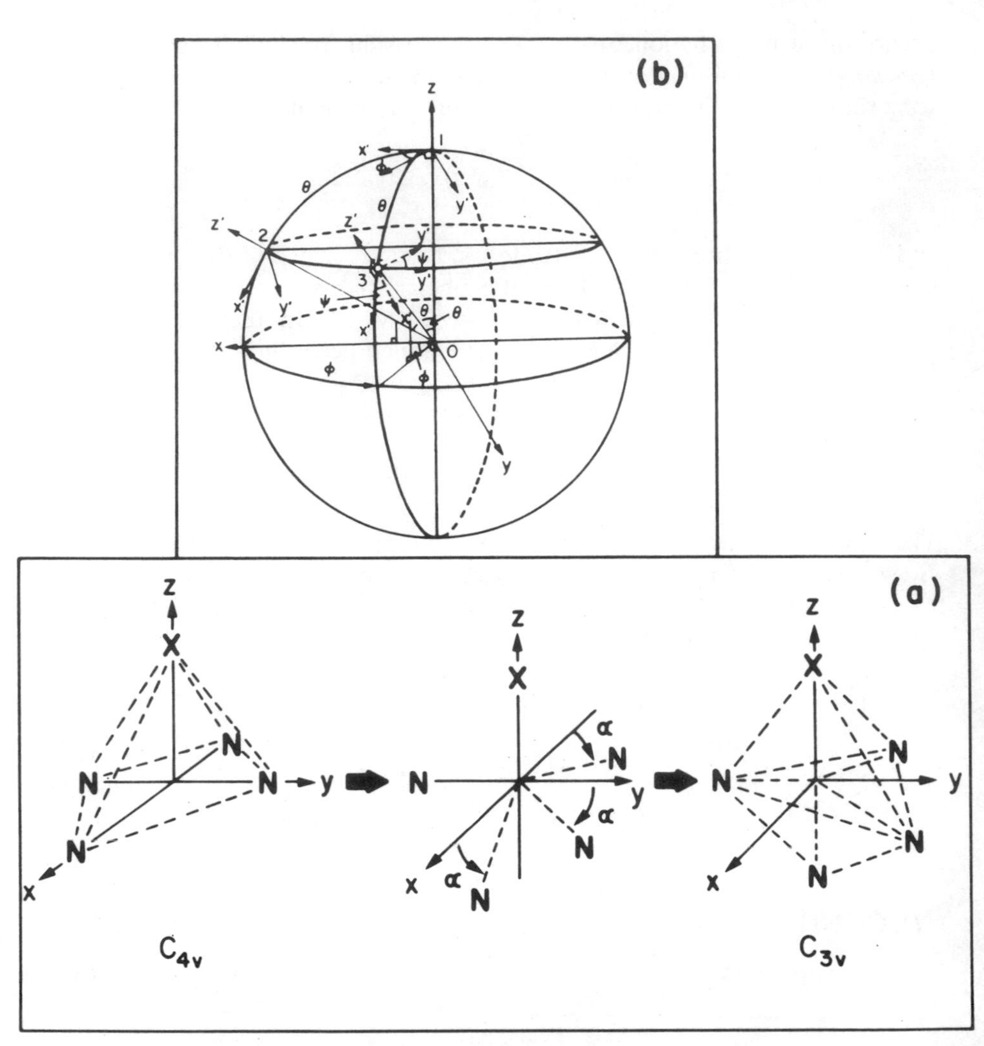

Figure 50. Illustration of the isomerization of a tetragonal pyramid (C_{4v} symmetry) into a trigonal bipyramid (C_{3v} symmetry) by a series of rotations about the center of coordinates (*a*). A general rotation in the three-dimensional space is illustrated in (*b*). In such rotations, we are concerned with two coordinate systems, (*XYZ*) and ($X'Y'Z'$), both centered at 0. The (*XYZ*) system is fixed in space, whereas ($X'Y'Z'$) is a system floating with respect to (*XYZ*). In this regard a transformation of coordinates is achieved with the Euler angles: the new basis is obtained by the rotation through Ψ about the Z axis, a rotation through θ about the newly generated Y'' axis, and a rotation through Ψ about the Z' axis until the ($X'Y'Z'$) system is reached.

Inspection of the bond energies in eqs. 31–33 shows that NCS must be the leaving ligand, a prediction that is supported by experimental results (Table 11). The agreement between the predictions of the model based on the excited state bond energies $I^*(M\text{–}L)$ and experimental results can be extended to compounds, namely *trans*-$Cr(en)_2F_+^2$, whose photochemistry is wrongly predicted by the Adamson's rules.

Kirk has summarized the stereomobility in the photosubstitution reactions of Cr(III) complexes as follows: "The entering ligand will stereospecifically occupy a position corresponding to entry into the coordination sphere trans to the leaving ligand."[287] Such stereomobility in most of the *trans*-diacidotetraamines or *trans*-diacidotetraammines means that only products with a *cis* configuration are produced in the photosubstitution reactions of Cr(III). There are, however, exceptions to the rule of total stereomobility, for example,

$$\begin{aligned} &trans\text{-}Cr(en)_2FCl^+ \xrightarrow[(-Cl^-)]{h\nu} (90\%)\,cis\text{-}Cr(en)_2(OH_2)F^{2+} \\ &\quad + (10\%)\,trans\text{-}Cr(en)_2(OH_2)F^{2+} \end{aligned} \tag{34}$$

$$trans\text{-}Cr(cyclam)Cl_2^+ \xrightarrow[(-Cl^-)]{h\nu} trans\text{-}Cr(cyclam)(OH_2)Cl^{2+} \tag{35}$$

which must be predicted by any mechanism with a general validity. In the approach of Vanquickenborne and Coulemans, the photodissociation of the excited state in a pentacoordinate intermediate (discussed above in terms of the AOM) is followed by two further steps: the isomerization of the intermediate and the association of the isomerized intermediate with the entering ligand.[283] The strategy of the model, therefore, is to construct state correlation diagrams, in the spirit of Woodward–Hoffman and Longuet–Higgins methodology, that reflect changes in the energies of various states as a function of a given reaction coordinate. In this regard, the isomerization of the tetragonal pyramid into a secondary trigonal bipyramid only involves changes of ligand–metal–ligand angles that describe the movement of the ligands on a spherical surface of fixed radius (Fig. 50). Furthermore, the association step progresses through changes of the metal–entering ligand distance and two ligand–metal–ligand angles to effect the rearrangement from the pentacoordinate bipyramid into the hexacoordinated octahedrum (Fig. 50). The variation of the radial part of the wave functions, that is, the variation associated with changes in the metal–entering ligand distance, are correlated with changes in the parameters σ_S and π_S of the entering ligand S. One further assumption that is required for the complete definition of the reaction coordinate is the constancy of the ratio σ_S/π_S at all values of the metal–entering ligand distance while the ligand–metal–ligand angles undergo identical and simultaneous changes that are concerted with the approach of the ligand S to the metal. Such variation of σ_S and π_S are incorporated in the AOM by means of

$$\sigma_S = \sigma_S^0 \left(\frac{\alpha}{30} \right)^n \tag{36}$$

where $n = \frac{1}{2}$ or 1, σ_S^0 is the value of the parameter at the equilibrium distance of the metal–entering ligand bond in the octahedrum, and α represents the difference $120-\theta$ between the 120° angle in the bipyramid and the actual value θ of the ligand–metal–ligand angles, for example, $120° \geq 90°$, $0° \leq \alpha \leq 30°$ for $\infty \geq$ M–S $\geq$ equilibrium bond distance. The simple manner by which eq. 36 couples the angular and radial displacements stands in contrast with the complicated dependence of the σ and l parameters on integrals containing the radial portions of the metal and ligand wave functions. Nevertheless, eq. 36 leads to the right predictions about the stereomobility of photosubstitution reactions.

For the final analysis of excited state reactivity one must combine the AOM description of the energy levels with information from Woodward–Hoffman and Longuet–Higgins symmetry-correlated state diagrams to establish the existence of activation barriers along the reaction paths shown in Figure 51.

Endicott has carried out AOM-based calculations which suggest that there are several pentacoordinated intermediates with energies equal to or less than the vibrationally equilibrated excited states.[288] Hence, the direct formation of trigonal bipyramidal intermediates from the vibrationally equilibrated state is plausible and removes the need for an isomerization step. Some heptacoordinated species, produced by the association of a solvent molecule with the excited state, seem to have the correct energies and can also be primary products. Among heptacoordinated intermediates, the capped trigonal prismatic geometry seems most likely to lead to the observed stereomobility in chromium photochemistry. Endicott has also devised a method for the calculation of specific activation energies associated with reaction paths connecting the hexacoordinated reactant with given pentacoordinated or heptacoordinated intermediates. One basic premise of the method is the assumption of the adiabatic transformation of the excited state to the primary product, a process evolving along the same surface as depicted in Figure 52. Also, the nuclear reorganizational barriers to surface crossings between the excited state and a ground state intermediate are expected to be significant in establishing some of the features of the relaxation path. In this approach, the activation energy E_a of the dissociation process

$$CrL_6 \longrightarrow CrL_5 + L \tag{37}$$

can be calculated with the expression $E_a \sim (\lambda/4)(1 - [\Delta E/\lambda]^2$, where λ is the nuclear reorganizational parameter and ΔE is the energy gap between the initial and final states. The parameter λ can be expressed in terms of the intrinsic reorganizational parameters of the hexacoordinate λ_{hex} and pentacoordinate λ_{pen} species, namely $\lambda = (\lambda_{ex} + \lambda_{pen})/2$. The intrinsic reorgani-

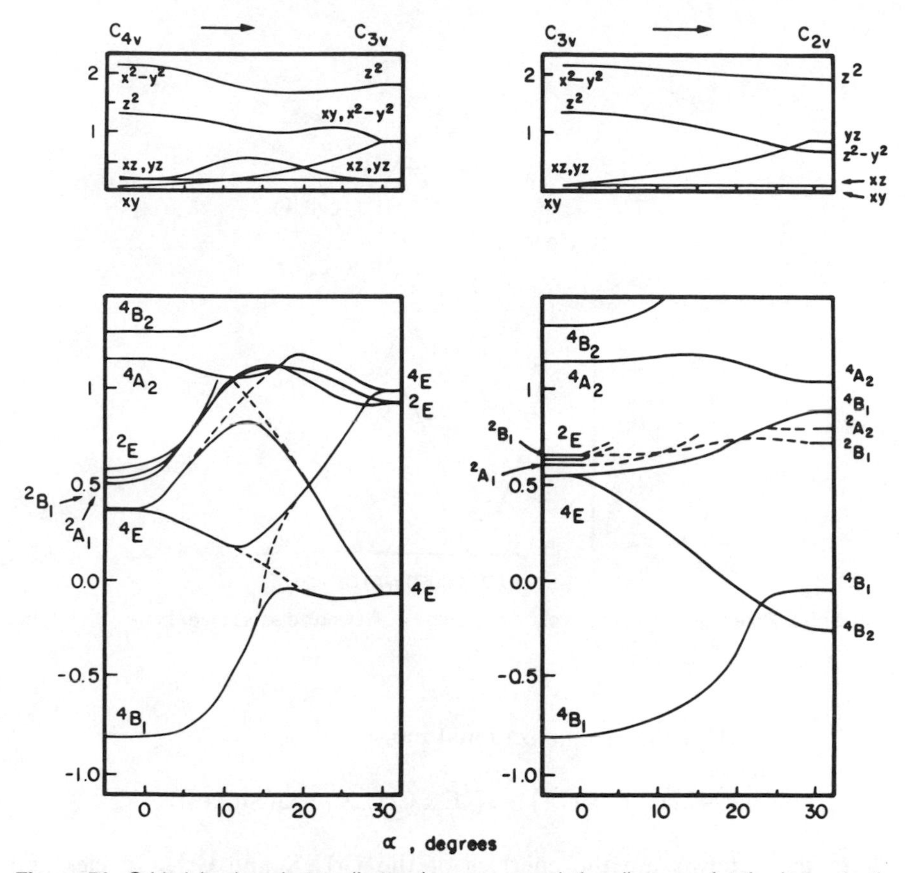

Figure 51. Orbital (top) and state (bottom) energy correlation diagrams for the isomerization of $Cr(NH_3)_4X^{2+}$ from tetragonal pyramid to trigonal bipyramid with X in an apical position (left) and an equatorial position (right).

zational parameters can be associated with the activation energies of the ligand exchange processes

$$CrL_6 + L' \rightleftharpoons CrL_5L' + L \tag{38}$$

$$CrL_5 + L' \rightleftharpoons CrL_4L' + L \tag{39}$$

Equations 38 and 39 describe reactions with a standard free energy $\Delta G^\circ \sim 0$ and activation energies $\Delta G_{\mathrm{i}}^{\ddagger} = \phi l_i/4$ where $\lambda_i = \lambda_{\mathrm{hex}}$ or λ_{pen}. Figure 52 shows that for a limiting dissociative behavior and small values of λ_{pen} the value of $\lambda_{\mathrm{hex}}/4$ is equal to the heterolytic bond dissociation energy of CrL_6. Therefore, for the adiabatic transformation

$$({}^4E_0)Cr(NH_3)_5 \xrightarrow{X} ({}^4B_1)Cr(NH_3)_4X + NH_3 \tag{40}$$

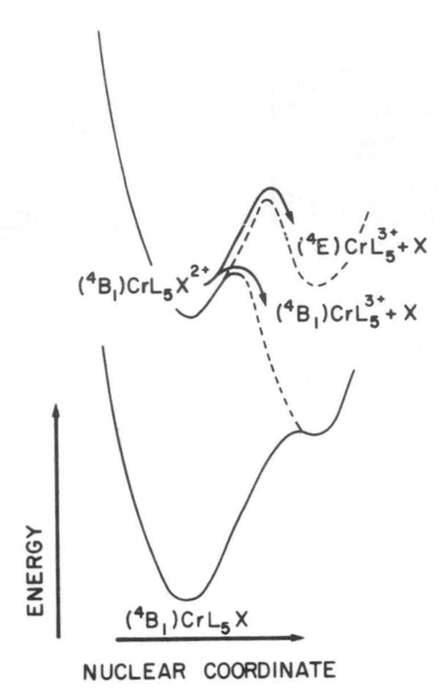

Figure 52. Potential energy curves describing a limiting dissociative behavior for the heterolytic dissociation of CrL_5X^{2+}.

($X = Cl^-$ or NH_3) the reorganizational parameter is

$$\lambda = 2\{\epsilon(^4E; CrL_5) - \epsilon(^4E_0; CrL_5X) + \Delta(\text{Stokes})\}$$

where the ϵ terms are the energies of the CrL_5X and CrL_5 species and Δ(Stokes) is the Stokes shift of the $(^4B_1)CrL_5 \rightarrow (^4E)CrL_5$ transition between square pyramidal species. If the Stokes shifts of $(^4B_1)CrL_5 \rightarrow (^4E)CrL_5X$ and $(^4B_1)CrL_5 \rightarrow (^4E)CrL_5$ are assumed to be the same, the activation energies ($E_a \sim 0$ for $X = Cl^-$ and $E_a \sim 16.9$ kJ for $X = NH_3$) suggest that the dissociation of the Cr–Cl bond takes place along this path, a result that contradicts experimental observations. However, eq. 40 represents an electronic forbidden transition, and the small value of the electronic matrix element selects against this path in favor of electronically allowed processes, such as those involving trigonal bipyramidal species.

6-3 PHOTOSUBSTITUTION REACTIONS OF d^6 METAL IONS

The metal-centered excited states of low-spin d^6 metal ion complexes involve the population of antibonding orbitals, for example, states associated with a transition $(t_{2g})^5(e_g)^1 \leftarrow (t_{2g})^6$. In this context, the geometries of the thermally equilibrated excited states associated with a $(t_{2g})^5(e_g)^1$ configuration are expected to be different to the ground state as a consequence of moderate enlargements (~10 pm) of the metal–ligand bonds in the excited states. For example, the $Co(CN)_6^{3-}$ ion exhibits a broad emission centered at $\lambda \sim 740$ nm which has been assigned to the $^3T_{1g} \rightarrow {}^1A_{1g}$ transition. Band shape analysis of such an emission was carried out with a function that represented a superposition of all possible vibronic transitions originating in the lowest vibrational state of the lowest excited electronic level. Such analysis suggests that the equilibrium nuclear configuration of the excited state is approximately octahedral with Co–C bonds that are ~10 pm longer than in the ground state.[289] It seems more reasonable, however, that the Stokes shift is related to Jahn–Teller-induced distortions in the excited state. The Franck–Condon least-squares fit of the emission, based on a t_{1u} odd-parity vibronic origin, lends support to the proposition of a tetragonally distorted excited state.[290,291] Magnetic field-induced change in the emission lifetime and photochemistry have been attributed to the Zeeman splitting of the tetragonally distorted excited state. This splitting is manifested, at room temperature and fluid solutions, as a dramatic dependence of the CN^- photolabilization yield (eq. 8) on the magnetic field intensity.[292]

There seem to be multiple patterns of photosubstitution reactivity among the low-spin d^6 metal complexes. The variety of photochemical behavior spans from the photoinert Co(III) amines and ammines, $\phi \leq 10^{-3}$, to the largely photoreactive, $\phi \sim 10^{-1}$, amines or ammines of Rh(III). One important difference between these compounds is that for Rh(III) compounds and $Co(CN)_6^{3-}$ the lowest lying excited state has an energy in excess of the energy required for the thermal (ground state) substitution, while for the Co(III), the lowest excited state energy is expected to be less than the energetic requirements for the ground state substitution reaction. Therefore, such a poor energetics can be regarded as the main cause of photoinertness in Co(III) amines or ammines in relationship to their photoreactive analogs, Co(III) cyano complexes, and compounds of Ru(II) and Rh(III).[227]

Information about the mechanism of photosubstitution reactions can be derived from the threshold energies for such photochemical reactions, namely the energies obtained by extrapolation to $\phi = 0$ in plots of the quantum yield versus photonic energy (action spectrum). These energies in the photochemistry of $Co(NH_3)_5Cl^{2+}$ are $E_{th}(NH_3) = 167$ kJ/mol for ammonia photolabilization and $E_{th}(Cl) = 109$ kJ/mol for chloride photolabilization, while the thermal aquation of chloride exhibits an activation energy $E_a = 100$ kJ/mol, comparable to $E_{th}(Cl)$.[277,293] It is possible to equate the larger threshold energy for ammonia photolabilization with an activation energy for the thermal hydrolysis of ammonia that is larger than the corresponding activation energy for chloride hydrolysis.

The photochemically reactive state of the Co(III) cyano complex is believed to be the lowest lying triplet state, a state that seems to be efficiently populated from upper states.[294] Such a population from upper states is considered to be fast, leading to constant values, $\phi \sim 0.3$ for

$254 \leq \lambda_{excit} \leq 365$ nm, of the photosolvation yields.[254] Moreover, nanosecond-lived intermediates have been assigned as the primary pentacoordinate products of the photosolvolysis, for example, species that are similar to the intermediates produced in thermal hydrolyses.[295] For amine or ammine Co(III) complexes, the lives of d metal-centered excited states seem to be restricted to a subnanosecond time domain. Transients with lifetimes $\tau = 140$ ps and $\tau = 40$ ps, assigned as dd excited states, have been detected following the flash excitation of *trans*-$Co(en)_2(NO_2)_2^+$ and *cis*-$Co(en)_2(NCS)Cl^+$, respectively.[296]

For Rh(III) amine complexes, the excited state luminescence is broad and exhibits two components in aqueous solutions at room temperature.[297,298] The high-energy component has been attributed to prompt fluorescence ($\tau \sim 10^{-10}$ s) from upper singlet states populated by the absorption light. A lower energy component ($\lambda_{max} \sim 700$ nm and $\tau \sim 10^{-8}$ s) has been assigned as phosphorescence from the lowest excited state, which is expected to be a triplet in accord with the ordering of energy levels for a d^6 metal ion in a strong field (Figs. 48 and 51). The lifetime of such a phosphorescence exhibits a temperature dependence (10^3–10^4 times longer lived at 77 K than in fluid solutions at room temperature) that, in the high-temperature level, is likely to be a consequence of the excited state activation-controlled radiationless relaxation. Such activation energy can be associated with the relative displacement of the ground and excited state potential surface discussed above in connection with the population of antibonding orbitals in the ligand field states.

The photolabilization in Rh(III) complexes is considered a dissociative process that leads to pentacoordinate primary products,[269,299,300]

$$({}^1A_2)Rh^{III}(NH_3)_4XY \xrightarrow{h\nu} (S_1)Rh^{III}(NH_3)_4XY \qquad (41)$$

$$(X, Y = NH_3, OH_2, \text{halogens})$$

$$(S_1)Rh^{III}(NH_3)_4XY \xrightarrow[\phi_{isc}]{k_{isc}} (T_1)Rh^{III}(NH_3)_4XY \qquad (42)$$

$$(T_1)Rh^{III}(NH_3)_4XY \xrightarrow[\phi_{nr}]{k_{nr}} ({}^1A_2)Rh^{III}(NH_3)_4XY \qquad (43)$$

$$(T_1)Rh^{III}(NH_3)_4XY \xrightarrow[\phi x]{kx} I + X \qquad (44)$$

$$(T_1)Rh^{III}(NH_3)_4XY \xrightarrow[\phi_y]{k_y} I' + Y \qquad (45)$$

$$I + H_2O \xrightarrow{\phi_{cis}} cis\text{-}({}^1A_1)Rh^{III}(NH_3)_4(OH_2)X \qquad (46)$$

$$I + H_2O \xrightarrow{\phi_{trans}} trans\text{-}({}^1A_1)Rh^{III}(NH_3)_4(OH_2)X \qquad (47)$$

$$I' + H_2O \xrightarrow{\phi'_{cis}} cis\text{-}({}^1A_1)Rh^{III}(NH_3)_4(OH_2)Y \quad (48)$$

$$I' + H_2O \xrightarrow{\phi'_{trans}} trans\text{-}({}^1A_1)Rh^{III}(NH_3)_4(OH_2)Y \quad (49)$$

The distribution of photoproducts in photolyses of various isotope-labeled Rh(III) amines, namely *cis*/*trans*-$Rh(NH_3)_4Cl_2^+$ and *cis*/*trans*-$Rh(NH_3)_4(OH_2)Cl^{2+}$, points to the formation of a common pentacoordinate intermediate, namely $Rh(NH_3)_4Cl^{2+}$, a behavior that is in accord with the finding of positive activation volumes for the photolabilization process.[301,302] Moreover, several studies of the Rh(III) photolabilization reactions have shown that intersystem crossing from singlet states (populated by the absorption of light) to the lowest triplet is a very efficient process, $\phi_{isc} \sim 1$ in eq. 42.[268] It seems likely that such a large intersystem crossing rate is related to an intense spin–orbit coupling perturbation that mixes states with nested (only slightly displaced) potential surfaces.

Questions have been raised about the assumptions made in proposing the mechanism of ligand and photolabilization for d^6 metal ion complexes. The partial quenching of the Rh(III) photosubstitution reactions by OH^- could signal, in principle, the participation of more than one excited state in the photochemistry of these complexes. It is possible, however, that OH^- is not an innocent quencher and forms adducts or amido-Rh(III) excited species with intrinsic chemical reactivities. Doubts have also been raised with regard to the prediction of the photochemical properties of d^6 metal ion complexes by using spectroscopic ligand field or ground state radial overlap parameters as in the case of the Cr(III) photosubstitution reactions. Insofar as there are appreciable differences in the metal–ligand bond lengths in the excited and ground states of d^6 metal ion complexes, there is no proof of the absolute validity of the model-defined parameters for the determination of excited state reactivities.[227]

For Ru(II) complexes, such as $Ru(NH_3)_5L$, the experimental evidence signals that large photolabilization yields (Table 12) are characteristic of compounds where the lowest energy, thermally equilibrated, excited state is ligand field instead of charge transfer.[271]

6-4 EXCEPTIONAL PHOTOREACTIVITY OF LIGAND FIELD EXCITED STATES

Most of the ruthenium(II) complexes undergo ligand photolabilization reactions which are attributed to the population of ligand field states, even when ligand field absorptions are hidden by strong charge transfer transitions. This is the case, for example, of $Ru(bipy)_3^{2+}$ where the ligand field state is thermally populated from the underlying MLCT charge transfer state.[271,303–305] In this context, it has been demonstrated that only Ru(II) complexes with low-lying ligand field states undergo photosubstitution reactions with large yields. For example, the photochemical transformation of $Ru(bipy)_3^{2+}$ in *trans*-$Ru(bipy)_2(X)_2$ complexes can be interpreted in terms of the reactivity of electronically excited pentacoordinate species similar to those described for d^6 and d^3 metal ion complexes

$$Ru(bipy)_3^{2+} + 2X^- \xrightarrow{h\nu} trans\text{-}Ru(bipy)_2(X)_2 + bipy \qquad (50)$$

Although the preceding sections suggest that photolabilization is the only photoreaction induced in ligand field states, examples have been found where these states participate in bimolecular electron transfer reactions. A well-documented example of such a behavior is provided by Cr(III) polypyridyl complexes, such as $Cr(bipy)_3^{3+}$, where outer sphere electron transfer reactions yield Cr(II) products

$$^*Cr(bipy)_3^{3+} + Fe^{2+} \longrightarrow Cr(bipy)_3^{2+} + Fe^{3+} \qquad (51)$$

In eq. 51, $^*Cr(bipy)_3^{3+}$ represents the thermal equilibrium distribution between ^2E and $^2T_{1g}$ states in O_h symmetry of ^2E, ^2E, ^2A in D_{3d} symmetry.[221] In such electron transfer reactions the excited state behaves similarly to that of the species in the ground state in the sense that one can use similar parametrizations based, for example, on Marcus–Hush theoretical treatments.[179–183,306]

A more striking reactivity has been exhibited by $Ir(bipy)_3^{3+}$ where the excitation induces linkage isomerization into an Ir–C ligated species[307–312]

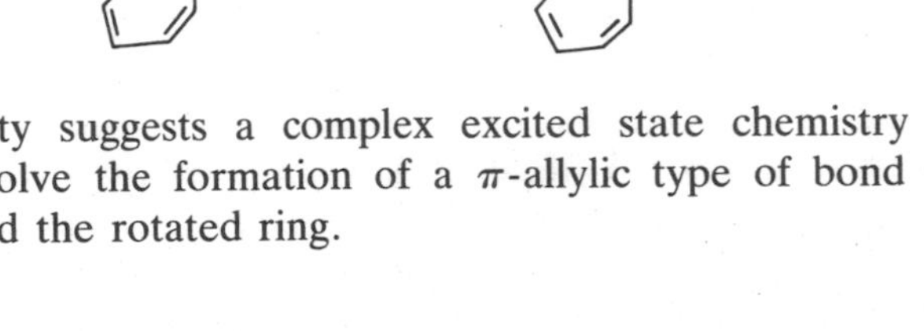

(52)

This reactivity suggests a complex excited state chemistry that can in principle involve the formation of a π-allylic type of bond between the excited Ir and the rotated ring.

7

ELEMENTS OF ORGANOMETALLIC PHOTOCHEMISTRY

7-1 EXCITED STATES IN ORGANOMETALLIC COMPOUNDS

The classification of the excited states in organometallic compounds as ligand field, charge transfer, ligand-centered, and metal–metal is based on the assumption that molecular orbitals in these complexes have very large contributions from either metal or ligand orbitals, as we would expect for highly ionic complexes.[313] Although such a classification can be applied to Werner-type complexes of the first and second row transition metals, organometallics are highly covalent compounds and the "metal-centered" d orbitals have considerable ligand character. The mixing of ligand character in the d orbitals has an important effect on the transition probabilities: ligand field transitions have extinction coefficients in the range of thousands. In other aspects, the pseudo ligand field states of organometallic compounds behave in a manner that bears strong resemblance to those of more simple coordination complexes.[314] For example, substitution reactions are induced in ligand field states of carbonyl compounds, a reactivity that was discussed in Chapter 4.

Insofar as organometallics have metals with small oxidation potentials and ligands whose empty π orbitals are placed at low energies, the CTTL excited states become readily available in these compounds. The CTTM states, however, are less important in determining the photochemical reactivity of organometallics than in Werner-type coordination complexes. It has been proposed that the complexes $Fe(\eta^5\text{-}C_5H_5)_2^+$, $V(CO)_6^-$, and $Re(CO)_5CH_3$ possess low-lying CTTM states that can in principle be involved in their photochemical reactions.[313,315–317]

In binuclear or cluster metallo complexes some orbitals are localized mainly over the metals centers. Transitions between metal-centered orbitals lead to the rupture or formation of metal–metal bonds. An example of this type of transition is found in the spectrum of $Mn_2(CO)_{10}$ which exhibits an intense absorption band ($\lambda_{max} \sim 375$ nm and $\epsilon_{max} > 10^4$ M^{-1} cm^{-1}) that is absent in mononuclear manganese carbonyl complexes, such as $Mn(CO)_5L$.[318] The transition has been assigned as $\sigma^* \leftarrow d(\pi)$, that is, between two molecular orbitals that are localized over the metal centers.

Charge transfer to solvent (CTTS) excited states do not seem to be common in organometallics. An absorption band, $\lambda_{max} \sim 307$ nm, appears in the spectrum of ferrocene, $Fe(\eta^5\text{-}C_5H_5)_2^+$, in halocarbon solvents but is not observed in solvents that are poorer electron acceptors, such as alcohols and cyclohexane. The new features observed in halocarbons has been assigned to a CTTS transition.[319,320]

Ligand-centered or intraligand transitions have been assigned, for example, to the lowest energy bands of fac-$ReCl(CO)_3L_2$ (L = *trans*-3- and *trans*-4-styrylpyridine) where the vibrational structure in these bands, at low temperature, is similar to those in the spectra of the free ligands.[321] Although we can expect these states to exhibit the same reactivity of the free ligand, intrinsic structural restraints in the complexes may induce new reactions.

7-2 METAL CARBONYLS

The bonding between CO and open-shell d^1–d^9 transition metal ions involves both a large π back bonding, associated with the delocalization of $d(\pi)$ metal electrons into empty antibonding π orbitals of the CO, and a large σ donation from filled orbitals of the CO that interact with empty metal orbitals of appropriate symmetry. Such electronic effects make the carbonyls very covalent compounds.[322,323] Nevertheless, the absorption bands in the electronic spectrum have been classified as LF, CTTL, or CTTM for transitions that involve the CO and other ligands, and ligand-centered.

Luminescence of carbonyl complexes has been associated with LF and CT states. The emission lifetimes of a number of Cr, Mo, and W carbonyls in glassy solutions or solids at 77 K are in the range of 10^{-5}–10^{-6} s[316] For low symmetry carbonyls (for example C_{2v}) of Cr, Mo, or W with derivatives of pyridine, such as 4-cyanopyridine, 4-benzoylpyridine, or bipyridine, it is possible that CTTL is the lowest and emissive state.[324] Rhenium carbonyls, such as $Re(CO)_3XL$ or $Re(CO)_3XL_2$ where X = Cl, Br, which have in addition electron acceptor ligands, for example, L = 1,10-phenanthroline, exhibit luminescence from CTTL states.[326] However, of the two emissions found with $Re(CO)_5X$ (3-benzoylpyridine)$_2$ in glassy solutions at 77 K, the short-lived (τ in the microsecond time range) has been assigned as emission from a CT state, while the long-lived (τ in the millisecond time range) has been assigned as emission from a ligand-centered state, $n\pi^*$. That the CT and $n\pi^*$ states are uncoupled is signaled by the two definite lifetimes associated with each state.

The large number of electronic excited states that are available upon excitation of transition metals carbonyls translates in a very rich photochemistry. It is not surprising, therefore, that photosubstitutions are important processes in the photochemistry of mononuclear carbonyls

$$M(CO)_{6-n}X_n \xrightarrow{h\nu} M(CO)_{5-n}X_nY + CO \quad (1)$$

$$M(CO)_{6-n}X_n \xrightarrow[Y]{} M(CO)_{6-n}X_{n-1}Y + X \quad (2)$$

The mechanism of these reactions has been investigated with scavengers and by flashphotolysis. Indeed, photolysis of $M(CO)_6$ at 77 K in methylcyclohexane glasses yields a long-lived intermediate that has been assigned as a coordinatively unsaturated species $M(CO)_5$ with C_{4v} symmetry. Infrared spectral changes observed at temperatures above 77 K have been interpreted as the isomerization of the primary species with C_{4v} symmetry into one of D_{3h} symmetry.[328–333] The $M(CO)_5$ intermediate has also been isolated in matrices, and from these studies it can be concluded that the species is extremely reactive, an observation that has been verified by flash photolysis.[334,335] Indeed, studies with picosecond time resolution show that the spectrum of the species is extremely dependent on the medium. This medium dependence has been attributed to the occupation of the sixth coordination position by molecules of the medium, that is, noble gasses, saturated hydrocarbons, and so on. In this context, the simplest mechanism for the photosubstitution reactions of carbonyls is

$$M(CO)_6 \xrightarrow{h\nu} [M(CO)_5, CO] \quad (3)$$

$$[M(CO)_5, CO] \xrightarrow[S]{} M(CO)_5S + CO \quad (4)$$

$$[M(CO)_5, CO] \longrightarrow M(CO)_6 \quad (5)$$

$$M(CO)_5S + L \longrightarrow M(CO)_5L + S \quad (6)$$

where the species in brackets are caged species, S is a very poor ligating molecule from the medium, for example, Ar in an Ar matrix, and L is a ligand capable of forming a stable product.

The photoreactivity described above is not different from the photosubstitution reactions of the Werner-type complexes and it is therefore possible to apply to the carbonyl photosubstitutions the same models discussed in Chapter 6. For carbonyls, the yields of these reactions depend on the excitation wavelength; that is; the yields increase with excitation energy.

One justification of this effect invokes the existence of two photoreactive LF states (Fig. 53) and assumes that thermal equilibration in each state is fast in comparison with bond dissociation.[324,336–338] The photochemistry of $M(CO)_5L$ ($L =$ aliphatic amine or NH_3) lends support to this mechanism. Indeed, these mixed ligand complexes exhibit photolabilization processes (eqs. 1 and 2) that can be assigned to orbital-specific effects. The population of ^1E or ^3E, LF states, induces the loss of L, a reactivity associated with an increase of the $\sigma^*(d_{z^2})$ character in the excited state bonding system of these complexes. In addition, the more efficient substitution of CO can be correlated with the population of LF excited states with higher energies than ^1E or ^3E, namely states presumably related to the population of the metal $d_{x^2-y^2}$.

In addition to photosubstitution reactions, the irradiation of metal carbonyls induces reactions in ligands coordinated to the metal. For example in the reaction

$$C_7H_7C(O)Mn(CO)_5 \xrightarrow{h\nu} (C_7H_7)Mn(CO)_3 + 3CO \qquad (7)$$

the reorganization of the bonds in the hydrocarbon takes place with photoelimination of carbon monoxide.[339] Reductive addition has also been observed with arene–carbonyl mixed ligand complexes[340]

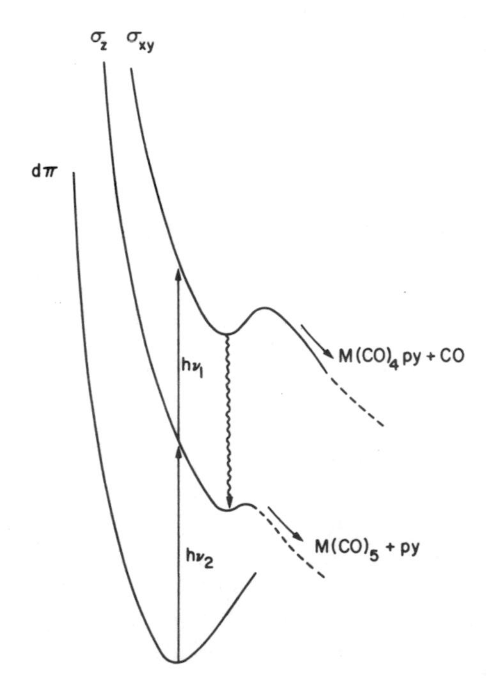

Figure 53. Potential curves describing the photodissociation process in $Mn(CO)_5Py$: labilization of equatorial CO and py.

$$Mn(\eta^5\text{-}C_5H_5)(CO)_3 \xrightarrow[SiCl_3H]{h\nu} Cl_3Si-Mn(CO)_3(\eta^6\text{-arene})(H) \tag{8}$$

Photolysis of dinuclear carbonyl complexes induces, in many cases, cleavage of the metal–metal bond. The intermediates are very reactive species that react with the solvent or other complexes[341]

$$Re_2(CO)_{10} \xrightarrow{h\nu}\ \xrightarrow{CCl_4} 2Re(CO)_5Cl \tag{9}$$

$$Re_2(CO)_{10} \xrightarrow{h\nu}\ \xrightarrow{Mn_2(CO)_{10}} 2MnRe(CO)_{10} \tag{10}$$

7-3 COMPOUNDS WITH σ (ALKYLIC) OR π (OLEFINIC) CARBON–METAL BONDS

Photolyses of some olefin complexes in the presence of hydrogen induce the hydrogenation of the olefin.[342–349] The photocatalytic process is observed with $M(CO)_6$ complexes, where the photoreaction involves the formation of an active catalyst. Indeed, the first step in the photoreaction with $M(CO)_6$ is the photosubstitution of CO

$$Cr(CO)_6 + 1,3\text{-diene} \xrightarrow{h\nu} Cr(1,3\text{-diene})(CO)_4 + 2CO \tag{11}$$

The $Cr(1,3\text{-diene})(CO)_4$ must be the receptor of a second photon in order to form the stable products. Most of the reported evidence suggests that $Cr(diene)(CO)_3$ is the main intermediate in the overall hydrogenation[350]

$$Cr(diene)(CO)_4 \xrightarrow{h\nu} Cr(diene)(CO)_3 + CO$$

$$Cr(diene)(CO)_3 \xrightarrow{+H_2} Cr(diene)(CO)_3H_2 \tag{12}$$

Olefin isomerization reactions are photoassisted by $M(CO)_6$ ($M = Mo$, W) in photoprocesses that involve the formation of olefin–carbonyl complexes.[351] Since prolonged photolysis leads to the total loss of the metal carbonyl, the proposed mechanism is multiphotonic in nature[351]

$$M(CO)_6 \xrightarrow[olefin]{h\nu} M(olefin)(CO)_5 + CO \tag{13}$$

$$M(olefin)(CO)_5 \xrightarrow[olefin]{h\nu} M(olefin)_2(CO)_4 + CO \tag{14}$$

$$M(olefin)(CO)_5 \xrightleftharpoons{h\nu,\ thermal} M(olefin')(CO)_5 \tag{15}$$

$$M(olefin)_2(CO)_4 \xrightleftharpoons{h\nu,\ thermal} M(olefin)(olefin')(CO)_4 \tag{16}$$

In such a process, olefin and olefin′ represent two cis/trans isomers. The mechanism of the photoisomerization (eq. 16) is believed to involve the formation of π-allyl hydrides as intermediates

R_1, R_2, R_3, R_4, $M(CO)_n$ —hν→ R, R, R_3, R_4, $H-M(CO)_{n-1}$ + CO (17)

A second possibility is the formation of "diradical species," which can be either excited states or short-lived intermediates

$M(CO)_n$ —hν→ [$M(CO)_n$]* → $\cdot M(CO)_n$ (18)

The formation of diradical species is well documented in the Cu(ethylene)$^{+}$ photoinduced polymerization of ethylene, where the process is initiated by the addition of ethylene to diradicals form in the primary process[352]

Cu^{+} ⇌(hν) Cu^{+} → Cu^{+} (19)

Dimerization reactions have also been induced in photolysis of carbonyl–olefin complexes of iron. For example, photolysis of $Fe(COT)(CO)_3$, COT = cyclooctatetraene, leads to dimeric species, which seem to originate in reactions of the primary species, $Fe(COT)(CO)_2$[353]

$Fe(COT)(CO)_3$ —hν, −CO→ $Fe(COT)(CO)_2$ —$Fe(COT)(CO)_3$→ $Fe(CO)_3$ / $Fe(CO)_3$ (20)

→ Fe—Fe, OC, OC, C O, CO, CO (21)

In addition to the dimerization, the reorganization of the cyclooctatetraene bonds generates a reactive intermediate that undergoes addition reactions with free hydrocarbons[354,355]

$Fe(COT)(CO)_3$ —hν→ $Fe(CO)_3$ —C_8H_8→ $Fe(CO)_3$ (22)

→ $Fe(CO)_3$ (23)

Copper olefin complexes exhibit a photochemistry extremely rich in valence isomerization and dimerization reactions.[356–359] For example, the CuOTF photoassisted dimerization of norbornene has been associated with the formation of reactive carbonions (eqs. 24 and 25)[357,358] However, the nature of the metal participation in the reaction is not clear. The metal can either facilitate the absorption of light and act as a template or have a direct participation through the population of specific excited states, such as CT

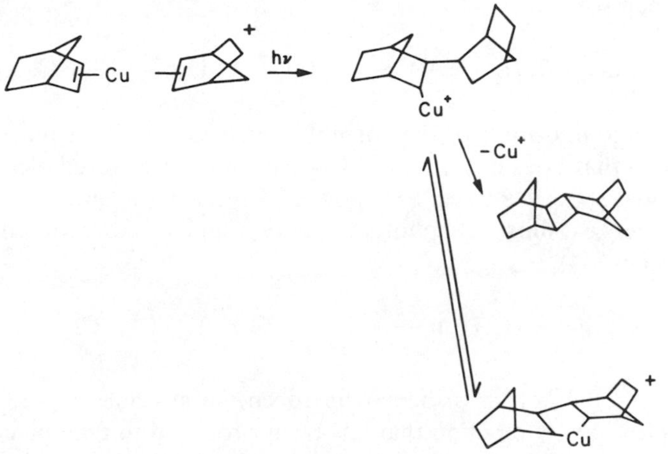

(24)

(25)

Some of these reactions, for example, the reorganization of the norbornene bonds, can also proceed by forming the diradical intermediates discussed above in connection with the polymerization of olefins (eq. 19). Nevertheless the copper(I) photoassisted valence isomerization of norbornadiene is proposed to involve an intermediate where a drastic bond reorganization has taken place[359]

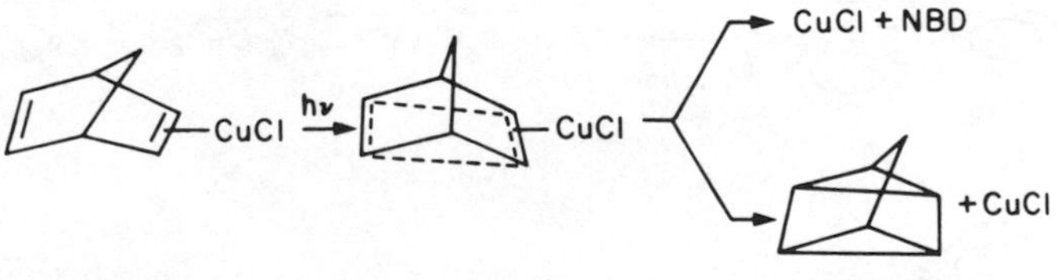

CuCl + NBD (26)

+ CuCl (27)

Metal to ligand charge transfer states could be responsible for much of the photoreactivity exhibited by the Cu(I)–olefin complexes. However, some copper(I) complexes can photocatalyze the valence isomerization of norbornadiene without forming Cu–norbornadiene complexes.[360] In these cases, energy transfer from the excited complex to a low-lying $\pi\pi^*$ excited state of the hydrocarbon is the most probable mechanism

$$Cu(BH_4)(Pph_3)_2 \overset{h\nu}{\rightleftharpoons} {}^*Cu(BH_4)(Pph_3)_2 \qquad (28)$$

$${}^*Cu(BH_4)(Pph_3)_2 + NBD \longrightarrow Cu(BH_4)(Pph_3)_2 + {}^*NBD \qquad (29)$$

$$^{*}\text{NBD} \longrightarrow \text{quadricyclene} \qquad (30)$$

For cyclopentadienyl complexes, the photochemical; features can be attributed to the population of CTTL and CTTM states. The photochemistry of ruthenocene in halocarbons, however, has been ascribed to the population of CTTS states that appear at appropriate energies in these solvents. Indeed, the compound is photoinert in hydrocarbons, namely cyclohexane, while undergoing a smooth oxidation to ruthenicenium in chlorocarbon solvents[361–363]

$$\text{Ru}(\eta^5\text{–C}_5\text{H}_5)_2 + \text{RCl} \xrightarrow{h\nu} \text{Ru}(\nu^5\text{–C}_5\text{H}_5)_2^+ + \text{R}^{\cdot} + \text{Cl}^- \qquad (31)$$

The photodecomposition of the solvent is associated with a reaction of the excited state that does not involve the generation of solvated electrons; that is the excited state itself acts as a powerful reducing agent.

Ferrocene exhibits a photoreactivity similar to ruthenocene in chlorocarbons[364,365]

$$\text{Fe}(\eta^5\text{–C}_5\text{H}_5)_2 \xrightarrow[\text{CCl}_4]{h\nu} [\text{Fe}(\eta^5\text{–C}_5\text{H}_5)_2]\text{Cl} + \text{CCl}_3^{\cdot} \qquad (32)$$

Moreover, the photolysis of benzoylferrocene in alcohols, eq. 33, produces solvated electrons, a reaction that has been proposed to take place in CTTS states lying at lower energies than other excited states (intramolecular states) of the complex.[366] Such ligand-centered states are photoreactive in compounds derived from ferrocene by substitution of organic groups on the cyclopentadienyl ring. For example, photolysis of benzoylferrocene (eq. 34) in polar solvents, for example, DMSO or DMF, induces photosolvation via cleavage of ring–metal bonds[367]

$$\text{Benzoylferrocene} \xrightarrow{\text{Alcohol}} \text{Fe}(\eta^5\text{-C}_5\text{H}_5)(\eta^5\text{-C}_5\text{H}_4\text{-}\overset{\text{O}}{\overset{\|}{\text{C}}}\text{-C}_6\text{H}_5)^+ + e^-_{solv} \qquad (33)$$

$$\text{Benzoylferrocene} \xrightarrow{\text{DMSO or DMF}} [\text{Fe}(\eta^5\text{-C}_5\text{H}_5)(\text{solvent})_2]\text{OCOPh} + \text{C}_5\text{H}_6 \qquad (34)$$

The alkyl complexes constitute another family of photoreactive compounds with metal–carbon bonds. Although earlier studies on these compounds regarded the alkyl–transition metal bond as similar to bonds in Grignard-type compounds, for example, CH_3MgCl, the considerable stability and characteristic reactivity of this family of transition metal complexes have led to the belief that the bond has a strong covalent character. The thermal stabilities of these compounds span a very broad range and the corresponding lifetimes in the dark, vary from a few nanoseconds to several years. A number of unstable alkyl complexes have been found as intermediates in reactions of alkyl radicals with Cu(II) and Cu(I) ions[368,369]

$$CH_3^{\cdot} \xrightarrow{Cu_{ag}^{2+}} CuCH_3^{2+} \xrightarrow{+H_2O} Cu_{aq}^{+} + CH_3OH \quad (35)$$

$$CH_3^{\cdot} \xrightarrow{Cu_{ag}^{+}} CuCH_3^{+} \xrightarrow{+H^+} Cu_{aq}^{2+} + CH_4 \quad (36)$$

Numerous photochemical studies signal that the photogeneration of reduced metallofragments and alkyl radicals is a rather common photoreaction.[370,371] The photolysis of benzylpentacyanocobaltatc(III),[372] for example, produces bibenzyl

$$2Co(CN)_5(CH_2C_6H_5) \xrightarrow{h\nu} 2Co(CN)_5^{3-} + (CH_2C_6H_5)_2 \quad (37)$$

This photobehavior has been equated to a favorable competition of the benzyl radical dimerization (eq. 38) versus the association of the radical to the Co(II) fragment (eq. 39)

$$C_6H_5CH_2^{\cdot} + C_6H_5CH_2 \longrightarrow (CH_2C_6H_5)_2 \quad (38)$$

$$C_6H_5CH_2^{\cdot} + Co(CN)_5^{3-} \longrightarrow Co(CN)_5CH_2C_6H_5 \quad (39)$$

Interception of the radical with O_2 is believed to lead to the formation of a peroxo complex (eqs. 40–42), which undergoes spontaneous decomposition into aquopentacyanocobaltate(III) and benzaaldehyde (eq. 43)

$$Co(CN)_5CH_2C_6H_5 \xrightarrow{h\nu} Co(CN)_5^{3-} + C_6H_5CH_2^{\cdot} \quad (40)$$

$$C_6H_5CH_2^{\cdot} + O_2 \longrightarrow C_6H_5CH_2O_2^{\cdot} \quad (41)$$

$$C_6H_5CH_2O_2^{\cdot} + Co(CN)_5^{3-} \longrightarrow Co(CN)_5(O_2CH_2C_6H_5)^{3-} \quad (42)$$

$$Co(CN)_5(O_2CH_2C_6H_5)^{3-} \xrightarrow[H_2O]{} Co(CN)_5OH_2^{2-} + C_6H_5CHO + HO^- \quad (43)$$

The similarity of the yields for the disappearance of the benzyl complex in the absence of O_2, $\phi \sim 0.13$, and in the presence of O_2, $\phi \sim 0.15$, provides strong support to the idea that the primary photoreaction is, in either case, a photohomolytic dissociation of the cobalt–carbon bond.

A large number of alkyl complexes that contain cobalt(III) coordinated to a macrocyclic ligand have been found to be extremely photosensitive.[167,373–375] The photohomolytic dissociation of the alkyl–cobalt bond has a high yield, on the order of 10^{-1}, that is independent of the excitation energy, and in consequence the threshold energy for photochemical reactivity seems to be placed below the energy of the lowest spectroscopic transitions. This photochemical behavior contrasts with the one of acidopentaammino complexes, $Co(NH_3)_5X^{2+}$ with $X = Cl^-$, Br^-, N_3^-, NO_2^-, as shown in Figure 54. Inspection of the absorption spectrum, however, shows that there is little difference between the spectra of acidopentaamminocobalt(III) complexes and alkyl–cobalt(III) complexes. Despite this fact, the lowest energy band, $\lambda_{max} \sim 547$ nm, in the spectrum of $Co([14]aneN_4)(OH_2)CH_3^{2+}$ has been assigned as a transition with CTTM character that involves the promotion of an electron from an alkyl–cobalt bonding orbital to a metal-centered, $d_{x^2-y^2}$ antibonding orbital.[373,376] Various considerations on the energetics of the photochemical reaction are in accord with this spectroscopic assignment. An alternative mechanism considers that a LF excited state could have the appropriate electronic distribution and sufficient energy for inducing the homolytic bond dissociation. The differences beween the two mechanisms are illustrated in Figure 55. One could expect that homologous complexes of d^6 metal ions with LF states

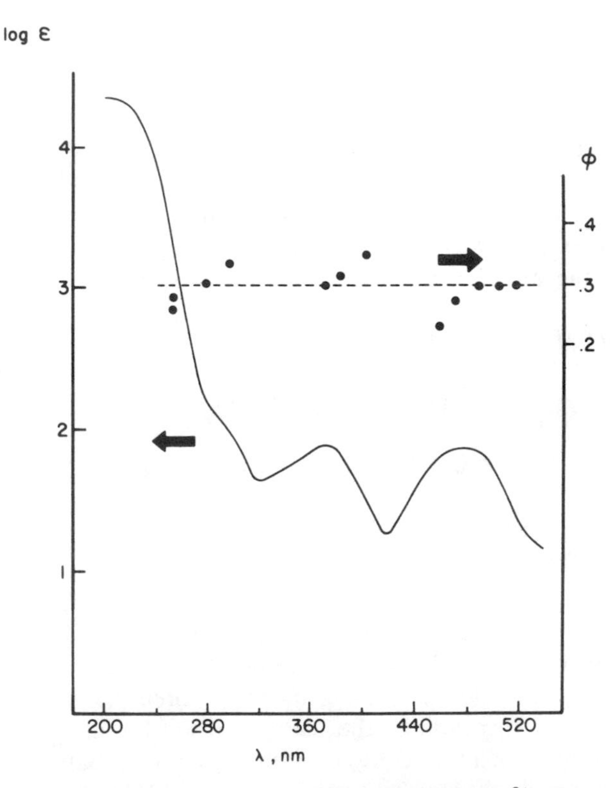

Figure 54. Absorption and action spectra of Co ([14]aneN$_4$)(OH$_2$)CH$_3^{2+}$. The yields correspond to the Co–alkyl photohomolysis.

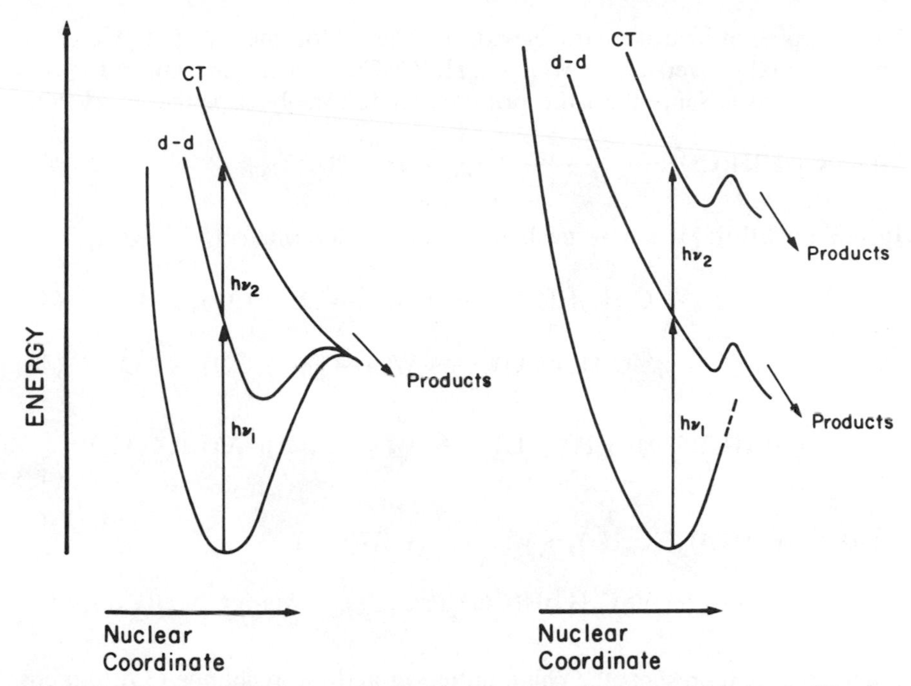

Figure 55. Possible potential curves for the photohomolysis of the Co–C bond in Co(III)–alkyl complexes.

placed at higher energies will show quantum yields independent of the excitation energy according to the second mechanism. The fact that the quantum yield for the homolytic dissociation of the Rh–C bond in photolysis of $Rh(NH_3)_4(OH_2)C_2H_5^{2+}$ depends on the excitation energy (eq. 44) lends some support to the proposition of a primary photoreaction initiated in a state with a high degree of charge transfer character[377]

$$Rh(NH_3)_4(OH_2)C_2H_5^{2+} \xrightarrow{h\nu} Rh(NH_3)_4^{2+} + C_2H_5^{\cdot} \tag{44}$$

7-4 HYDRIDE COMPLEXES

The hydride complexes exhibit an interesting photochemistry with potential practical applications to the storage of solar energy and carbon dioxide fixation.

A photoinduced chain mechanism has been proposed for the photosubstitution reaction of $W(\eta^5\text{–}C_5H_5)H(CO)_3$, where the yields for 311-nm excitations are in the range of 6 to 30[378,379]

$$W(\eta^5\text{–}C_5H_5)H(CO)_3 \xrightarrow[P(C_4H_9)_3]{h\nu} W(\eta^5\text{–}C_5H_5)H(CO)_2P(C_4H_9)_3 + CO \tag{45}$$

The trapping of hydrogen atoms with triphenylchloromethane, $(C_6H_5)_3CCl$, and the total conversion to $W(\eta^5{-}C_5H_5)Cl(CO)_3$ in the presence of such a scavenger gives support to the primary photohomolysis of the W–H bond

$$W(\eta^5{-}C_5H_5)H(CO)_3 \xrightarrow[(C_6H_5)_3CCl]{h\nu} W(\eta^5{-}C_5H_5)Cl(CO)_3 \tag{46}$$

In this regard the reaction mechanism can be described by

$$W(\eta^5 - C_5H_5)H(CO)_3 \xrightarrow{h\nu} W(\eta^5{-}C_5H_5)(CO)_3 + H^{\cdot} \tag{47}$$

$$W(\eta^5{-}C_5H_5)(CO)_3 \longrightarrow W(\eta^5{-}C_5H_5)(CO)_2 + CO \tag{48}$$

$$W(\eta^5{-}C_5H_5)(CO)_2 + P(C_4H_9)_3 \longrightarrow W(\eta^5{-}C_5H_5)(CO)_2P(C_4H_9)_3 \tag{49}$$

$$W(\eta^5{-}C_5H_5)(CO)_2P(C_4H_9)_3 + W(\eta^5{-}C_5H_5)H(CO)$$
$$\longrightarrow W(\eta^5{-}C_5H_5)H(CO)_2P(C_4H_9)_3 + W(\eta^5{-}C_5H_5)(CO)_3 \tag{50}$$

where eq. 47 represents the chain initiation and the reactions shown in eqs. 48–50 may lead to a chain length of about 2000.

The photoelimination of hydrogen does not necessarily proceed through radicals and can also be highly reversible. For example, the photoelimination of hydrogen from $CoH_2(ppy)(PR_3)_2$, ppy = phen or bipy and PR_3 = $P(C_2H_5)_3$, $P(C_2H_5)_2(C_6H_5)$, can be prevented in an atmosphere of H_2[380]

$$CoH_2(ppy)(PR_3)_2^+ \xrightleftharpoons{h\nu} Co(ppy)(PR)_3)_2^+ + H_2 \tag{51}$$

A reversible photoelimination of hydrogen is also found in the family of Ir complexes.[381,382] Indeed, 360-nm photolysis of $IrH_2Cl(P(C_6H_5)_3)_3$ induces elimination of hydrogen, a reaction that reverses in an H_2 atmosphere. This reversibility is associated by the trapping of a very reactive species, $IrCl(P(C_6H_5)_3)_3$, by hydrogen (eqs. 52 and 53). In the absence of hydrogen or in prolonged photolysis the intermediate product undergoes a reversible *ortho*-metallation reaction

$$IrClH_2(P(C_6H_5)_3)_3 \xrightleftharpoons{h\nu} IrCl(P(C_6H_5)_3)_3 + H_2 \tag{52}$$

$$IrCl(P(C_6H_5)_3)_3 \xrightleftharpoons{h\nu \text{ or thermal}} \text{[Ir: H, Cl, } (C_6H_5)_3P, P(C_6H_5)_2 \text{ (ortho-metallated } C_6H_4), P(C_6H_5)_3] \tag{53}$$

A similar *ortho*-metallation reaction is observed with $IrH_3(P(C_6H_5)_3)_3$

$$IrH_3(P(C_6H_5)_3)_3 \xrightarrow{h\nu} H_2 + IrH(P(C_6H_5)_3)_3 \tag{54}$$

$$IrH(P(C_6H_5)_3)_3 \longrightarrow \text{(H)(H)Ir(C}_6\text{H}_4\text{)(P(C}_6\text{H}_5)_3)_2\text{(P(C}_6\text{H}_5)_2) \tag{55}$$

The reactions shown in eqs. 52–55 do not require radical formation, that is, they do not proceed through photohomolysis, and it is unlikely that they involve the formation of H^+ or H^- ions. In these contexts, the photoelimination of hydrogen can be regarded as a concerted path that involves metal–hydrogen bond breaking as well as hydrogen–hydrogen bond formation[382]

$$\text{M(H)}_2 \xrightarrow{h\nu} {}^*[\text{M}\cdots\text{H}_2] \longrightarrow [\text{M} + H_2] \tag{56}$$

The photochemistry of other hydrides is in general very complex due to the generation of very reactive intermediates, which can lead to a variety of products.[383] For example, ReH_3LL where $LL = (C_6H_5)_2P{-}CH_2CH_2{-}P(C_6H_5)_2$, photoeliminates hydrogen and forms a very reactive species that is able to coordinate N_2, CO, C_2H_4, CO_2[384]

$$ReH_3(LL)_2 \xrightarrow{h\nu} ReH(LL)_2 + H_2 \tag{57}$$

$$ReH(LL)_2 \xrightarrow{N_2} Re(N_2)H(LL)_2 \tag{58}$$

$$ReH(LL)_2 \xrightarrow{CO} ReCOH(LL)_2 \tag{59}$$

$$ReH(LL)_2 \xrightarrow{C_2H_4} Re(C_2H_4)H(LL)_2 \tag{60}$$

$$ReH(LL)_2 \xrightarrow{CO_2} Re(HCO_2)(LL)_2 \tag{61}$$

The reaction with CO_2 (eq. 61) produces a formato complex and causes in consequence the fixation of carbon dioxide. Similar reactions of Re complexes have been used for this purpose.[385]

APPENDIX I

PHYSICAL CONSTANTS AND CONVERSION FACTORS FOR PHOTOCHEMICAL WORK

General Constants

Avogadro's number	$N = 6.02252 \times 10^{23}$
Boltzmann's constant	$k = 1.38054 \times 10^{-16}$ erg/deg
Bohr radius	$a_0 = 0.52917$ Å
Bohr magneton	$eh/2mc = 9.2732 \times 10^{-21}$ erg/G
Electronic charge	$e = 4.80298 \times 10^{-10}$ esu
Electronic rest mass	$m_e = 9.10908 \times 10^{-28}$ g
Electron/proton moment ratio	$g_e\beta_e/g_N\beta_N = 658.230$
Faraday constant	$F = 9.6487 \times 10^4$ C/gew
Gas constant	$R = 1.9872$ cal/deg mol
	$= 8.3143$ J/deg mol
	$= 8.205 \times 10^{-2}$ L atm/deg mol
g value of free electron	$g_e = 2.002322$
g value of proton	$g_H = 5.58536$
Gyromagnetic ratio of proton	$g_H = 2.6752 \times 10^4$ rad/s G
Molar volume (ideal STP)	2.2414×10^4 cm^3/mol
Nuclear magneton	$\beta_N = eh/2M_pc = 5.0505 \times 10^{-24}$ erg/G
Plank's constant	$h = 6.62559 \times 10^{-27}$ erg s
Proton mass	$M_H = 1.67252 \times 10^{-24}$ g

Energy Conversion Factors

cm^{-1}/erg	1.1963×10^8
cm^{-1}/cal	2.8591
cm^{-1}/eV	1.2398×10^{-4}

cal/eV	4.3363×10^{-5}
J/cal	0.2390

Photonic Energy

For photonic energies $E_{h\nu}$ (in kJ)

$$E_{h\nu} = 1.1917 \times 10^{5}/\lambda = 1.1917 \times 10^{-2}\tilde{\nu}$$

where λ is in nm, and $\tilde{\nu}$ in cm^{-1}.

APPENDIX II

CHARACTER TABLES FOR SYMMETRY GROUPS

The Nonaxial Groups

C_1	E
A	1

C_8	E	σ_h		
A′	1	1	T_x, T_y, R_z	x^2, y^2, z^2, xy
A″	1	−1	T_z, R_x, R_y	yz, xz

C_i	E	i		
A_g	1	1	R_x, R_y, R_z	$x^2, y^2, z^2,$
A_u	1	−1	x, y, z	xy, xz, yz

The C_n Groups

C_2	E	C_2		
A	1	1	z, R_z	x^2, y^2, z^2, xy
B	1	−1	x, y, R_x, R_y	yz, xz

C_3	E	C_3	C_3^2		$\epsilon = \exp(2\pi i/3)$
A	1	1	1	z, R_z	$x^2 + y^2, z^2$
E	$\begin{Bmatrix} 1 \\ 1 \end{Bmatrix}$	$\begin{matrix} \epsilon \\ \epsilon^* \end{matrix}$	$\begin{matrix} \epsilon^* \\ \epsilon \end{matrix}$	$(x, y)(R_x, R_y)$	$(x^2 - y^2, xy)(yz, xz)$

C_4	E	C_4	C_2	C_4^3		
A	1	1	1	1	z, R_z	x^2+y^2, z^2
B	1	−1	1	−1		x^2-y^2, xy
E	1	i	−1	$-i$	$(x, y)(R_x, R_y)$	(yz, xz)
	1	$-i$	−1	i		

C_5	E	C_5	C_5^2	C_5^3	C_5^4		$\epsilon = \exp(2\pi i/5)$
A	1	1	1	1	1	z, R_z	x^2+y^2, z^2
E	1	ϵ	ϵ^2	ϵ^{2*}	ϵ^*	$(x, y)(R_x, R_y)$	(yz, xz)
	1	ϵ^*	ϵ^{2*}	ϵ^2	ϵ		
E_2	1	ϵ^2	ϵ^*	ϵ	ϵ^{2*}		(x^2-y^2, xy)
	1	ϵ^{2*}	ϵ	ϵ^*	ϵ^2		

C_6	E	C_6	C_3	C_2	C_3^2	C_6^5		$\epsilon = \exp(2\pi i/6)$
A	1	1	1	1	1	1	z, R_z	x^2+y^2, z^2
B	1	−1	1	−1	1	−1		
E_1	1	ϵ	$-\epsilon^*$	−1	$-\epsilon$	ϵ^*	(x, y)	(xz, yz)
	1	ϵ^*	$-\epsilon$	−1	$-\epsilon$	ϵ	(R_z, R_y)	
E_2	1	$-\epsilon^*$	$-\epsilon$	1	$-\epsilon^*$	$-\epsilon$		(x^2-y^2, xy)
	1	$-\epsilon$	$-\epsilon^*$	1	$-\epsilon$	$-\epsilon^*$		

C_7	E	C_7	C_7^2	C_7^3	C_7^4	C_7^5	C_7^6		$\epsilon = \exp(2\pi i/7)$
A	1	1	1	1	1	1	1	z, R_z	x^2+y^2, z^2
E_1	1	ϵ	ϵ^2	ϵ^3	ϵ^{3*}	ϵ^{2*}	ϵ^*	(x, y)	(xz, yz)
	1	ϵ^*	ϵ^{2*}	ϵ^{3*}	ϵ^3	ϵ^2	ϵ	(R_x, R_y)	
E_2	1	ϵ^2	ϵ^{3*}	ϵ^*	ϵ	ϵ^3	ϵ^{2*}		(x^2-y^2, xy)
	1	ϵ^{2*}	ϵ^3	ϵ	ϵ^*	ϵ^{3*}	ϵ^2		
E_3	1	ϵ^3	ϵ^*	ϵ^2	ϵ^{2*}	ϵ	ϵ^{3*}		
	1	ϵ^{3*}	ϵ	ϵ^{2*}	ϵ^2	ϵ^*	ϵ^3		

C_8	E	C_8	C_4	C_2	C_4^3	C_8^3	C_8^5	C_8^7		$\epsilon = \exp(2\pi i/8)$
A	1	1	1	1	1	1	1	1	z, R_z	x^2+y^2, z^2
B	1	−1	1	1	1	−1	−1	−1		
E_1	1	ϵ	i	−1	$-i$	$-\epsilon^*$	$-\epsilon$	ϵ^*	(x, y)	(xz, yz)
	1	ϵ^*	$-i$	−1	i	$-\epsilon$	$-\epsilon^*$	ϵ	(R_x, R_y)	
E_2	1	i	−1	1	−1	$-i$	i	$-i$		(x^2-y^2, xy)
	1	$-i$	−1	1	−1	i	$-i$	i		
E_3	1	$-\epsilon$	i	−1	$-i$	ϵ^*	ϵ	$-\epsilon$		
	1	$-\epsilon^*$	$-i$	−1	i	ϵ	ϵ^*	$-\epsilon$		

The D_n Groups

D_2	E	$C_2(z)$	$C_2(y)$	$C_2(x)$		
A	1	1	1	1		x^2, y^2, z^2
B_1	1	1	−1	−1	z, R_z	xy
B_2	1	−1	1	−1	y, R_y	xz
B_3	1	−1	−1	1	x, R_x	yz

D_3	E	$2C_3$	$3C_2$		
A_1	1	1	1		$x^2 + y^2, z^2$
A_2	1	1	−1	z, r_z	
E	2	−1	0	$(x, y)(R_x, R_y)$	$(x^2 - y^2, xy)(xz, yz)$

D_4	E	$2C_4$	$C_2(=C_4^2)$	$2C_2'$	$2C_2''$		
A_1	1	1	1	1	1		$x^2 + y^2, z^2$
A_2	1	1	1	−1	−1	z, R_z	
B_1	1	−1	1	1	−1		$x^2 - y^2$
B_2	1	−1	1	−1	1		xy
E	2	0	−2	0	0	$(x, y)(R_x, R_y)$	(xz, yz)

D_5	E	$2C_5$	$2C_5^2$	$5C_2$		
A_1	1	1	1	1		$x^2 + y^2, z^2$
A_2	1	1	1	−1	z, R_z	
E_1	2	2 cos 72°	2 cos 144°	0	$(x, y)(R_x, R_y)$	(xz, yz)
E_2	2	2 cos 144°	2 cos 72°	1		$(x^2 - y^2, xy)$

D_6	E	$2C_6$	$2C_3$	C_2	$3C_2'$	$3C_2''$		
A_1	1	1	1	1	1	1		$x^2 + y^2, z^2$
A_2	1	1	1	1	−1	−1	z, R_z	
B_1	1	−1	1	−1	1	−1		
B_2	1	−1	1	−1	−1	1		
E_1	2	1	−1	−2	0	0	$(x, y)(R_x, R_y)$	(xz, yz)
E_2	2	−1	−1	2	0	0		$(x^2 - y^2, xy)$

The C_{nv} Groups

C_{2v}	E	C_2	$\sigma_v(xz)$	$\sigma_v'(yz)$		
A_1	1	1	1	1	z	x^2, y^2, z^2
A_2	1	1	−1	−1	R_z	xy
B_1	1	−1	1	−1	x, R_y	xz
B_2	1	−1	−1	1	y, R_x	yz

C_{3v}	E	$2C_3$	$3\sigma_v$		
A_1	1	1	1	z	$x^2 + y^2, z^2$
A_2	1	1	−1	R_z	
E	2	−1	0	$(x, y)(R_x, R_y)$	$(x^2 - y^2, xy)(xz, yz)$

C_{4v}	E	$2C_4$	C_2	$2\sigma_v$	$2\sigma_d$		
A_1	1	1	1	1	1	z	$x^2 + y^2, z^2$
A_2	1	1	1	−1	−1	R_z	
B_1	1	−1	1	1	−1		$x^2 - y^2$
B_2	1	−1	1	−1	1		xy
E	2	0	−2	0	0	$(x, y)(R_x, R_y)$	(xz, yz)

C_{5v}	E	$2C_5$	$2C_5^2$	$5\sigma_v$		
A_1	1	1	1	1	z	$x^2 + y^2, z^2$
A_2	1	1	1	−1	R_z	
E_1	2	2 cos 72°	2 cos 144°	0	$(x, y)(R_x, R_y)$	(xz, yz)
E_2	2	2 cos 144°	2 cos 72°	0		$(x^2 - y^2, xy)$

C_{6v}	E	$2C_6$	$2C_3$	C_2	$3\sigma_v$	$3\sigma_d$		
A_1	1	1	1	1	1	1	z	$x^2 - y^2, z^2$
A_2	1	1	1	1	−1	−1	R_z	
B_1	1	−1	1	−1	1	−1		
B_2	1	−1	1	−1	−1	1		
E_1	2	1	−1	−2	0	0	$(x, y)R_x, R_y)$	(xz, yz)
E_2	2	−1	−1	2	0	0		$(x^2 - y^2, xy)$

The C_{nh} Groups

C_{2h}	E	C_2	i	σ_h		
A_g	1	1	1	1	R_z	x^2, y^2, z^2, xy
B_g	1	−1	1	−1	R_x, R_y	xz, yz
A_u	1	1	−1	−1	z	
B_u	1	−1	−1	1	x, y	

C_{3h}	E	C_3	C_3^2	σ_h	S_3	S_3^5		$\epsilon = \exp(2\pi i/3)$
A′	1	1	1	1	1	1	R_z	$x^2 + y^2, z^2$
E′	$\begin{cases} 1 \\ 1 \end{cases}$	ϵ ϵ^*	ϵ^* ϵ	1 1	ϵ ϵ^*	ϵ^* ϵ	(x, y)	$(x^2 - y^2, xy)$
A″	1	1	1	−1	−1	−1	z	
E″	$\begin{cases} 1 \\ 1 \end{cases}$	ϵ ϵ^*	ϵ^* ϵ	−1 −1	$-\epsilon$ $-\epsilon^*$	$-\epsilon^*$ $-\epsilon$	(R_x, R_y)	(xy, yz)

C_{4h}	E	C_4	C_2	C_4^3	i	S_4^3	σ_h	S_4		
A_g	1	1	1	1	1	1	1	1	R_z	$x^2 + y^2, z^2$
B_g	1	−1	1	−1	1	−1	1	−1		$x^2 - y^2, xy$
E_g	$\begin{cases} 1 \\ 1 \end{cases}$	i $-i$	−1 −1	$-i$ i	1 1	i $-i$	−1 −1	$-i$ i	(R_x, R_y)	(xz, yz)
A_u	1	1	1	1	−1	−1	−1	−1	z	
B_u	1	−1	1	−1	−1	1	−1	1		
E_u	$\begin{cases} 1 \\ 1 \end{cases}$	i $-i$	−1 −1	$-i$ i	−1 −1	$-i$ i	1 1	i $-i$	(x, y)	

C_{5h}	E	C_5	C_5^2	C_5^3	C_5^4	σ_h	S_5	S_5^7	S_5^3	S_5^9		$\epsilon = \exp(2\pi i/5)$
A′	1	1	1	1	1	1	1	1	1	1	R_z	$x^2 + y^2, z^2$
E_1'	1	ϵ	ϵ^2	ϵ^{2*}	ϵ^*	1	ϵ	ϵ^2	ϵ^{2*}	ϵ^*	(x, y)	
	1	ϵ^*	ϵ^{2*}	ϵ^2	ϵ	1	ϵ^*	ϵ^{2*}	ϵ^2	ϵ		
E_2'	1	ϵ^2	ϵ^*	ϵ	ϵ^{2*}	1	ϵ^2	ϵ^2	ϵ^*	ϵ	ϵ^{2*}	$(x^2 - y^2, xy)$
	1	e^{2*}	ϵ	ϵ^*	ϵ^2	1	ϵ^{2*}	ϵ	ϵ^*	ϵ^2		
A″	1	1	1	1	1	−1	−1	−1	−1	−1	z	
E_1''	1	ϵ	ϵ^2	ϵ^{2*}	ϵ^*	−1	$-\epsilon$	$-\epsilon^2$	$-\epsilon^{2*}$	$-\epsilon^*$	(R_x, R_y)	(xz, yz)
	1	ϵ^*	ϵ^{2*}	ϵ^2	ϵ	−1	$-\epsilon^*$	$-\epsilon^{2*}$	$-\epsilon^2$	$-\epsilon$		
E_2''	1	ϵ^2	ϵ^*	ϵ	ϵ^{2*}	−1	$-\epsilon^2$	$-\epsilon^*$	$-\epsilon$	$-\epsilon^{2*}$		
	1	ϵ^{2*}	ϵ	ϵ^*	ϵ^2	−1	$-\epsilon^{2*}$	$-\epsilon$	$-e^*$	$-\epsilon^2$		

C_{6h}	E	C_6	C_3	C_2	C_3^2	C_6^5	i	S_3^5	S_6^5	σ_h	S_6	S_3		$\epsilon = \exp(2\pi i/6)$
A_g	1	1	1	1	1	1	1	1	1	1	1	1	R_z	$x^2 + y^2, z^2$
B_g	1	−1	1	−1	1	−1	1	−1	1	−1	1	−1		
E_{1g}	1	ϵ	$-\epsilon^*$	−1	$-\epsilon$	ϵ^*	1	ϵ	$-\epsilon^*$	−1	$-\epsilon$	ϵ^*	(R_x, R_y)	(xz, yz)
	1	ϵ^*	$-\epsilon$	−1	$-\epsilon^*$	ϵ	1	ϵ^*	$-\epsilon$	−1	$-\epsilon^*$	ϵ		
E_{2g}	1	$-\epsilon^*$	$-\epsilon$	1	$-\epsilon^*$	$-\epsilon$	1	$-\epsilon^*$	$-\epsilon$	1	$-\epsilon^*$	$-\epsilon$		$(x^2 - y^2, z^2)$
	1	$-\epsilon$	$-\epsilon^*$	1	$-\epsilon$	$-\epsilon^*$	1	$-\epsilon$	$-\epsilon^*$	1	$-\epsilon$	$-\epsilon^*$		
A_u	1	1	1	1	1	1	−1	−1	−1	−1	−1	−1	z	
B_u	1	−1	1	−1	1	−1	−1	1	−1	1	−1	1		
E_{1u}	1	ϵ	$-\epsilon^*$	−1	$-\epsilon$	ϵ^*	−1	$-\epsilon$	ϵ^*	1	ϵ	$-\epsilon^*$	(x, y)	
	1	ϵ^*	$-\epsilon$	−1	$-\epsilon^*$	ϵ	−1	$-\epsilon^*$	ϵ	1	ϵ^*	$-\epsilon$		
E_{2u}	1	$-\epsilon^*$	$-\epsilon$	1	$-\epsilon^*$	$-\epsilon$	−1	ϵ^*	ϵ	−1	ϵ^*	ϵ		
	1	$-\epsilon$	$-\epsilon^*$	1	$-\epsilon$	$-\epsilon^*$	−1	ϵ	ϵ^*	−1	ϵ	ϵ^*		

The D_{nh} Groups

D_{2h}	E	$C_2(z)$	$C_2(y)$	$C_2(x)$	i	$\sigma(xy)$	$\sigma(xz)$	$\sigma(yz)$		
A_g	1	1	1	1	1	1	1	1		x^2, y^2, z^2
B_{1g}	1	1	−1	−1	1	1	−1	−1	R_z	xy
B_{2g}	1	−1	1	−1	1	−1	1	−1	R_y	xz
B_{3g}	1	−1	−1	1	1	−1	−1	1	R_x	yz
A_u	1	1	1	1	−1	−1	−1	−1		
B_{1u}	1	1	−1	−1	−1	−1	1	1	z	
B_{2u}	1	−1	1	−1	−1	1	−1	1	y	
B_{3u}	1	−1	−1	1	−1	1	1	−1	x	

D_{3h}	E	$2C_3$	$3C_2$	σ_h	$2S_3$	$3\sigma_v$		
A_1'	1	1	1	1	1	1		$x^2 + y^2, z^2$
A_2'	1	1	−1	1	1	−1	R_z	
E'	2	−1	0	2	−1	0	(x, y)	$(x^2 - y^2, xy)$
A_1''	1	1	1	−1	−1	−1		
A_2''	1	1	−1	−1	−1	1	z	
E'	2	−1	0	−2	1	0	(R_x, R_y)	(xz, yz)

D_{4h}	E	$2C_4$	C_2	$2C_2'$	$2C_2''$	i	$2S_4$	σ_h	$2\sigma_v$	$2\sigma_d$		
A_{1g}	1	1	1	1	1	1	1	1	1	1		$x^2 + y^2, z^2$
A_{2g}	1	1	1	−1	−1	1	1	1	−1	−1	R_z	
B_{1g}	1	−1	1	1	−1	1	−1	1	1	−1		$x^2 - y^2$
B_{2g}	1	−1	1	−1	1	1	−1	1	−1	1		xy
E_g	2	0	−2	0	0	2	0	−2	0	0	(R_z, R_y)	(xz, yz)
A_{1u}	1	1	1	1	1	−1	−1	−1	−1	−1		
A_{2u}	1	1	1	−1	−1	−1	−1	−1	1	1	z	
B_{1u}	1	−1	1	1	−1	−1	1	−1	−1	1		
B_{2u}	1	−1	1	−1	1	−1	1	−1	1	−1		
E_u	2	0	−2	0	0	−2	0	2	0	0	(x, y)	

D_{5h}	E	$2C_5$	$2C_5^2$	$5C_2$	σ_h	$2S_5$	$2S_5^3$	$5\sigma_v$		
A_1'	1	1	1	1	1	1	−1	1		x^2+y^2, z^2
A_2'	1	1	1	−1	1	1	−1	−1	R_z	
E_1'	2	2 cos 72°	2 cos 144°	0	2	2 cos 72°	2 cos 144°	0	(x, y)	
E_2'	2	2 cos 144°	2 cos 72°	0	2	2 cos 144°	2 cos 72°	0		(x^2-y^2, xy)
A_1''	1	1	1	1	−1	−1	−1	−1		
A_2''	1	1	1	−1	−1	−1	−1	1	z	
E_1''	2	2 cos 72°	2 cos 144°	0	−2	−2 cos 72°	−2 cos 144°	0	(R_x, R_y)	(xz, yz)
E_2''	2	2 cos 144°	2 cos 72°	0	−2	−2 cos 144°	−2 cos 72°	0		

D_{6h}	E	$2C_6$	$2C_3$	C_2	$3C_2'$	$3C_2''$	i	$2S_3$	$2S_6$	σ_h	$3\sigma_d$	$3\sigma_v$		
A_{1g}	1	1	1	1	1	1	1	1	1	1	1	1		x^2+y^2, z^2
A_{2g}	1	1	1	1	−1	−1	1	1	1	1	−1	−1	R_z	
B_{1g}	1	−1	1	−1	1	−1	1	−1	1	−1	1	−1		
B_{2g}	1	−1	1	−1	−1	1	1	−1	1	−1	−1	1		
E_{1g}	2	1	−1	−2	0	0	2	1	−1	−2	0	0	(R_x, R_y)	(xz, yz)
E_{2g}	2	−1	−1	2	0	0	2	−1	−1	2	0	0		(x^2-y^2, xy)
A_{1u}	1	1	1	1	1	1	−1	−1	−1	−1	−1	−1		
A_{2u}	1	1	1	1	−1	−1	−1	−1	−1	−1	1	1	z	
B_{1u}	1	−1	1	−1	1	−1	−1	1	−1	1	−1	1		
B_{2u}	1	−1	1	−1	−1	1	−1	1	−1	1	1	−1		
E_{1u}	2	1	−1	−2	0	0	−2	−1	1	2	0	0	(x, y)	
E_{2u}	2	−1	−1	2	0	0	−2	1	1	−2	0	0		

The D_{nd} Groups

D_{2d}	E	$2S_4$	C_2	$2C_2'$	$2\sigma_d$		
A_1	1	1	1	1	1		x^2+y^2, z^2
A_2	1	1	1	−1	−1	R_z	
B_1	1	−1	1	1	−1		x^2-y^2
B_2	1	−1	1	−1	1	z	xy
E	2	0	−2	0	0	$(x, y); (R_x, R_y)$	(xz, yz)

D_{3d}	E	$2C_3$	$3C_2$	i	$2S_6$	$3\sigma_d$		
A_{1g}	1	1	1	1	1	1		x^2+y^2, z^2
A_{2g}	1	1	−1	1	1	−1	R_z	
E_g	2	−1	0	2	−1	0	(R_x, R_y)	(x^2-y^2, xy), (xz, yz)
A_{1u}	1	1	1	−1	−1	−1		
A_{2u}	1	1	−1	−1	−1	1	z	
E_u	2	−1	0	−2	1	0	(x, y)	

D_{4d}	E	$2S_8$	$2C_4$	$2S_8^3$	C_2	$4C_2'$	$4\sigma_d$		
A_1	1	1	1	1	1	1	1		x^2+y^2, z^2
A_2	1	1	1	1	1	−1	−1	R_z	
B_1	1	−1	1	−1	1	1	−1		
B_2	1	−1	1	−1	1	−1	1	z	
E_1	2	$\sqrt{2}$	0	$-\sqrt{2}$	−2	0	0	(x, y)	
E_2	2	0	−2	0	2	0	0		(x^2-y^2, xy)
E_3	2	$-\sqrt{2}$	0	$\sqrt{2}$	−2	0	0	(R_x, R_y)	(xz, yz)

D_{5d}	E	$2C_5$	$2C_5^2$	$5C_2$	i	$2S_{10}^3$	$2S_{10}$	$5\sigma_d$		
A_{1g}	1	1	1	1	1	1	1	1		$x^2 + y^2, z^2$
A_{2g}	1	1	1	−1	1	1	1	−1	R_z	
E_{1g}	2	2 cos 72°	2 cos 144°	0	2	2 cos 72°	2 cos 144°	0	(R_x, R_y)	(xz, yz)
E_{2g}	2	2 cos 144°	2 cos 72°	0	2	2 cos 144°	2 cos 72°	0		$(x^2 - y^2, xy)$
A_{1u}	1	1	1	1	−1	−1	−1	−1		
A_{2u}	1	1	1	−1	−1	−1	−1	1	z	
E_{1u}	2	2 cos 72°	2 cos 144°	0	−2	−2 cos 72°	−2 cos 144°	0	(x, y)	
E_{2u}	2	2 cos 144°	2 cos 72°	0	−2	−2 cos 144°	−2 cos 72°	0		

D_{6d}	E	$2S_{12}$	$2C_6$	$2S_4$	$2C_3$	$2S_{12}^5$	C_2	$6C_2'$	$6\sigma_d$		
A_1	1	1	1	1	1	1	1	1	1		x^2+y^2, z^2
A_2	1	1	1	1	1	1	1	−1	−1	R_z	
B_1	1	−1	1	−1	1	−1	1	1	−1		
B_2	1	−1	1	−1	1	−1	1	−1	1	z	
E_1	2	$\sqrt{3}$	1	0	−1	$-\sqrt{3}$	−2	0	0	(x, y)	
E_2	2	1	−1	−2	−1	1	2	0	0		(x^2-y^2, xy)
E_3	2	0	−2	0	2	0	−2	0	0		
E_4	2	−1	−1	2	−1	−1	2	0	0		
E_5	2	$-\sqrt{3}$	1	0	−1	$\sqrt{3}$	−2	0	0	(R_x, R_y)	(xz, yz)

The S_n Groups

S_4	E	S_4	C_2	S_4^3		
A	1	1	1		R_z	x^2+y^2, z^2
B	1	−1	1	−1	z	x^2-y^2, xy
E	1	i	−1	$-i$	$(x, y); (R_x, R_y)$	(xz, yz)
	1	$-i$	−1	i		

S_6	E	C_3	C_3^2	i	S_6^5	S_6		$\epsilon = \exp(2\pi i/3)$
A_g	1	1	1	1	1	1	R_z	x^2+y^2, z^2
E_g	1	ϵ	ϵ^*	1	ϵ	ϵ^*	(R_x, R_y)	(x^2-y^2, xy)
	1	ϵ^*	ϵ	1	ϵ^*	ϵ		(xz, yz)
A_u	1	1	1	−1	−1	−1	z	
E_u	1	ϵ	ϵ^*	−1	$-\epsilon$	$-\epsilon^*$	(x, y)	
	1	ϵ^*	ϵ	−1	$-\epsilon^*$	$-\epsilon$		

S_8	E	S_8	C_4	S_8^3	C_2	S_8^5	C_4^3	S_8^7		$\epsilon = \exp(2\pi i/8)$
A	1	1	1	1	1	1	1	1	R_z	x^2+y^2, z^2
B	1	−1	1	−1	1	−1	1	−1	z	
E_1	1	ϵ	i	$-\epsilon^*$	−1	$-\epsilon$	$-i$	ϵ^*	$(x, y);$	
	1	ϵ^*	$-i$	$-\epsilon$	−1	$-\epsilon^*$	i	ϵ	(R_x, R_y)	
E_2	1	i	−1	$-i$	1	i	−1	$-i$		(x^2-y^2, xy)
	1	−1	−1	i	1	$-i$	−1	i		
E_3	1	$-\epsilon^*$	$-i$	ϵ	−1	ϵ^*	i	$-\epsilon$		(xz, yz)
	1	$-\epsilon$	i	ϵ^*	−1	ϵ	$-i$	$-\epsilon^*$		

The Cubic Groups

T	E	$4C_3$	$4C_3^2$	$3C_2$		$\epsilon = \exp(2\pi i/3)$
A	1	1	1	1		$x^2 + y^2 + z^2$
E	1	ϵ	ϵ^*	1		$(2z^2 - x^2 - y^2,$
	1	ϵ^*	ϵ	1		$x^2 - y^2)$
T	3	0	0	−1	(R_x, R_y, R_z); (x, y, z)	(xy, xz, yz)

T_h	E	$4C_3$	$4C_3^2$	$3C_2$	i	$4S_6$	$4S_6^5$	$3\sigma_n$		$\epsilon = \exp(2\pi i/3)$
A_g	1	1	1	1	1	1	1	1		$x^2 + y^2 + z^2$
A_u	1	1	1	1	−1	−1	−1	−1		
E_g	1	ϵ	ϵ^*	1	1	ϵ	ϵ^*	1		$(2z^2 - x^2 - y^2,$
	1	ϵ^*	ϵ	1	1	ϵ^*	ϵ	1		$x^2 - y^2)$
E_u	1	ϵ	ϵ^*	1	−1	$-\epsilon$	$-\epsilon^*$	−1		
	1	ϵ^*	ϵ	1	−1	$-\epsilon^*$	$-\epsilon$	−1		
T_g	3	0	0	−1	1	0	0	−1	(R_x, R_y, R_z)	(xz, yz, xy)
T_u	3	0	0	−1	−1	0	0	1	(x, y, z)	

T_d	E	$8C_3$	$3C_2$	$6S_4$	$6\sigma_d$		
A_1	1	1	1	1	1		$x^2 + y^2 + z^2$
A_2	1	1	1	−1	−1		
E	2	−1	2	0	0		$(2z^2 - x^2 - y^2,$ $x^2 - y^2)$
T_1	3	0	−1	1	−1	(R_x, R_y, R_z)	
T_2	3	0	−1	−1	1	(x, y, z)	(xy, xz, yz)

O	E	$6C_4$	$3C_2(=C_4^2)$	$8C_3$	$6C_2$		
A_1	1	1	1	1	1		$x^2 + y^2 + z^2$
A_2	1	−1	1	1	−1		
E	2	0	2	−1	0		$(2z^2 - x^2 - y^2,$ $x^2 - y^2)$
T_1	3	1	−1	0	−1	(R_x, R_y, R_z) (x, y, z)	
T_2	3	−1	−1	0	1		(xy, xz, yz)

O_h	E	$8C_3$	$6C_2$	$6C_4$	$3C_2(=C_4^2)$	i	$6S_4$	$8S_6$	$3\sigma_h$	$6\sigma_d$		
A_{1g}	1	1	1	1	1	1	1	1	1	1		$x^2+y^2+z^2$
A_{2g}	1	1	−1	−1	1	1	−1	1	1	−1		
E_g	2	−1	0	0	2	2	0	−1	2	0		$(2z^2-x^2-y^2, x^2-y^2)$
T_{1g}	3	0	−1	1	−1	3	1	0	−1	−1	(R_x, R_y, R_z)	
T_{2g}	3	0	1	−1	−1	3	−1	0	−1	1		(xz, yz, xy)
A_{1u}	1	1	1	1	1	−1	−1	−1	−1	−1		
A_{2u}	1	1	−1	−1	1	−1	1	−1	−1	1		
E_u	2	−1	0	0	2	−2	0	1	−2	0		
T_{1u}	3	0	−1	1	−1	−3	−1	0	1	1	(x, y, z)	
T_{2u}	3	0	1	−1	−1	−3	1	0	1	−1		

The Icosahedral Group

I_h	E	$12C_5$	$12C_5^2$	$20C_3$	$15C_2$	i	$12S_{10}$	$12S_{10}^3$	$20S_6$	15σ		
A_g	1	1	1	1	1	1	1	1	1	1		$x^2+y^2+z^2$
T_{1g}	3	$(1+\sqrt{5})/2$	$(1-\sqrt{5})/2$	0	−1	3	$(1-\sqrt{5})/2$	$(1+\sqrt{5})/2$	0	−1	(R_x, R_y, R_z)	
T_{2g}	3	$(1-\sqrt{5})/2$	$(1+\sqrt{5})/2$	0	−1	3	$(1+\sqrt{5})/2$	$(1-\sqrt{5})/2$	0	−1		
G_g	4	−1	−1	1	0	4	−1	−1	1	0		
H_g	5	0	0	−1	1	5	0	0	−1	1		$(2z^2-x^2-y^2,$ $x^2-y^2,$ $xy, yz, xz)$
A_u	1	1	1	1	1	−1	−1	−1	−1	−1		
T_{1u}	3	$(1+\sqrt{5})/2$	$(1-\sqrt{5})/2$	0	−1	−3	$-(1-\sqrt{5})/2$	$-(1+\sqrt{5})/2$	0	1	(x, y, z)	
T_{2u}	3	$(1-\sqrt{5})/2$	$(1+\sqrt{5})/2$	0	−1	−3	$-(1+\sqrt{5})/2$	$-(1-\sqrt{5})/2$	0	1		
G_u	4	−1	−1	1	0	−4	1	1	−1	0		
H_u	5	0	0	−1	1	−5	0	0	1	−1		

The Groups $C_{\infty v}$ and $D_{\infty h}$ for Linear Molecules

$C_{\infty v}$	E	$2C_{\infty}^{\Phi}$	$\cdots$	$\infty\sigma_v$		
$A_1 \equiv \Sigma^+$	1	1		1	z	x^2+y^2, z^2
$A_2 \equiv \Sigma^-$	1	1		−1	R_z	
$E_1 \equiv \Pi$	2	$2\cos\Phi$		0	$(x, y); (R_x, R_y)$	(xz, yz)
$E_2 \equiv \Delta$	2	$2\cos 2\Phi$		0		(x^2-y^2, xy)
$E_3 \equiv \Phi$	2	$2\cos 3\Phi$		0		
$\vdots$	$\vdots$	$\vdots$		$\vdots$		

$D_{\infty h}$	E	$2C_{\infty}^{\Phi}$	$\cdots$	$\infty\sigma_l$	i	$2S_{\infty}^{\Phi}$	$\cdots$	∞C_2		
Σ_g^+	1	1		1	1	1		1		x^2+y^2, z^2
Σ_g^-	1	1		−1	1	1		−1	R_z	
Π_g	2	$2\cos\Phi$		0	2	$-2\cos\Phi$		0	(R_x, R_y)	(xz, yz)
Δ_g	2	$2\cos 2\Phi$		0	2	$-2\cos 2\Phi$		0		(x^2-y^2, xy)
$\vdots$	$\vdots$	$\vdots$		$\vdots$	$\vdots$	$\vdots$		$\vdots$		
Σ_u^+	1	1		1	−1	−1		−1	z	
Σ_u^-	1	1		−1	−1	−1		1		
Π_u	2	$2\cos\Phi$		0	−2	$2\cos\Phi$		0	(x, y)	
Δ_u	2	$2\cos 2\Phi$		0	−2	$-2\cos 2\Phi$		0		
$\vdots$	$\vdots$	$\vdots$		$\vdots$	$\vdots$	$\vdots$		$\vdots$		

Character Tables for Double Groups

D_2'	E	R	C_2^X	C_2^Y	C_2^Z	Bethe's
			$C_2^X R$	$C_2^Y R$	$C_2^Z R$	nomenclature
A′	1	1	1	1	1	Γ_1
B_1'	1	1	1	−1	−1	Γ_2
B_2'	1	1	−1	1	−1	Γ_3
B_3'	1	1	−1	−1	1	Γ_4
E′	2	−2	0	0	0	Γ_5

D_3'	E	R	C_3 C_3^2R	C_3^2 C_3R	$3C_2$	$3C_2R$	Bethe's nomenclature
A_1'	1	1	1	1	1	1	Γ_1
A_2'	1	1	1	1	−1	−1	Γ_2
E_1'	1	−1	−1	1	i	−1	
	1	−1	−1	1	$-i$	i	Γ_3
E_2'	2	−2	1	−1	0	0	Γ_4
E_3'	2	2	−1	−1	0	0	Γ_5

D_4'	E	R	C_4 C_4^3R	C_4R C_4^3	C_2 C_2R	$2C_2'$ $2C_2'R$	$2C_2''$ $2C_2''R$	Bethe's nomenclature
A_1'	1	1	1	1	1	1	1	Γ_1
A_2'	1	1	1	1	1	−1	−1	Γ_2
B_1'	1	1	−1	−1	1	1	−1	Γ_3
B_2'	1	1	−1	−1	1	−1	1	Γ_4
E_1'	2	2	0	0	−2	0	0	Γ_5
E_2'	2	−2	$\sqrt{2}$	$-\sqrt{2}$	0	0	0	Γ_6
E_3'	2	−2	$-\sqrt{2}$	$\sqrt{2}$	0	0	0	Γ_7

D_6'	E	R	C_2 C_2R	C_3 C_3^2R	C_3^2 C_3R	C_6 C_6^5R	C_6^5 C_6R	$3C_2'$ $3C_2'R$	$3C_2''$ $3C_2''R$	Bethe's nomenclature
A_1'	1	1	1	1	1	1	1	1	1	Γ_1
A_2'	1	1	1	1	1	1	1	−1	−1	Γ_2
B_1'	1	1	−1	1	1	−1	−1	1	−1	Γ_3
B_2'	1	1	−1	1	1	−1	−1	−1	1	Γ_4
E_1'	2	−2	0	1	−1	$\sqrt{3}$	$-\sqrt{3}$	0	0	Γ_5
E_2'	2	−2	0	1	−1	$-\sqrt{3}$	$\sqrt{3}$	0	0	Γ_6
E_3'	2	−2	0	−2	2	0	0	0	0	Γ_7
E_4'	2	2	−2	−1	−1	−1	−1	0	0	Γ_8
E_5'	2	2	2	−1	−1	−1	−1	0	0	Γ_9

T'	E	R	$4C_3$	$4C_3^2$	$4C_3R$	$4C_3^2R$	$3C_2$ $3C_2R$	Bethe's nomenclature
A_1'	1	1	1	1	1	1	1	Γ_1
	1	1	ϵ	ϵ^2	ϵ	ϵ^2	1	
E'	1	1	ϵ^2	ϵ	ϵ^2	ϵ	1	Γ_2
E''	2	−2	1	−1	−1	1	0	Γ_3
F	3	3	0	0	0	0	−1	Γ_4
	2	−2	ϵ	$-\epsilon^2$	$-\epsilon$	ϵ^2	0	
G	2	−2	ϵ^2	ϵ	$-\epsilon^2$	ϵ	0	Γ_5

O'	E	R	$4C_3$ $4C_3^2R$	$4C_3R$ $4C_3^2$	$3C_2$ $3C_2R$	$3C_4$ $3C_4^3R$	$3C_4R$ $3C_4^3$	$6C_2'$ $6C_2'R$	Bethe's nomenclature
A_1'	1	1	1	1	1	1	1	1	Γ_1
A_2'	1	1	1	1	1	−1	−1	−1	Γ_2
E_1'	2	2	−1	−1	2	0	0	0	Γ_3
T_1'	3	3	0	0	−1	1	1	−1	Γ_4
T_2'	3	3	0	0	−1	−1	−1	1	Γ_5
E_2'	2	−2	1	−1	0	$\sqrt{2}$	$-\sqrt{2}$	0	Γ_6
E_3'	2	−2	1	−1	0	$-\sqrt{2}$	$\sqrt{2}$	0	Γ_7
G'	4	−4	−1	1	0	0	0	0	Γ_8

APPENDIX III

VIBRATIONAL MOTIONS

The points about nuclear motions discussed in Chapter 3 will be illustrated by analyzing the vibrational displacements in simple molecules.[66,67] For example, the trigonal complex MX_3 belongs to a point group D_3h. Vectorial instantaneous displacements can be expressed as the resultant of three-dimensional bases vectors. These bases are Cartesian axes mounted on each atom, and the atoms are placed at the nuclear equilibrium positions (Fig. 56). Operation with the symmetry transformations upon the 12 vectors of MX_3, that is, vectors representing the molecule's $3N$ degrees of motional freedom, leads to the following reducible representation

	E	$2C_3$	$3C_2$	σ_h	$2S_3$	$3\sigma_v$
Γ_T	12	0	−2	4	−2	2

It is possible to decompose this representation in the species of the D_{3h} point group by making use of the orthogonality of irreducible representations

$$\sum_k \chi_i(R_k)\chi_j(R_k) = h\,\delta_{ij}$$

where the subscripts i and j indicate two given irreducible representations, the sum spans all the elements R_k of the group, and h is the group's order. With the same notation, the character $\chi(R)$ of a matrix corresponding to a reducible representation can be expressed as a linear combination of the characters $\chi_i(R)$ of all the irreducible representations in the group

$$\chi(R) = \sum_i a_i \chi_i(R)$$

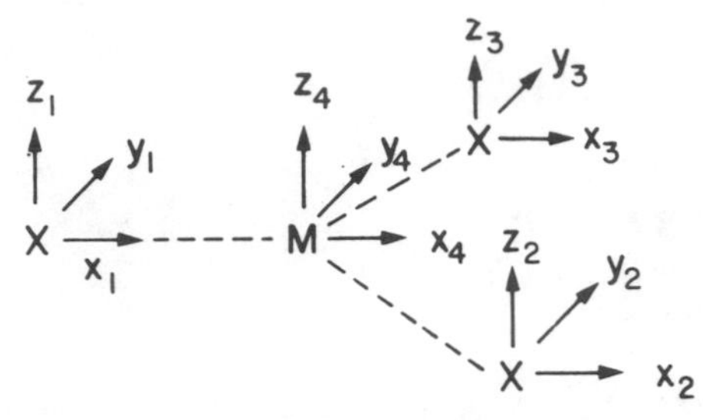

Figure 56. Cartesian coordinates for the study of vibrational motions in a planar MX_3 (D_{3h} symmetry) molecule.

Combination of these two expressions leads to the relationship

$$a_i = \frac{1}{h} \sum_R \chi(R)\chi_i(R)$$

which gives the number of times a_i that the irreducible representation ith is contained in the reducible representation. This expansion of the reducible representation Γ gives

$$\Gamma_T = A_1' + A_2' + 3E' + 2A_2'' + E''$$

where on the right side of the equality are irreducible representations in D_{3h}.

Rigid body displacements in addition to rotations around the molecule's center of mass correspond to six degrees of freedom, which span the irreducible representations

$$\Gamma_{t+\Gamma} = A_2' + E' + A_2'' + E''$$

However, motions leading to rigid body displacements or rotations around the center of mass are considered spurious motions whose species must be removed from Γ_T to restrict the expansion to only those species of the vibrations. The species $\Gamma_{t+\Gamma}$ of such spurious motions are usually indicated in one side of the character tables of the point groups (Appendix II). They can also be investigated by studying the symmetry transformations of unitary vectors placed along Cartesian axes pinned on the center of mass. The unitary vectors signaling the displacement of the center of mass are pointed in the direction of the x, y, and z axes, while each rotation is represented by a unitary vector parallel to the rotation axes and pointed in the direction of the forward movement of a screw whose turning follows the rotational movement. Subtraction of $\Gamma_{t+\Gamma}$ from Γ_T leads to the irreducible representations

$$\Gamma_{vib} = A_1' + 2E' + A_2'$$

for the normal vibrations.

One can use linear combinations of the rectangular coordinates, for example, vectors x_i, y_i, z_i with $i = 1$–4 in Figure 56, whose symmetry transformations are bases for the irreducible representations in r_{vib}. The symmetry-adapted linear combinations of Cartesian vectors (or scalar functions) can be obtained by applying to them the projection operator

$$\hat{\mathcal{P}}_j = \frac{l_j}{h} \sum_R \chi_j(R)\hat{R}$$

where the projection takes place over the jth irreducible representation with dimension l_j in a group of order h. For example, the transformations of the bases vectors shown in Table 15 can be used with the projection operator for calculating a normal coordinate,

$$\hat{\mathcal{P}}_{A_2'} z_i = z_4 - (z_1 + z_2 + z_3)$$

that spans the irreducible representation A_2'.

Although it is possible to use linear combinations of the rectangular coordinates for the expression of normal coordinates, it is more convenient to use internal coordinates, that is, increments of bond lengths and angles, since the corresponding force constants have a clear physical meaning. The number of internal coordinates minus the number of equations restricting the nuclear displacements must equal the number of internal coordinates. For the trigonal molecule, the nuclear displacements can be represented by increments of three different MX bonds, the three bending angles X–M–X, and an out-of-plane deformation angle. Insofar as $3N - 6$ coordinates are expected to be linearly independent, there is a redundant internal coordinate within the seven coordinates described above that must be expressed in terms of the other six. The condition $2\pi = \Sigma_{i=1}^{3} \alpha_i$, establishing that the sum of the angles around the central atom must remain equal to 2π, leads to

$$0 = \sum_{i=1}^{3} \Delta\alpha_i$$

which restricts their increments. The application of group theory to the three M–X bonds and XMX angular increments gives the following reducible representations

	E	$2C_3$	$3C_2$	σ_h	$2S_3$	$3\sigma_v$
Γ_{MX}	3	0	1	3	0	1
Γ_{XMX}	3	0	1	3	0	1

Decomposition of Γ_{MX} and Γ_{XMX} into the species of the D_3h point group gives

$$\Gamma_{MX} = A_1' + E'$$

TABLE 15 Matrices for the Transformations of the x_1, y_1, z_i ($i = 1, 2, 3$) Vectors under the Symmetry Operations of the D_{3h} Point Group[a]

$\hat{E}$	$\hat{C}_3$	$\hat{C}_2$	$\hat{\sigma}_h$	$\hat{\sigma}_v$	$\hat{S}_3$
		$\begin{pmatrix} 1 & 0 \\ 0 & -1 \end{pmatrix}\begin{pmatrix} x_1 \\ y_1 \end{pmatrix}$	$\begin{pmatrix} 1 & 0 \\ 0 & 1 \end{pmatrix}\begin{pmatrix} x_1 \\ y_1 \end{pmatrix}$	$\begin{pmatrix} 1 & 0 \\ 0 & -1 \end{pmatrix}\begin{pmatrix} x_1 \\ y_1 \end{pmatrix}$	b
$\begin{pmatrix} 1 & 0 \\ 0 & 1 \end{pmatrix}\begin{pmatrix} x_1 \\ y_1 \end{pmatrix}$	$\begin{pmatrix} -\cos 60 & \sin 60 \\ -\sin 60 & -\cos 60 \end{pmatrix}\begin{pmatrix} x_2 \\ y_2 \end{pmatrix}$	$\begin{pmatrix} \cos 60 & \sin 60 \\ \sin 60 & \cos 60 \end{pmatrix}\begin{pmatrix} x_2 \\ y_2 \end{pmatrix}$		$\begin{pmatrix} \cos 60 & \sin 60 \\ \sin 60 & \cos 60 \end{pmatrix}\begin{pmatrix} x_2 \\ y_2 \end{pmatrix}$	b
	$\begin{pmatrix} -\cos 60 & -\sin 60 \\ \sin 60 & -\cos 60 \end{pmatrix}\begin{pmatrix} x_3 \\ y_3 \end{pmatrix}$	$\begin{pmatrix} -\cos 60 & -\sin 60 \\ -\sin 60 & -\cos 60 \end{pmatrix}\begin{pmatrix} x_3 \\ y_3 \end{pmatrix}$		$\begin{pmatrix} -\cos 60 & -\sin 60 \\ -\sin 60 & \cos 60 \end{pmatrix}\begin{pmatrix} x_3 \\ y_3 \end{pmatrix}$	b
$\begin{pmatrix} 1 & 0 & 0 \\ 0 & 1 & 0 \\ 0 & 0 & 1 \end{pmatrix}\begin{pmatrix} z_1 \\ z_2 \\ z_3 \end{pmatrix}$	$\begin{pmatrix} 0 & 0 & 1 \\ 1 & 0 & 0 \\ 0 & 1 & 0 \end{pmatrix}\begin{pmatrix} z_1 \\ z_2 \\ z_3 \end{pmatrix}$	$\begin{pmatrix} -1 & 0 & 0 \\ 0 & 0 & -1 \\ 0 & -1 & 0 \end{pmatrix}\begin{pmatrix} z_1 \\ z_2 \\ z_3 \end{pmatrix}$	$\begin{pmatrix} -1 & 0 & 0 \\ 0 & -1 & 0 \\ 0 & 0 & -1 \end{pmatrix}\begin{pmatrix} z_1 \\ z_2 \\ z_3 \end{pmatrix}$	c	d
	$\begin{pmatrix} 0 & 1 & 0 \\ 0 & 0 & 1 \\ 1 & 0 & 0 \end{pmatrix}\begin{pmatrix} z_1 \\ z_2 \\ z_3 \end{pmatrix}$	$\begin{pmatrix} 0 & 0 & -1 \\ 0 & -1 & 0 \\ -1 & 0 & 0 \end{pmatrix}\begin{pmatrix} z_1 \\ z_2 \\ z_3 \end{pmatrix}$			
		$\begin{pmatrix} 0 & -1 & 0 \\ -1 & 0 & 0 \\ 0 & 0 & -1 \end{pmatrix}\begin{pmatrix} z_1 \\ z_2 \\ z_3 \end{pmatrix}$			

a For the relative orientations of the vectors see Figure 56.
b Similar to the matrices for $\hat{C}_3$.
c Similar to the matrices for $\hat{C}_2$.
d Similar to the matrices for $\hat{C}_3$ times the scalar -1.

$$\Gamma_{XMX} = A'_1 + E'$$

It can be demonstrated that A'_1 represents an impossible condition where the three angles must increase or decrease in the same amount, and that the distortion from planarity spans the irreducible representation A'_2. The symmetry coordinates can now be expressed as linear combinations of the internal coordinates by using the projection operator described above. Such combinations are

$$\hat{\mathscr{P}}_{A'_1} d_1 = d_1 + d_2 + d_3$$

$$\hat{\mathscr{P}}_{E'} d_2 = 2d_2 - d_1 - d_3$$

$$\hat{\mathscr{P}}_{E'} d_3 = 2d_3 - d_1 - d_2$$

where d_i $(i = 1, 2, 3)$ represents increments of the three M–X bond distances, and the linear combinations

$$\hat{\mathscr{P}}_{E'} \Delta\alpha_{12} = 2\Delta\alpha_{12} - \Delta\alpha_{13} - \Delta\alpha_{23}$$

$$\hat{\mathscr{P}}_{E'} \Delta\alpha_{13} = 2\Delta\alpha_{13} - \Delta\alpha_{12} - \Delta\alpha_{23}$$

of the $\Delta\alpha_{ij} (ij = 12, 13, 23)$, namely the angular increments of X_i–M–X_j angles. The potential energy of the molecule may be written

$$2V = \sum_{i,k} f_{ik} s_i s_k$$

where s_i and s_k are changes of the internal coordinates. Terms with $i = k$ represent contributions from a given mode while terms with $i \neq k$ correspond to interactions between different vibrational modes. These internal coordinates are related to the symmetry coordinates S_i by orthogonal transformations, $S = Us$ in a matrix form. The transformation matrix U has a reciprocal matrix that is equal to the corresponding transpose $U^{-1} = U^{t}$. In this regard, the potential energy can be written

$$2V = \sum_{i,k} f_{ik} s_i s_k = \sum_{i,k} F_{ik} S_i S_k$$

where F_{ik} is the force constant for the symmetry coordinates S_i and S_k, and in the form of matrices

$$2V = \tilde{s} f s = \tilde{S} F S$$

where $\tilde{s}$ and $\tilde{S}$ are row matrices obtained by transposing s and S, respectively. The rectangular coordinates matrix Q_{xyz} can also be related to the internal coordinates by an orthogonal transformation

$$s = BQ_{xyz}$$

which can also be used for the definition of the G matrix

$$G = BM^{-1}B^{t}$$

The diagonal matrix M^{-1}

$$M^{-1} = \begin{vmatrix} \mu_1 & 0 & 0 & \cdots & 0 \\ 0 & \mu_2 & 0 & \cdots & 0 \\ 0 & 0 & \mu_3 & \cdots & 0 \\ & & & \cdot & \\ & & & \cdot & \\ & & & \cdot & \end{vmatrix}$$

has the reciprocal of the atomic masses $\mu_i = M_i^{-1}$ as the elements of the diagonal while the diagonal elements are zeros. The usefulness of the G matrix is evident in expressing the kinetic energy as an operation between matrices

$$2T = \tilde{\dot{s}}G^{-1}\dot{s}$$

with the row matrix $\tilde{\dot{s}}$ and the column matrix $\dot{s}$ containing the time derivatives of the internal coordinates. Newton's equation of motion

$$\frac{\partial}{\partial t}\left(\frac{\partial T}{\partial \dot{s}_k}\right) + \frac{\partial V}{\partial s_k} = 0$$

can be used with the matrice definitions of T and V for establishing a secular equation

$$|F - G^{-1}4\pi^2C^2\tilde{\nu}^2| = 0$$

which can be reorganized into

$$|GF - Ea| = 0$$

where $a = 4\pi^2C^2\tilde{\nu}^2$ and E is the unity matrix. Vibrational frequencies can be obtained by solving the secular. For this purpose we can use an F matrix constructed with a suitable set of force constants.

The nuclear displacements in a chemical reaction can be expressed in terms of the normal vibrations. Like vibratory motions, the nuclear displacements in a chemical reaction must obey constraints that prevent rigid body displacements or rotations around the center of mass, for example

$$\sum_i m_i \Delta g_i = 0$$

where m_i is the mass and Δg_i the displacement vector of the ith atom. For example, in a tetragonally distorted hexacoordinate complex MX_6, the dissociation of one of the elongated bonds can be described by the set of displacement vectors shown in Figure 57. These displacements span the reducible representation

	E	$2C_4$	$C_2(=C_4^2)$	$2C_2'$	$2C_2''$
Γ	7	6	3	−6	−2

in the subgroup D_4 of the molecule's point group D_{4h}. The decomposition in irreducible representations gives

$$\Gamma = A_1 + 3A_2 + B_2 + E$$

which are normal vibrations of the molecule. The net movement leading to the dissociation, therefore, spans the irreducible representation E.

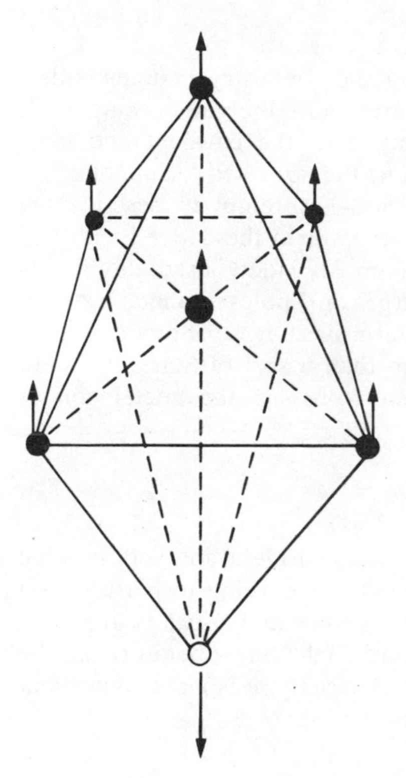

Figure 57. Nuclear displacements bringing about the dissociation of one bond (ligand represented by an open circle).

APPENDIX IV

DESCRIPTION OF THE CHEMICAL BONDING IN COORDINATION COMPLEXES

Bonding in coordination compounds of organic chemistry (amine–oxides, amine–boranes, benzene–iodine adduct, etc.) and inorganic compounds (transition metal complexes) is so different from the covalent and ionic bondings that it requires a new classification, the so-called "semipolar" or "dative" bond. A number of theoretical models attempt to describe the migration of charge from the donor to the acceptor in the dative bond, and the crystal field method, still one of the most widely used despite its limitations, considers the effect of point charges or dipoles on metal orbitals. Such an effect, the perturbation of the metal wave functions by the overall electrostatic field of the ligands, is no more than a sort of Stark effect. In this model, the Hamiltonian for the electrons in the metal ion is formulated[386]

$$\hat{\mathcal{H}} = \hat{\mathcal{H}}^0 + \mathcal{U}$$

where $\hat{\mathcal{H}}^0$ is the Hamiltonian for the free (gaseous) metal ion and $\mathcal{U}$ is the electrostatic potential generated by the ligands. According to perturbation theory, the zero-order wave function of $\hat{\mathcal{H}}^0$ is known and $\mathcal{U}$ can be regarded as a first-order perturbation operator. Although this operator can be expressed in terms of the basic principles of electricity, it is more convenient to expand $\mathcal{U}$ in a series of spherical harmonics

$$\mathcal{U} = \sum_i \sum_l \sum_m Y_l^m(\theta_i, \varphi_i) R_{nl}(r_i)$$

where the summation over i comprises the electrons of the metal ion.[386]

Moreover, the appropriate values of l and m become evident when one applies to v the condition that makes it transform as the totally symmetric irreducible representation in the point group of the molecule. It must be noticed that the operator $\mathcal{U}$ described in this manner is a one-electron operator.

Insofar as the overlap between the electronic clouds of the metal and the ligand have been ignored, the crystal field model is strictly valid in the limit of highly ionic coordination compounds with the lowest members of the nephelauxetic series. The approach of the ligand field model is to introduce a number of semiempirical parameters to correct the deficiencies of the crystal field model. Interelectronic repulsion energies $\langle \Psi_i | e^2/r_{ij} | \Psi_j \rangle$, where i, j represent a pair of electrons, are described in terms of the Racah parameters A, B, C or the Slater–Condon integrals F_k.[387,388] The relationship between the two sets of parameters is

$$A = F_0 + 49F_4$$

$$B = F_2 - 5F_4$$

$$C = 35F_4$$

and the Slater–Condon integrals are the same as those used to calculate the Coulomb and exchange integrals in the theory of atomic spectra.[95]

In more general terms, the crystal and ligand field models described above can be regarded as first-order approximations that neglect the overlap integrals. For a second-order approximation, we must specify the effective Hamiltonian

$$\hat{\mathcal{H}}_{\text{eff}} = -\tfrac{1}{2}\nabla^2 + \mathcal{U}_{\text{M}}(i) + \mathcal{U}_{\text{X}}(i)$$

where the potential felt by the electron i has been split into metal $\mathcal{U}_{\text{M}}(i)$ and ligands $\mathcal{U}_{\text{X}}(i)$ contributions corresponding to metal $\phi_{\text{M}\alpha}$ and ligand $\phi_{\text{g}\alpha}$ orbitals of a second-order wave function

$$\Psi_{\pm} = c_{\text{M}}\phi_{\text{M}\alpha} \pm c_{\text{X}}\phi_{\text{g}\alpha}$$

If the electron i is under the field of M or X alone, the corresponding energies are

$$E_{\text{M}}^0 = \langle \phi_{\text{M}\alpha} | -\tfrac{1}{2}\nabla^2 + \mathcal{U}_{\text{M}}(i) | \phi_{\text{M}\alpha} \rangle$$

and

$$E_{\text{X}}^0 = \langle \phi_{\text{g}\alpha} | -\tfrac{1}{2}\nabla^2 + \mathcal{U}_{\text{X}}(i) | \phi_{\text{g}\alpha} \rangle$$

The secular equation

$$(H_{MM} - E)(H_{XX} - E) - (H_{MX} - ES)(H_{XM} - ES) = 0$$

obtained when the wave function $\Psi_{\pm}$ is operated with the effective Hamiltonian $\mathscr{H}_{eff}$ gives the energy levels E. The various elements of the secular equation can be considered as functions of the energy E, that is, $W_{AB} = H_{AB} - ES_{AB}$. These expressions can be introduced into the equations of the coefficients c_M and c_X

$$c_M W_{MM} + c_X W_{MX} = 0 \qquad c_M W_{XM} + c_X W_{XX} = 0$$

Reorganization of the wave functions $\Psi_{\pm}$ according to

$$\Psi_{\pm} = c_M \phi_{M\alpha} \pm c_X \phi_{g\alpha} = N_{\pm}(\phi_{M\alpha} \pm \mu_{\pm} \phi_{g\alpha})$$

leads to normalization factors

$$N_{\pm} = (1 \pm 2\mu_{\pm} S + \mu_{\pm}^2)^{-1/2}$$

and the orthogonality between Ψ_+ and Ψ_- imposes a further relationship

$$\mu_- = \frac{\mu_+ + S}{1 + \mu_+ S}$$

The standard crystal field model is recovered by making $\mu_+ = 0$ and $\mu_- = S$, that is, $\Psi_- = (1 - S^2)^{-1/2} (\phi_{M\alpha} - S\phi_{g\alpha})$. Such an approximation gives

$$c_M = (1 - S^2)^{-1/2}$$

$$c_X = -S(1 - S^2)^{-1/2}$$

for the coefficients, and

$$H_{MM} = E_M^0 + \langle \phi_{M\alpha} | \mathscr{U}_X | \phi_{M\alpha} \rangle$$

$$H_{MX} = E_M^0 S + \langle \phi_{M\alpha} | \mathscr{U}_X | \phi_g \alpha \rangle$$

for the matrix elements. Moreover, the use of the H_{MM} and H_{MX} values in the solution of the secular equation

$$E = (1 - S^2)^{-1}(H_{MM} - SH_{MX})$$

gives

$$E = E_M^0 + (1 - S^2)^{-1}(\langle \phi_{M\alpha} | \mathscr{U}_X | \phi_{M\alpha} \rangle - S\langle \phi_{M\alpha} | \mathscr{U}_X | \phi_{g\alpha} \rangle)$$

The last integral can be expanded by using the Mulliken–Ruedenberg approximation, which gives the value of the function $\phi_{M\alpha}$ in the neighborhood of the ligand nuclei as a product of the overlap integral and the ligand orbital, $\phi_{M\alpha} \sim S\phi_{g\alpha}$. This approximation and the expansion of the $(1 - S^2)^{-1}$ in terms of S^2 leads to

$$E = E^0_M + \langle \phi_{M\alpha} | \mathcal{U}_X | \phi_{M\alpha} \rangle + S^2(\langle \phi_{M\alpha} | \mathcal{U}_X | \phi_{g\alpha} \rangle - \langle \phi_{g\alpha} | \mathcal{U}_X | \phi_{g\alpha} \rangle) \cdots$$

$$\sim E^0_M + E^1_{CF} + S^2(E^1_{CF} - E^1_{LF})$$

where

$$E^1_{LF} = \langle \phi_{g\alpha} | \mathcal{U}_X | \phi_{g\alpha} \rangle$$

is an intrinsic ligand field contribution.

A second-order ligand field model, called the angular overlap model, was initially developed by Jørgensen under the assumption that $|H_{MX}| < |H_{MM} - H_{XX}|$.[389] For this model, E^0_M and $\langle \phi_{M\alpha} | \mathcal{U}_X | \phi_{M\alpha} \rangle$ are regarded as the first and second terms of the perturbatory expansion. The numerator and denominator for the second term can be calculated from the expression for the elements of the secular equation

$$H_{MX} - ES = \langle \phi_{M\alpha} | \mathcal{U}_X | \phi_{M\alpha} \rangle$$

$$H_{XX} - E = E^0_X - E^0_M$$

where $E^0_X = \langle \phi_{g\alpha} | -\frac{1}{2}\nabla^2 + \mathcal{U}_X | \phi_{g\alpha} \rangle$. Hence, the energy of the electron in a ligand field is

$$E = E^0_M + \langle \phi_{M\alpha} | \mathcal{U}_X | \phi_{M\alpha} \rangle + |\langle \phi_{M\alpha} | \mathcal{U}_X | \phi_{g\alpha} \rangle|^2 / (E^0_X - E^0_M)$$

with the first term corresponding to the orbital energy of the isolated atom, the second term representing the crystal field correction, and the third term providing the ligand field correction which is proportional to the square of the overlap integral. The group overlap integral S can be expressed as a product of a radial group integral S_{rad} and an angular factor

$$S = S_{rad} \mathcal{F}_\lambda(\chi_M, X_1, X_2, \ldots, X_N)$$

where the subscript λ signals the irreducible representations of the chemical bonding; for d orbitals $\lambda = 0, 1, 2$ for σ, π, and δ bonds, respectively. Moreover, the angular factors are written

$$\mathcal{F}_\lambda(\chi_M, X_i) = \mathcal{N}_\lambda F_{M,\lambda}(\theta_i, \varphi_i)$$

where θ_i, φ_i represent the angular coordinates of the ligand X_i and $\mathcal{N}_\lambda$ is a proportionality factor, namely $\mathcal{N}_\lambda$ is $(2l+1)^{1/2}$, $[2(2l+1)l(l+1)]^{1/2}$ and $[1/2(2l+1)-(l-1)(l+1)(l+2)]^{1/2}$ for σ, π, and δ bonding, respectively, with a metal orbital of angular quantum number l. Incorporation of $\mathcal{N}$ into the radial part gives

$$S = F_{\mathrm{M},\lambda}(\theta_i, \varphi_i)\tilde{S}_{\mathrm{MX}_i}$$

where the sign over the overlap integral S_{MX_i} signals the absence of angular dependences in the atomic functions. The expression of the potential $\mathcal{U}_\mathrm{X}$ in terms of ligand contributions

$$\mathcal{U}_\mathrm{X} = \sum_i \mathcal{U}_{\mathrm{X}_i}$$

and the factorization of the ligand field integrals

$$\langle \phi_{\mathrm{M}\alpha} | \sum_i \mathcal{U}_{\mathrm{X}_i} | \phi_{\mathrm{M}\alpha} \rangle = \sum_i F^2_{\mathrm{M},\lambda}(\theta_i, \varphi_i)\tilde{\mathcal{U}}_{\mathrm{MM}}$$

$$\langle \phi_{\mathrm{M}\alpha} | \sum_i \mathcal{U}_{\mathrm{X}_i} | \phi_{\mathrm{g}\alpha} \rangle \approx \langle \phi_{\mathrm{M}\alpha} | \sum_i \mathcal{U}_{\mathrm{X}_i} \chi_{\mathrm{X}_i} \rangle$$

$$\langle \phi_{\mathrm{M}\alpha} | \sum_i \mathcal{U}_{\mathrm{X}_i} \chi_{\mathrm{X}_i} \rangle = \sum_i F_{\mathrm{M},\lambda}(\theta_i, \varphi_i)\tilde{\mathcal{U}}_{\mathrm{MX}_i}$$

gives

$$\Delta E_\mathrm{M} = \sum_i F_{\mathrm{M},\lambda}(\theta_i, \varphi_i)[\tilde{\mathcal{U}}_{\mathrm{MM}} + \tilde{\mathcal{U}}^2_{\mathrm{MX}}/(E^0_\mathrm{M} - E^0_\mathrm{X})] = \sum_i F_{\mathrm{M},\lambda}(\theta_i, \varphi_i)\sigma_{\lambda,\mathrm{X}_i}$$

for the energy change from the energy level E^0_M. The expression in brackets is a parameter $\sigma_{\lambda,\mathrm{X}_i}$ characteristic of the M–X_i bond and independent of the other ligands and particular symmetry. Examples of this treatment are given later in this appendix.

The previous treatments are mainly electrostatic models that neglect the fact that ligands form bonds with the metal, a problem that is fully considered in various molecular orbital treatments. To apply the LCAO–MO treatment to transition metal complexes, we must find the linear combination of one-electron atomic orbitals χ_p.[390–399] Such linear combinations

$$\phi_i = \sum_{p=1}^{m} C_{p,i}\chi_p$$

are adapted to the symmetry of the molecule, and the atomic orbitals are limited to those in the valence shell. The coefficients $C_{p,i}$ are defined by the set of homogeneous equations

$$\sum_{q=1}^{m} C_{q,i}(H_{pq} - S_{pq}) = 0 \qquad (p = 1, 2, \ldots, m)$$

The overlap integrals

$$S_{pq} = \langle \chi_p | \chi_q \rangle$$

are evaluated from the explicit algebraic expressions of the functions χ. The integrals

$$H_{pq} = \langle \chi_p | \mathcal{H}_{\text{eff}} | \chi_q \rangle$$

are the matrix elements of the effective Hamiltonian for the movement of a given electron in the average field of the other electrons. In the Wolfsberg–Helmholtz method, the diagonal terms H_{pp} are assumed to be equal to the energy required for the extraction of an electron from the shell *nl* of the atom *p*. Such energies, known as valence orbital ionization potentials, or in short VOIP, are used to calculate nondiagonal elements H_{pq} according to recommendations by various authors. For example, the mathematical average rule

$$H_{pq} = kS_{pq}(H_{pp} + H_{qq})/2$$

was initially proposed by Wolfsberg–Helmholtz; [391] the geometrical average rule

$$H_{pq} = kS_{pq}(H_{pp} H_{qq})^{1/2}$$

was proposed by Balhausen and Gray;[392] and the reciprocal mean rule

$$H_{pq} = kS_{pq} \frac{2H_{pp}H_{qq}}{(H_{pp} + H_{qq})}$$

was proposed by Yeranos and Hasman.[399] Some semiempirical LCAO–MO treatments add to the previous features a variable number of charge iteration cycles which bring into the models the electroneutrality principle and the fact that the VOIP changes with the charges on the atoms involved in forming the chemical bond. Bach's method starts the calculation of the VOIP, $I_{\text{VO}}(q)$, with a relationship

$$I_{\text{VO}}(q) = I_{\text{O}}(q) + E_{\text{Av}}(p^q) + E_{\text{Av}}(p^{q+1})$$

where $I_{\text{O}}(q)$ is the ionization potential of the ion p^q, formal electric charge q, in its spectroscopic ground state, and $E_{\text{Av}}(p^q)$, $E_{\text{Av}}(p^{q+1})$ are the baricenters of the configurations involved in the ionization process, measured with respect to their corresponding ground states.[392,395,396] For variations of $I_{VO}(q)$ with successive cycles of the interaction, the potential is represented by a quadratic expression in q

$$I_{VO}(q) = a + bq + cq^2$$

However, the charge iteration is not always the recommended procedure for the description of a dative bonding. In many instances, calculations with a more complete set of wave functions (double-zeta functions, contracted Gaussian functions) and without charge iteration cycles yield better results than calculations with reduced basis and charge iteration.

An interesting semiempirical molecular orbital method was developed by McLure and Yamatera to derive chemical information from electronic spectra.[400,401] The method is better illustrated by considering a complex ML_5X, where a ligand X is placed on the positive z axis (Fig. 58). In a zero-order approximation, the complex can be regarded as octahedral. The symmetry-adapted linear combinations (Appendix III) of σ and π ligand orbitals (Fig. 58) can be used in the calculation of the charge densities that determine the metal orbital energies. For example, the energy of the orbital $x^2 - y^2$ can be written as $4\epsilon_\sigma$ since there are contributions from four equivalent σ bonds. The energy of the orbital z^2 can be obtained in relationship to the energy of the orbital $x^2 - y^2$. For this purpose, one must consider that the relative charge densities along the Cartesian axes are the square of the coefficients for each ligand divided by the sum of the squared orbital coefficients. In this regard, the electronic density is

$$\rho_z(z^2) = 2^2/[2^2 + 2^2 + (-1)^2 + (-1)^2 + (-1)^2 + (-1)^2] = 1/3$$

per ligand along the z axis, and

$$\rho_{xy}(z^2) = (-1)^2/[2^2 + 2^2 + (-1)^2 + (-1)^2 + (-1)^2 + (-1)^2 = 1/12$$

per ligand along the xy plane. The densities relative to the orbital $x^2 - y^2$ are

$$\frac{\rho_z(z^2)}{\rho(x^2 - y^2)} = \frac{1/3}{1/4} = \frac{4}{3}$$

Since the energy per bond for the orbital $x^2 - y^2$ is ϵ_σ, the energy of the orbital z^2 may be written

$$\epsilon_{z^2} = 4(1/3)\epsilon_\sigma + (4/3)\epsilon'_\sigma + (4/3)\epsilon_\sigma = (4/3)(2\epsilon_\sigma + \epsilon'_\sigma)$$

where the energies ϵ_σ and ϵ'_σ reflect differences in the bonds that L and X form with the metal in the ML_5X complex. Moreover, the introduction of parameters reflecting the differences between equatorial and axial bondings can be used to express the orbital energies in the ML_5X complexes as

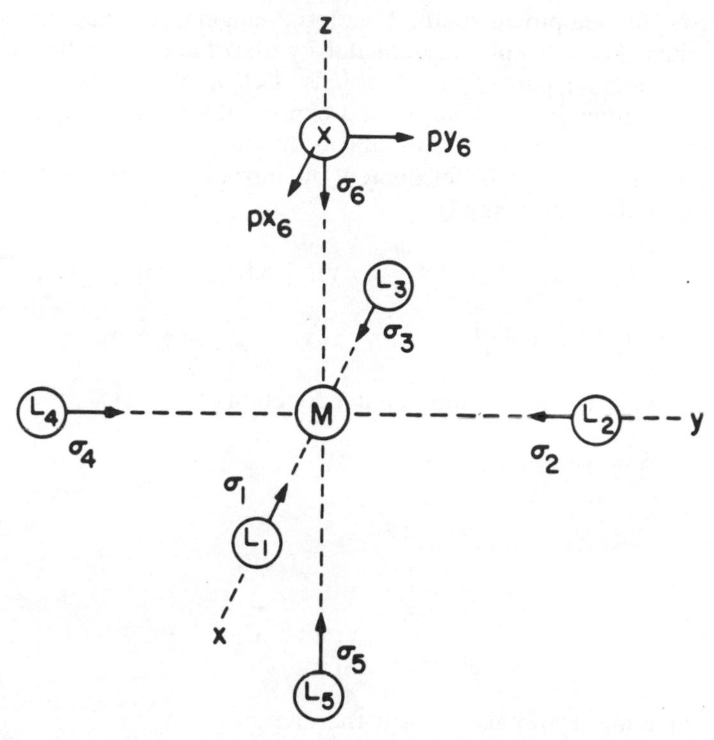

Figure 58. Ligand orbitals with σ and π symmetry for an ML_5X complex. Linear combinations of given symmetries are:

X	Metal	σ	π
a_{1g}	s	$(1/\sqrt{6})(\sigma_1+\sigma_2+\sigma_3+\sigma_4+\sigma_5+\sigma_6)$	
e_g	x^2-y^2	$(1/\sqrt{2})(\sigma_1-\sigma_2+\sigma_3-\sigma_4)$	
	z^2	$\frac{1}{2}\sqrt{3}(2\sigma_5+2\sigma_6-\sigma_1-\sigma_2-\sigma_3-\sigma_4)$	
t_{1u}	px	$(1/\sqrt{2})(\sigma_1-\sigma_3)$	$-px_6$
	py	$(1/\sqrt{2})(\sigma_2-\sigma_4)$	py_6
	pz	$(1\sqrt{2})(\sigma_5-\sigma_6)$	
t_{2g}	xz		$+px_6$
	yz		$-py_6$
	xy		

perturbations of the orbital energies ML_6. For example, if we use the energy difference $\delta\epsilon=\epsilon'_\sigma-\epsilon_\sigma$ between Γ bonds formed by X and L, the energy of the orbital z^2 in ML_5X can be recast as

$$\epsilon_{z^2}=4\epsilon_\sigma+(4/3)\delta\epsilon$$

where the first-order term $4\epsilon_\sigma$ is the energy of the orbital z^2 in ML_6.

McLure's semiempirical method and the angular overlap model give similar results. For example, the angular overlap factors can be calculated by using the angular part of the d orbitals. Let σ_L and σ_X be the intrinsic ligand contributions from L and X, and express the angular dependence of the z^2 orbital in terms of the Cartesian coordinates, $z^2 - (1/2)(x^2 + y^2)$. The angular factors $F_{M,\lambda}$ can be evaluated by introducing the corresponding ligand coordinates, for example

$$\epsilon_{z^2} = [F_\sigma^2(x) + F_\sigma^2(-x) + F_\sigma^2(y) + F_\sigma^2(-y)\sigma_L + [F_\sigma^2(-z)]\sigma_L + [F_\sigma^2(z^2)]\sigma_X$$

$$= [(1/4) + (1/4) + (1/4) + (1/4)]\sigma_L + \sigma_L + \sigma_X = 2\sigma_L + \sigma_X$$

In a similar way, the use of the angular functions

$$d_{x^2-y^2} \propto (3^{1/2}/2)(x^2 - y^2)$$

$$d_{xy} \propto xy$$

$$d_{yz} \propto yz$$

$$d_{xz} \propto xz$$

for the remaining d orbitals leads to the energies

$$\epsilon_{x^2-y^2} = 3\sigma_L$$

$$\epsilon_{xy} = 0$$

$$\epsilon_{xz} = \epsilon_{yz} = (1/4)\pi_X$$

where π_X represents an intrinsic π contribution of the ligand X along the planes xz and yz. It is easy to relate these energies to those obtained with McLure's procedure by using

$$\epsilon_\sigma = (3/4)\sigma_L$$

$$\epsilon'_\sigma = (3/4)\sigma_X$$

APPENDIX V

CHARGE TRANSFER TRANSITIONS

The AOM expression for the energies of the metal orbitals in the coordination complex (Appendix IV) can be recast in the form

$$E_{\mathrm{M}} = E_{\mathrm{M}}^{0} - \mathcal{U}_{\mathrm{M}}^{0} - \sum_{\alpha} F_{\mathrm{M},\lambda}(\theta_{\alpha}, \varphi_{\alpha})\sigma_{\lambda,\mathrm{X}_{\alpha}}$$

where the term $\mathcal{U}_{\mathrm{M}}^{0}$ represents an electrostatic perturbation of spherical symmetry upon the metal orbitals. If $E_{\mathrm{X}_{\alpha}}^{0}$ is the energy of a given ligand orbital, the transition energy between one-electron energy levels are

$$\Delta E = (E_{\mathrm{X}}^{0} - [E_{\mathrm{M}}^{0} - \mathcal{U}_{\mathrm{M}}^{0}]) + \sum_{\alpha} F_{\mathrm{M},\lambda}(\theta_{\alpha}, \varphi_{\alpha})\sigma_{\lambda,\mathrm{X}_{2}}$$

It is possible to equate the first term with a difference in optical electronegativities, that is, $(E_{\mathrm{X}}^{0} - [E_{\mathrm{M}}^{0} - \mathcal{U}_{M}^{0}]) \propto \chi_{\mathrm{X}} - \chi_{\mathrm{M}}$. Moreover, the addition of a term, δSPE, correcting for the difference between the interelectronic energies of the ground and excited state configurations gives the energy of the electronic transition E_{CT}, where charge shifts from the ligand to the metal or vice versa. In this regard, the complete expression for the energy of a charge transfer transition is

$$E_{\mathrm{CT}} = (\chi_{\mathrm{X}} - \chi_{\mathrm{M}}) + \sum_{\alpha} F_{\mathrm{M},\lambda}(\theta_{\alpha}, \varphi_{\alpha})\sigma_{\lambda,\mathrm{X}_{2}} + \delta\mathrm{SPE} + \cdots$$

where the E_{CT} energy is given in kilojoules. For a vertical transition in an octahedral complex, the expression can easily be reduced to eq. 9 in Chapter 5. Moreover, the angular contributions introduced by the factors $\sigma_{\mathrm{M},\lambda}$, and the radial contributions of the factors $\sigma_{\lambda,\mathrm{X}_{\alpha}}$ can be used to calculate the energy gap between a CT excited state and a ground state with different nuclear configurations.

The change in bond energies associated with population of a CT state reflect the destabilization brought about in a given bond. For example, McLure's procedure gives the excited state bond energies

$$*I_{MX} = -(\epsilon_d^0 + 3\epsilon_X^0(\pi)) + e_\sigma(x) + 2e'_\sigma(L) - 12e'_\pi(X)$$

$$*I_{ML} = -(\epsilon_d^0 + 2\epsilon_L^0(\sigma)) + e_\sigma(X)$$

for a $\pi_X \rightarrow d_{z^2}$ charge transfer transition in an ML_5X complex. The changes in bond energies

$$\Delta I_{MX} = *I_{ML} - I_{ML} = -\epsilon_d^0 - \epsilon_X^0(\pi) + 4e'_\pi(X) + e_\sigma(X) + 2e_\sigma(L)$$

$$\Delta I_{ML} = *I_{ML} - I_{ML} = -\epsilon_d^0 + e'_\sigma(X) + 2e'_\sigma(L)$$

signal a considerable destabilization of the MX bond that gets more pronounced with the difference $\epsilon_d^0 - \epsilon_X^0(\pi)$. A similar destabilization is experienced by the MX bond in a $\pi_X \rightarrow d_{x^2-y^2}$ charge transfer transition. Although these observations suggest that the ML_5X complex must undergo homolysis of the MX bond, we must consider that the formation of the primary products is dynamical in nature and usually involves competition between various processes.

APPENDIX **VI**

BORN CYCLES RELATED TO CHARGE TRANSFER PROCESSES

The calculation of the primary products formation enthalpy

$$M^{III}L_5X \xrightarrow{\Delta H_{CT}} [M^{II}L_5, X^{\cdot}]$$

where $[M^{II}L_5, X^{\cdot}]$ represents a radical pair, can be carried out with the Born cycles shown in Figure 59.[402,403] George and McLure established a thermodynamic cycle for estimating the standard free energy of a redox process. Several cycles have also been proposed for the calculation of other thermochemical properties in coordination complexes.[403] For example, scheme I in Figure 59 is used in connection with the standard enthalpy of hydration hydration

$$\Delta H_h^0 = \Delta H_F^0 - L_v - \sum_{n=1}^{3} I_n - na$$

where L_v is the latent heat of sublimitation, $\Sigma_{n=1}^{3} I_n$ is the sum of the first n ionization ethalpies corrected from the ionization potentials, ΔH_h^0 is the enthalpy of the process

$$M^{+n}(g) \longrightarrow M^{+n}(aq)$$

and a is the enthalpy of the process

$$H^+(aq) + e^- \longrightarrow \tfrac{1}{2}H_2(g)$$

A thermodynamic cycle for the formation of a complex from an aqueous ion

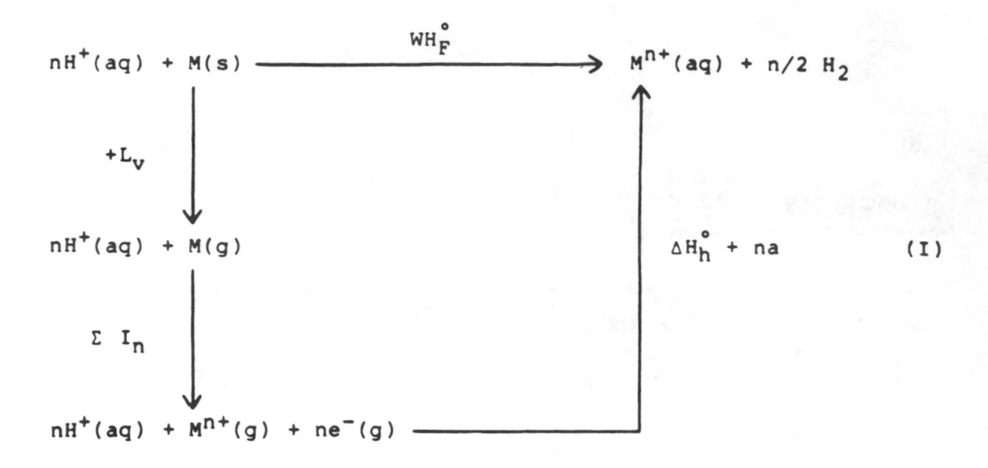

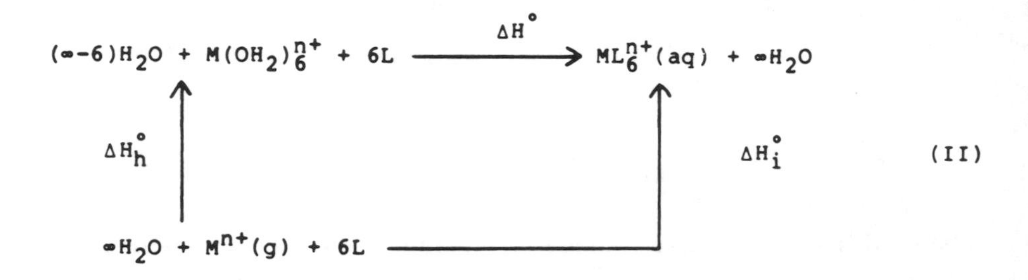

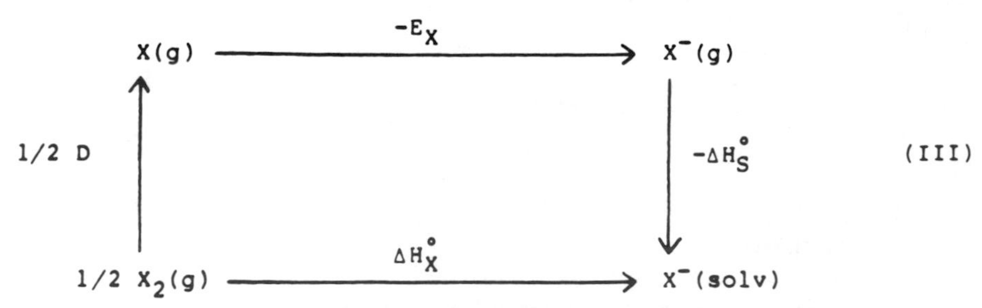

Figure 59. Born cycles for the calculation of thermochemical quantities associated with charge transfer transitions.

in aqueous solution is shown in scheme II (Fig. 59). The cycle shows that the enthalpy of immersion of the gaseous ion in a very dilute solution of the ligand with formation of the hydrated complex ML_6^{3+} is

$$\Delta H_i^0 = \Delta H_{CF}^0 + \Delta H_h^0$$

where ΔH_h^0 has been defined in connection with scheme I. Moreover, provided the difference in the spherical ion terms do not change radically, ΔH_{CF}^0 can be related to the differences in ligand field stabilization energies

$$\Delta H_{CF}^0 \sim (\alpha D_q - \alpha' D_q')$$

A similar cycle can be used to calculate the thermochemical propertics associated with the process

$$\tfrac{1}{2}H_2(g) + (ML_5^{III})_{aq} \underset{-e^-}{\overset{e^-}{\rightleftharpoons}} (M^{II}L_5)_{aq} + H^+(aq)$$

In this context, we find that the third ionization enthalpy I_3, the change in ligand field stabilization energies $(\alpha D_q - \alpha' D_q')$, and the solvation enthalpies ΔH_S^0 and ΔH_S^0 contribute to the reaction's enthalpy

$$\Delta H_M^0 = I_3 + (\alpha D_q - \alpha' D_q) + (\Delta H_s^0 - \Delta H_S^{0\prime})_M - a$$

Inasmuch as the reaction free energy is

$$\Delta G_M^0 = \Delta H_M^0 - T\Delta S_M^0$$

the calculation of ΔG_M^0 is limited by the possibility of estimating related entropy changes ΔS_M^0. However, we can use the relationship

$$\Delta H_M^0 \sim \Delta E_M^0$$

as a workable hypothesis that is based on the condition

$$P\,\Delta V_M^0 \ll \Delta E_M^0$$

A Born cycle (scheme III in Fig. 59) can also be used for the enthalpy of the ligand oxidation

$$\Delta H_X^0 = \tfrac{1}{2}D - \epsilon_X - \Delta H_S^0(X) + a$$

It is possible to relate the enthalpies ΔH_M^0 and ΔH_X^0 to the enthalpy of the charge transfer process. For example, a charge transfer process involving ion pairs can be described by means of the equations

$$ML_6^{3+} + e^- \rightleftarrows ML_6^{2+} \qquad \Delta H_M^0$$

$$X^{\cdot} + e^{-} \rightleftarrows X^{-} \qquad \Delta H^{0}_{X}$$

$$ML_6^{3+} + X^{-} \rightleftarrows [ML_6^{3+}, X^{-}] \qquad \Delta H^{0}_{ip}$$

$$ML_6^{2+} + X^{\cdot} \rightleftarrows [ML_6^{2+}, X^{\cdot}] \qquad \Delta H^{0}_{rp}$$

where the species in brackets represent ion and ion-radical pairs, while ΔH^{0}_{ip} and ΔH^{0}_{rp} represent the corresponding formation enthalpies. The enthalpy ΔH_{CT} associated with the charge transfer process

$$[ML_6^{3+}, X^{-}] \xrightarrow{\Delta H_{CT}} [ML_6^{2+}, X^{\cdot}]$$

can be expressed by

$$\Delta H_{CT} = \Delta H^{0}_{M} - \Delta H^{0}_{X} - \Delta H^{0}_{ip} + \Delta H^{0}_{rp}$$

if ML^{3+} and ML^{2+} are in their corresponding nuclear equilibrium configurations. To calculate the threshold enthalpy ΔH^{th}_{CT} we must consider that the nuclear configuration and spin state of the primary product ML_6^{2+} differ from those in the equilibrated ground state product. In this regard, the threshold enthalpy can be corrected from ΔH_{CT} by introducing the differences in Franck–Condon enthalpies ΔH_{FC} and spin pairing energies ΔSP in the expression of ΔH_{CT}

$$\Delta H^{th}_{CT} = \Delta H^{0}_{M} - \Delta H^{0}_{X} - \Delta H^{0}_{ip} + \Delta H^{0}_{rp} + \Delta H_{FC} + \Delta SP$$

REFERENCES

1. M. Planck, *Ann. Phys.* **1901**, *4*, 553.
2. A. Einstein, *Ann. Phys.* **1905**, *17*, 132.
3. B. Rossi, *Optics*, Reading, MA: Addison-Wesley, **1957**, pp. 312–343.
4. C. Davisson and L. Germer, *Phys. Rev.* **1927**, *30*, 705.
5. G. P. Thomson, *Proc. R. Soc. London* **1928**, *117*, 600.
6. W. Duane, *Proc. Natl. Acad. Sci.* **1923**, *9*, 158.
7. A. H. Compton, *Proc. Natl. Acad. Sci.* **1923**, *9*, 359.
8. W. Heisenberg, *Z. Phys.* **1927**, *43*, 172.
9. B. Rossi, *Optics*, Reading, MA: Addison-Wesley, **1957**, pp. 366–378.
10. J. G. Calvert and J. N. Pitts, Jr., *Photochemistry*, New York: Wiley **1967**, pp. 19–21.
11. J. Granifo and G. Ferraudi, *J. Phys. Chem.* **1985**, *89*, 1206.
12. C. G. Hatchard and C. A. Parker, *Proc. R. Soc. London Ser. A* **1956**, *235*, 518.
13. J. Lee and H. H. Seliger, *J. Chem. Phys.* **1964**, *40*, 519.
14. J. N. Demas, W. D. Bowman, E. F. Zalewski, and R. A. Velapoldi, *J. Phys. Chem.* **1981**, *85*, 2766.
15. C. A. Disher, P. F. Smith, I. Lippman, and R. Turse, *J. Phys. Chem.* **1963**, *67*, 2501.
16. W. G. Leighton and G. S. Forbes, *J. Am. Chem. Soc.* **1930**, *52*, 3139.
17. M. I. Christie and G. Porter, *Proc. R. Soc. London* **1952**, *212*, 390.
18. J. N. Pitts, Jr., J. D. Margerum, R. P. Taylor, and W. Brim, *J. Am. Chem. Soc.* **1955**, *77*, 5499.
19. K. Porter and D. H. Volman, *J. Am. Chem. Soc.* **1962**, *84*, 2011.
20. E. E. Wegner and A. Adamson, *J. Am. Chem. Soc.* **1966**, *88*, 394.
21. L. B. Thomas, *J. Am. Chem. Soc.* **1940**, *62*, 1879.
22. RCA Photomultiplier Manual, Technical Series PT-61, RCA Corporation, **1970**.
23. J. Franck and E. Rabinowitch, *Trans. Faraday Soc.* **1934**, *30*, 120.
24. R. M. Noyes, in *Progress in Reaction Kinetics*, Vol. 1 (G. Porter, Ed.), New York: Pergamon, **1961**, p. 129.
25. J. P. Lorand, *Progr. Inorg. Chem.* **1972**, *17*, 207.

26. D. H. Volman and J. C. Chen, *J. Am. Chem. Soc.* **1959**, *81*, 4141.
27. R. M. Noyes, *J. Am. Chem. Soc.* **1956**, *78*, 5486.
28. T. Koenig, *J. Am. Chem. Soc.* **1969**, *91*, 2558.
29. W. Braun, L. Rajbenbach, and F. R. Eirich, *J. Phys. Chem.* **1962**, *66*, 1591.
30. S. W. Benson, *Thermochemical Kinetics*, New York: Wiley, **1976**, pp. 204–207.
31. G. Herzberg, *Molecular Spectra and Molecular Structure*, Vol. 1, Princeton, NJ: Van Nostrand, **1966**, p. 315.
32. G. Porter and M. R. Wright, *Disc. Faraday Soc.* **1959**, *27*, 18.
33. N. J. Turro, *Modern Molecular Photochemistry*, Menlo Park, CA: Benjamin/Cummings, **1978**, pp. 160, 161.
34. A. J. Bard and L. R. Faulkner, *Electrochemical Methods*, New York: Wiley, **1980**, pp. 130–133.
35. M. von Smoluchowski, *Z. Phys. Chem.* **1917**, *92*, 129.
36. G. V. Buxton and R. M. Sellers, "Compilation of Rate Constants for the Reactions of Metal Ions in Unusual Oxidation States," Natl. Stand. Ref. Data Ser., National Bureau of Standards; NSRDS-NBS 62.
37. A. B. Ross and P. Neta, "Rate Constants for Reactions of Inorganic Radicals in Aqueous Solution," Natl. Stand. Ref. Data Ser., National Bureau of Standards; NSRDS-NBS 65.
38. A. Anbar, M. Bambenek and A. Ross, "Selected Specific Rates of Reactions of Transients from Water in Aqueous Solution. 1. Hydrated Electron," Natl. Stand. Ref. Data Ser., National bureau of Standards; NSRDS-NBS 43.
39. G. L. Hug, "Optical Spectra of Nonmetallic Inorganic Transient Species in Aqueous Solution," Natl. Stand. Ref. Data Ser., National Bureau of Standards; NSRDS-NBS 69.
40. A. Ross and P. Neta, "Rate Constants for Reactions of Aliphatic Carbon-Centered Radicals in Aqueous Solution," Natl. Stand. Ref. Data Ser., National Bureau of Standards; NSRDS-NBS 70.
41. M. S. Matheson, W. A. Mulac, J. L. Weeks, and J. Rabani, *J. Phys. Chem.* **1966**, *70*, 2092.
42. M. Schöneshöfer and A. Henglein, *Ber. Bunsenges. Phys. Chem.* **1969**, *73*, 289.
43. J. H. Baxendale, P. L. Bevan, P. L. Bevan, and A. D. Stott, *Trans. Faraday Soc.* **1968**, *64*, 2389.
44. M. Schöneshöfer and A. Henglein, *Ber. Bunsenses. Phys. Chem.* **1970**, *74*, 393.
45. M. Gratzel, A. Henglein, J. Lilie, and G. Beck, *Ber. Bunsenges. Phys. Chem.* **1969**, *73*, 646.
46. G. Stein, *Israel J. Chem.* **1970**, *8*, 691.
47. M. Shirom and M. Tomkiewicz, *J. Chem. Phys.* **1972**, *56*, 2731.
48. R. M. Noyes, *J. Am. Chem. Soc.* **1955**, *77*, 2042.
49. G. Porter, "Flash Photolysis," Technique of Organic Photochemistry, Vol. VIII, Part. II, 2nd ed. Interscience Publishers, NY, 1963.
50. S. Claesson and L. Lindquists, *Arkiv. Kemi* **1957**, *11*, 535.
51. S. Claesson and L. Lindquists, *Arkiv. Kemi* **1958**, *12*, 1.
52. C. Lewis, W. R. Ware, L. J. Doemeny, and T. L. Nemzek, *Rev. Sci. Instrum.* **1973**, *44*, 107.

53. D. L. Chapman and F. Briers, *J. Chem. Soc.* **1928**, 1802.
54. L. E. Hargrove, R. L. Fork, and M. A. Pollack, *Appl. Phys. Lett.* **1964**, *5*, 4.
55. P. M. Rentzepis, *Chem. Phys. Lett.* **1968**, *2*, 117.
56. D. Von der Linde, D. Bernecker, and W. Kaiser, *Optics Commun.* **1970**, *2*, 149.
57. M. Malley and P. M. Rentzepis, *Chem. Phys. Lett.* **1970**, *7*, 57.
58. M. R. Topp, P. M. Rentzepis, and R. P. Jones, *Chem. Phys. Lett.* **1971**, *9*, 1.
59. T. L. Netzel, P. M. Rentzepis, and J. S. Leigh, *Science* **1973**, *182*, 238.
60. P. M. Rentzepis, R. P. Jones, and J. Jortner, *J. Chem. Phys.* **1973**, *59*, 766.
61. G. A. Kenney-Wallace and D. C. Walker, *J. Chem. Phys.* **1971**, *55*, 447.
62. M. Born and J. R. Oppenheimer, *Ann. Phys.* **1927**, *84*, 457.
63. M. Born, *Gött. Nachr. Math. Phys. KI.* **1951**, 1.
64. M. Bixon and J. Jortner, *J. Chem. Phys.* **1968**, *48*, 715.
65. C. J. Ballhausen, *Molecular Electronic Structures of Transition Metal Complexes*, New York: McGraw-Hill, **1979**.
66. G. Herzberg, *Molecular Spectra and Molecular Structure, Infrared and Raman Spectral of Polyatomic Molecules, Vol. 2*, Princeton, NJ: Van Nostrand, **1945**.
67. E. B. Wilson, J. C. Decius, and P. C. Cross, *Molecular Vibrations*, New York: McGraw-Hill, **1955**.
68. L. Landau, *Phys. Z. Sowjetunion* **1932**, *2*, 46.
69. C. Zener, *Proc. R. Soc. London Sec. A* **1932**, *137*, 696.
70. C. Zener, *Proc. R. Soc. London Ser. A* **1932**, *140*, 660.
71. L. Landau and E. M. Lifshitz, *Quantum Mechanics*, Reading, MA: Addison-Wesley, **1958**, pp. 304–312.
72. G. S. Hammond, *Adv. Photochem.* **1964**, *7*, 373.
73. B. H. Hathaway, M. Douggan, A. Murphy, J. Mullane, C. Power, A. Walsh, and B. Walsh, *Coord. Chem. Rev.* **1981**, *36*, 267.
74. J. Gazo, I. B. Bersuker, J. Garaj, M. Kabesova, J. Kohout, R. Langerlderova, M. Melnik, M. Serator, and F. Valach, *Coord. Chem. Rev.* **1976**, *21*, 253.
75. H. A. Jahn and E. Teller, *Proc. R. Soc. London Ser. A* **1937**, *161*, 220.
76. H. C. Longuet-Higgins, *Adv. Spectrosc.* **1961**, *2*, 429.
77. R. Englman, *The Jahn Teller Effect in Molecules and Crystals*, New York: Wiley-Interscience, **1972**.
78. C. E. Schäffer and C. K. Jørgensen, *J. Inorg. Nucl. Chem.* **1958**, *8*, 143.
79. L. E. Orgel, *J. Chem. Soc.* **1952**, 4752, 4756.
80. R. J. P. Williams, *Disc. Faraday Soc.* **1958**, *26*, 180.
81. B. R. Hollebone, *Theor. Chim. Acta* **1980**, *56*, 45.
82. J. Donini and B. R. Hollebone, *Theor. Chim. Acta* **1976**, *42*, 97.
83. M. H. L. Pryce, *Proc. Phys. Soc. A* **1950**, *63*, 25.
84. H. Kamimura, *J. Phys. Soc. Japan* **1956**, *11*, 1171.
85. K. W. H. Stevens, *Proc. R. Soc. London Ser. A* **1953**, *219*, 542.
86. M. H. L. Pryce, *Proc. R. Soc. London Ser. A* **1950**, *63*, 25.
87. M. H. L. Pryce, *Nuovo Cimento*, *ser.* 10, **1957**, *6*, 817.

88. K. Ikeda and S. Maeda, *Inorg. Chem.* **1978**, *17*, 2698.
89. C. D. Flint and A. G. Paulusz, *Mol. Phys.* **1980**, *41*, 907.
90. R. Wernicke, G. Eyring, and H. Schmidtke, *Chem. Phys. Lett.* **1978**, *58*, 267.
91. H. Schmidtke and D. Strand, *Inorg. Chim. Acta* **1982**, *62*, 153.
92. B. A. Kozikowski and T. A. Keiderling, *Mol. Phys.* **1980**, *40*, 477.
93. T. Schonherr, R. Wernicke, and H. Schmidtke, *Spectro-chim. Acta Part A* **1982** *38*, 679.
94. A. Einstein, *Phys. Z.* **1917**, *18*, 121.
95. E. U. Condon and G. H. Shortley, *The Theory of Atomic Spectra*, Cambridge: Cambridge Univ. Press, **1935**, p. 6.
96. R. F. Fenske, *J. Am. Chem. Soc.* **1967**, *89*, 252.
97. M. T. Vala, C. J. Ballhausen, R. Dingle, and S. L. Holt, *Mol. Phys.* **1972**, *23*, 217.
98. T. G. Harrison, H. H. Patterson, and J. J. Godfrey, *Inorg. Chem.* **1976**, *15*, 1291.
99. L. E. Orgel, *J. Chem. Phys.* **1955**, *23*, 1824.
100. P. A. Ford and O. F. Hill, *Spectrochim. Acta* **1960**, *16*, 493.
101. M. L. H. Pryce and W. A. Runciman, *Disc. Faraday Soc.* **1958**, *26*, 34.
102. L. S. Forster, *Transition Metal Chem.* **1969**, *5*, 1.
103. N. A. P. Kane-Maguire and C. H. Langford, *Chem. Commun.* **1971**, 895.
104. C. D. Flint and A. P. Matthews, *J. Chem. Soc. Faraday Trans. 2* **1976**, *72*, 379.
105. R. B. Wilson and E. I. Solomon, *Inorg. Chem.* **1978**, *17*, 1729.
106. L. S. Forster, J. V. Rund, F. Castelli, and P. Adams, *J. Phys. Chem.* **1982**, *86*, 2395.
107. K. Kuhn, F. Wagestian, and H. Kupka, *J. Phys. Chem.* **1981**, *85*, 665.
108. A. Pfeil, *J. Am. Chem. Soc.* **1971**, *93*, 5395.
109. F. Seel and D. Meyer, *Z. Anorg. Allgem. Chem.* **1974**, *408*, 275.
110. E. Konig, E. Lindner, I. P. Lorenz, G. Ritter, and H. Gausman, *J. Inorg. Nucl. Chem.* **1971**, *33*, 3305.
111. R. A. Plane and J. P. Hunt, *J. Am. Chem. Soc.* **1957**, *79*, 3343.
112. R. Dingle, *J. Chem. Phys.* **1969**, *50*, 1952.
113. A. Pfeil, Ph.D. Thesis, University of British Columbia, **1971**.
114. W. H. Fonger and C. Struck, *Phys. Rev. B* **1975**, *11*, 3251.
115. P. C. Mitchell, *Quart. Rev.* **1966**, *20*, 103.
116. M. J. Reisfeld, N. A. Matwiyoff, and L. B. Asprey, *J. Mol. Spectrosc.* **1971**, *39*, 8.
117. S. L. Chodos, A. M. Black, and C. D. Flint, *J. Chem. Phys.* **1976**, *65*, 4816.
118. C. A. Parker, *Photoluminescence of Solutions*, Amsterdam: Elsevier, **1968**.
119. C. D. Flint and P. Greenough, *J. Chem. Soc. Faraday Trans. 2* **1974**, *70*, 130.
120. C. K. Jørgensen, *Adv. Chem. Phys.* **1965**, *8*, 47.
121. H. L. Schlafer, H. Gracesmann, H. F. Wagestian, and H. Zander, *Z. Phys. Chem.* **1966**, *51*, 274.
122. C. Conti and L. S. Forster, *J. Am. Chem. Soc.* **1977**, *99*, 613.

123. C. A. Parker and C. G. Hatchard, *Trans. Faraday Soc.* **1961**, *57*, 1894.
124. C. A. Parker and C. G. Hatchard, *Proc. Chem. Soc.* **1962**, 147.
125. C. A. Parker and C. G. Hatchard, *Proc. R. Soc. London Ser. A* **1962**, *269*, 574.
126. D. G. Whiten, J. K. Roy, and F. A. Carrol, "The Exciplex," (M. Gordon and W. R. Ware, Eds.), New York: Academic, **1975**.
127. Mark E. Frink and G. Ferraudi, *Chem. Phys. Lett.* **1986**, *124*, 576.
128. G. Ferraudi, M. E. Frink, and D. K. Geiger, *J. Phys. Chem.* **1986**, *90*, 1924.
129. G. Ferraudi and D. R. Prasad, *Inorg. Chem.* **1983**, *22*, 1672.
130. M. T. Indelli, R. Ballardini, C. A. Bignozzi, and F. Scandola, *J. Phys. Chem.* **1982**, *86*, 4284.
131. G. N. Lewis and M. Kasha, *J. Am. Chem. Soc.* **1945**, *67*, 994.
132. S. J. Stickler and R. A. Berg, *J. Chem. Phys.* **1962**, *37*, 814.
133. L. S. Forster, *Transition Metal Chemistry*, Vol. 1 (R. L. Carlin, Ed.), New York: Dekker, **1969**.
134. J. Jortner, S. A. Rice, and R. M. Hochstrasser, *Advances in Photochemistry*, Vol. 7 (J. N. Pitts, Jr., G. S. Hammond, and W. A. Noyes, Eds.), New York: Wiley-Interscience, **1969**, pp. 149–160, 184–192.
135. M. Bixon and J. Jortner, *J. Chem. Phys.* **1968**, *48*, 715.
136. J. Jortner and R. S. Berry, *J. Chem. Phys.* **1968**, *48*, 2757.
137. D. P. Chock, J. Jortner, and S. A. Rice, *J. Chem. Phys.* **1968**, *49*, 610.
138. K. F. Freed and J. Jortner, *J. Chem. Phys.* **1970**, *52*, 6272.
139. R. Englman and J. Jortner, *Mol. Phys.* **1970**, *18*, 145.
140. T. Ramasami, J. F. Endicott, and G. Brubaker, *J. Phys. Chem.* **1983**, *87*, 5057.
141. T. Förster, *Disc. Faraday Soc.* **1959**, *27*, 7.
142. T. Förster, *Naturwissenchaften* **1946**, *33*, 166.
143. D. L. Dexter, *J. Chem. Phys.* **1953**, *21*, 836.
144. V. Balzani, F. Bolleta, and F. Scandola, *J. Am. Chem. Soc.* **1980**, *102*, 2152.
145. N. Sutin, *Progr. Inorg. Chem.* **1983**, *30*, 441.
146. F. A. Kröger, *Some Aspects of the Luminescence of Solids*, Houston: Elsevier, **1948**.
147. J. F. Endicott, *Coord. Chem. Revs.* **1985**, *64*, 293.
148. J. F. Endicott, *ACS Symp. Ser.* **1982**, *198*, 227.
149. R. Tamilarasan and J. F. Endicott, *J. Phys. Chem.* **1986**, *90*, 1027.
150. R. S. Mulliken, *J. Phys. Chem.* **1952**, *74*, 811.
151. R. S. Mulliken, *J. Phys. Chem.* **1952**, *56*, 801.
152. C. K. Jørgensen, *Progr. Inorg. Chem.* **1962**, *4*, 73.
153. C. K. Jørgensen, *Orbitals in Atoms and Molecules* New York: Academic, **1962**.
154. C. K. Jørgensen, *Oxidation Numbers and Oxidation States*, New York: Springer, **1969**.
155. J. F. Endicott, *Concepts in Inorganic Photochemistry* (A. Adamson and P. Fleischauer, Eds.), New York: Wiley-Interscience, **1975**, pp. 81–86.
156. N. Bjerrum, *Kgl. Danske Videnskab. Selskab. Mat. fys. Medd.* **1926**, *9*, 7.
157. M. G. Evans and G. H. Nancollas, *Trans. Faraday Soc.* **1953**, *49*, 363.

158. J. A. Caton and J. E. Prue, *J. Chem. Soc.* **1956**, 671.
159. H. Taube and F. A. Posey, *J. Am. Chem. Soc.* **1953**, *75*, 1463.
160. F. A. Posey and H. Taube, *J. Am. Chem. Soc.* **1956**, *78*, 15.
161. H. Elsgernd and J. K. Beattie, *Inorg. Chem.* **1968**, *7*, 2468.
162. M. T. Beck, *Coord. Chem. Rev.* **1968**, *3*, 91.
163. M. F. Manfrin, G. Varani, L. Moggi, and V. Balzani, *Mol. Photochem.* **1969**, *1*, 387.
164. P. D. Fleischauer, A. W. Adamson, and G. Sartori, *Progr. Inorg. Chem.* **1972**, *17*, 1.
165. A. W. Adamson, W. L. Waltz, E. Zinalto, D. W. Watts, P. D. Fleischauer, and R. D. Lindholm, *Chem. Rev.* **1968**, *68*, 541.
166. A. W. Adamson, *Disc. Faraday Soc.* **1960**, *29*, 163.
167. J. F. Endicott and T. L. Netzel, *J. Am. Chem. Soc.* **1979**, *101*, 4000.
168. V. Balzani, R. Ballardini, N. Sabbatini, and L. Moggi, *Inorg. Chem.* **1968**, *7*, 1938.
169. F. Scandola, C. Bartocci, and M. A. Scandola, *J. Am. Chem. Soc.* **1973**, *95*, 7898.
170. J. F. Endicott and G. Ferraudi, *J. Am. Chem. Soc.* **1974**, *96*, 3681.
171. E. M. Kober, J. V. Caspar, R. S. Lumpkin, and T. J. Meyer, *J. Phys. Chem.* **1986**, *90*, 3722, and references therein.
172. R. Dallinger and W. Woodruff, *J. Am. Chem. Soc.* **1979**, *101*, 4391.
173. K. W. Hipps, *Inorg. Chem.* **1980**, *19*, 1390.
174. J. Ferguson, E. R. Krausz, and M. Maeder, *J. Phys. Chem.* **1985**, *89*, 1852.
175. J. V. Caspar, T. D. Westmoreland, G. H. Allen, P. G. Bradley, T. J. Meyer, and W. Woodruff, *J. Am. Chem. Soc.* **1984**, *106*, 3492.
176. J. F. Endicott, G. J. Ferraudi, and J. R. Barber, *J. Phys. Chem.* **1975**, *79*, 630.
177. J. F. Endicott, G. J. Ferraudi, and J. R. Barber, *J. Am. Chem. Soc.* **1975**, *97*, 219.
178. J. F. Endicott and G. Ferraudi, *Inorg. Chem.* **1975**, *14*, 3133.
179. R. A. Marcus, *J. Phys. Chem.* **1968**, *72*, 891.
180. R. A. Marcus, *J. Phys. Chem.* **1963**, *67*, 853.
181. R. A. Marcus, *Ann. Rev. Phys. Chem.* **1964**, *15*, 155.
182. N. Sutin, *Ann. Rev. Nucl. Sci.* **1962**, *12*, 285.
183. R. A. Marcus, *J. Chem. Phys.* **1956**, *24*, 966.
184. R. S. Berry, C. N. Reimann, and G. N. Spokes, *J. Chem. Phys.* **1962**, *37*, 2278.
185. E. Rabinowitch, *Rev. Mod. Phys.* **1942**, 112.
186. M. Smith and M. C. R. Symons, *J. Chem. Phys.* **1956**, *25*, 1084.
187. M. J. Blandamer and M. Fox, *Chem. Rev.* **1970**, *70*, 59.
188. K. L. Stevenson and D. D. Davis, *Inorg. Nucl. Chem. Lett.* **1976**, *12*, 905.
189. G. Ferraudi, *Inorg. Chem.* **1978**, *187*, 1741.
190. R. Platzman and J. Frank, *Z. Phys.* **1954**, *138*, 411.
191. J. Jortner, *J. Chem. Phys.* **1957**, *27*, 823.
192. J. Jortner, *J. Chem. Phys.* **1959**, *30*, 839.

193. J. Jortner, *Radiat. Res. Suppl.* **1964**, *4*, 24.
194. T. P. Das, *Adv. Chem. Phys.* **1962**, *4*, 338.
195. G. Stein and A. Treinin, *Trans. Faraday Soc.* **1959**, *55*, 1086.
196. A. Treinin, *J. Phys. Chem.* **1964**, *68*, 893.
197. M. Smith and M. C. R. Symons, *Trans. Faraday Soc.* **1958**, *54*, 338.
198. M. Smith and M. C. R. Symons, *Trans. Faraday Soc.* **1958**, *54*, 346.
199. M. Smith and M. C. R. Symons, *Disc. Faraday Soc.* **1957**, *24*, 206.
200. M. Fox, *Concepts in Inorganic Photochemistry* (A. Adamson and P. Fleischauer, Eds.), New York: Wiley-Interscience, **1975**, p. 338.
201. C. K. Jørgensen, *Solid State Phys.* **1962**, *13*, 375.
202. C. K. Jørgensen, *Adv. Chem. Phys.* **1963**, *5*, 33.
203. L. Doglioti and E. Hayon, *J. Phys. Chem.* **1968**, *72*, 1800.
204. A. Treinin and M. Yaacobi, *J. Phys. Chem.* **1963**, *68*, 2487.
205. G. Zimmerman, *J. Chem. Phys.* **1955**, *23*, 825.
206. M. Daniels, R. V. Meyers, and E. V. Belardo, *J. Phys. Chem.* **1968**, *72*, 389.
207. U. Shuali, M. Ottolenghi, J. Rabani, and Z. Jelin, *J. Phys. Chem.* **1969**, *73*, 3445.
208. G. Ferraudi, J. F. Endicott, and J. Barber, *J. Am. Chem. Soc.* **1975**, *97*, 6406.
209. C. D. Flint and D. J. Palacio, *J. Chem. Soc. Faraday Trans.* **1979**, *75*, 1159.
210. N. J. Linck, S. J. Berens, M. Magde, and R. G. Linck, *J. Phys. Chem.* **1983**, *87*, 1733.
211. O. S. Mortensen, *J. Chem. Phys.* **1967**, *47*, 4215.
212. A. D. Kirk, *Coord. Chem. Revs.* **1981**, *39*, 225.
213. C. K. Jørgensen, *Absorption Spectra and Chemical Bonding in Complexes*, New York: Pergamon, **1962**.
214. R. T. Walters and A. W. Adamson, *Acta Chem. Scand. Ser. A* **1979**, *33*, 53.
215. W. Geiss and H. L. Schlafer, *Z. Phys. Chem.* **1969**, *63*, 107.
216. A. D. Kirk and G. B. Porter, *Inorg. Chem.* **1980**, *19*, 445.
217. A. D. Kirk, L. A. Frederick, and C. F. Wong, *Inorg. Chem.* **1979**, *18*, 448.
218. H. F. Wagestian, *Z. Phys. Chem.* **1969**, *67*, 3827.
219. C. D. Flint and P. Greenough, *J. Chem. Soc. Faraday Trans. 2* **1974**, *70*, 130.
220. A. Chiang and A. Adamson, *J. Phys. Chem.* **1968**, *72*, 3827.
221. M. A. Jamieson, N. Serpone, and M. Z. Hoffman, *Coord. Chem. Rev.* **1981**, *39*, 121.
222. N. A. P. Kane-Maguire and C. H. Langford, *J. Chem. Soc. Chem. Commun.* **1971**, 895.
223. D. Sandrini, M. Gandolfi, L. Moggi, and V. Balzani, *J. Am. Chem. Soc.* **1978**, *100*, 1463.
224. M. A. Jamieson, N. Serpone, M. S. Henry, and M. Z. Hoffman, *Inorg. Chem.* **1979**, *18*, 214.
225. K. DeArmond and L. S. Forster, *Spectrochim. Acta* **1963**, *19*, 1393.
226. E. Zinato, P. Riccieri, and M. Prelati, *Inorg. Chem.* **1981**, *20*, 1423.
227. J. F. Endicott, personal communication.

228. C. D. Flint and A. D. Mattheus, *J. Chem. Soc. Faraday Trans. 2* **1973**, *69*, 419.
229. J. R. Permureddi, *Coord. Chem. Rev.* **1969**, *4*, 73.
230. L. G. Vanquickenborne and A. Coulemans, *J. Am. Chem. Soc.* **1977**, *99*, 2208.
231. L. G. Vanquickenborne and A. Coulemans, *Coord. Chem. Rev.* **1983**, *100*, 157.
232. A. F. Fucaloro, L. S. Forster, J. V. Rund, and S. H. Lin, *J. Phys. Chem.* **1983**, *87*, 1796.
233. L. S. Forster, J. V. Rund, and A. F. Fucaloro, *J. Phys. Chem.* **1984**, *88*, 5012.
234. P. Riccieri and H. L. Schlafer, *Inorg. Chem.* **1970**, *9*, 727.
235. H. F. Wasgestian and H. L. Schlafer, *Z. Phys. Chem.* **1968**, *57*, 282.
236. H. F. Wasgestian and H. L. Schlafer, *Z. Phys. Chem.* **1968**, *62*, 127.
237. E. Zinato and P. Riccieri, *Inorg. Chem.* **1973**, *12*, 1451.
238. P. Riccieri and E. Zinato, *Inorg. Chem.* **1980**, *19*, 3279.
239. C. F. C. Wong and A. D. Kirk, *Inorg. Chem.* **1977**, *16*, 3148.
240. L. S. Garner and D. A. House, *Transition Metal Chem.* **1970**, *6*, 59.
241. P. Riccieri and E. Zinato, *J. Am. Chem. Soc.* **1975**, *97*, 6071.
242. C. D. Flint and A. P. Matthews, *J. Chem. Soc. Faraday Trans. 2* **1980**, *76*, 1381.
243. S. Yamada, *Coord. Chem. Revs.* **1967**, *2*, 83.
244. C. D. Flint and D. A. Matthews, *J. Chem. Soc. Faraday Trans. 2* **1974**, *70*, 1307.
245. L. Dubicki, M. A. Hitchman, and P. Day, *Inorg. Chem.* **1970**, *9*, 188.
246. S. C. Pyke and R. G. Linck, *Inorg. Chem.* **1980**, *19*, 2468.
247. M. F. Mantrin, M. Gandolfi, L. Moggi, and V. Balzani, *Gazz. Chem. Ital.* **1973**, *103*, 1189.
248. P. C. Ford, D. Wink, and J. Dibenedetto, *Progr. Inorg. Chem.* **1983**, *30*, 213.
249. E. Zinato, A. W. Adamson, J. L. Reed, J. P. Puaux, and P. Riccieri, *Inorg. Chem.* **1984**, *23*, 1138.
250. E. Zinato, P. Riccieri, and M. Prelati, *Inorg. Chem.* **1981**, *20*, 1423.
251. E. Zinato, P. Riccieri, and M. Prelati, *Inorg. Chem.* **1981**, *20*, 1432.
252. R. A. Pribush, C. K. Poon, C. M. Bruce, and A. W. Adamson, *J. Am. Chem. Soc.* **1974**, *96*, 3027.
253. M. Manfrini, G. Varani, L. Moggi, and V. Balzani, *Mol. Photochem.* **1969**, *1*, 387.
254. L. Moggi, F. Bolletta, V. Balzani, and F. Scandola, *J. Inorg. Nucl. Chem.* **1966**, *28*, 2589.
255. G. Emschwiller and J. Legros, *Compte Rendu* **1965**, *261*, 1535.
256. D. A. Sexton, L. H. Skibsted, D. Magde, and P. C. Ford, *Inorg. Chem.* **1984**, *23*, 4533.
257. L. H. Skibsted, D. Strauss, and P. C. Ford, *Inorg. Chem.* **1979**, *18*, 3171.
258. D. A. Sexton, P. C. Ford, and D. Magde, *J. Phys. Chem.* **1983**, *87*, 197.
259. M. Nishizawa, T. M. Suzuki, S. Prouse, R. J. Watts, and P. C. Ford, *Inorg. Chem.* **1984**, *23*, 1837.
260. R. Siram, Ph.D. Dissertation, University of Southern California, 1972.

261. J. F. Endicott and G. Ferraudi, *J. Phys. Chem.* **1976**, *80*, 949.
262. G. Ferraudi and J. F. Endicott, *J. Chem. Soc. Chem. Commun.* **1973**, 674.
263. A. W. Adamson and A. H. Sporer, *J. Am. Chem. Soc.* **1958**, *80*, 3865.
264. A. W. Adamson and A. H. Sporer, *J. Inorg. Nucl. Chem.* **1958**, *8*, 209.
265. A. W. Adamson, A. Chiang, and E. Zinato, *J. Am. Chem. Soc.* **1969**, *91*, 5467.
266. M. E. Frink, D. Magde, D. A. Sexton, and P. C. Ford, *Inorg. Chem.* **1984**, *25*, 238.
267. M. A. Bergkamp, J. Brannon, D. Magde, R. J. Watts, and P. C. Ford, *J. Am. Chem. Soc.* **1979**, *101*, 4549.
268. T. L. Kelly and J. F. Endicott, *J. Phys. Chem.* **1972**, *76*, 1937.
269. L. Mønsted and L. H. Skibsted, *Acta Chem. Scand. Ser. A* **1983**, *37*, 663.
270. C. Kutal and A. W. Adamson, *Inorg. Chem.* **1973**, *12*, 1454.
271. P. C. Ford, *Rev. Chem. Interm.* **1979**, *2*, 267.
272. P. Riccieri and E. Zinato, Proceedings of the XIV International Conference on Coordination Chemistry, International Union of Pure and Applied Chemistry, Toronto, Canada, 1972, p. 252.
273. A. W. Adamson, *J. Phys. Chem.* **1967**, *71*, 798.
274. H. L. Schlafer, *J. Phys. Chem.* **1965**, *69*, 2201.
275. N. A. P. Kane-Maguire, J. Conway, and C. H. Langford, *J. Chem. Soc. Chem. Commun.* **1974**, 801.
276. G. Ferraudi and M. Pacheco, *Chem. Phys. Lett.* **1984**, *112*, 187.
277. R. B. Woodward and R. Hoffmann, *Angew. Chem., Int. Ed. Engl.* **1969**, *8*, 781.
278. H. C. Longuet-Higgins and E. W. Abrahamson, *J. Am. Chem. Soc.* **1965**, *87*, 2045.
279. B. R. Hollebone, C. H. Langford, and N. Serpone, *Coord. Chem. Rev.* **1981**, *39*, 181.
280. J. I. Zink, *Inorg. Chem.* **1975**, *14*, 4467.
281. J. I. Zink, *J. Am. Chem. Soc.* **1974**, *96*, 4464.
282. J. I. Zink, *Inorg. Chem.* **1973**, *12*, 1018.
283. L. G. Vanquickenborne and A. Ceulenmans, *J. Am. Chem. Soc.* **1978**, *100*, 475.
284. L. G. Vanquickenborne and A. Ceulenmans, *J. Am. Chem. Soc.* **1977**, *99*, 2208.
285. L. G. Vanquickenborne and A. Ceulenmans, *Inorg. Chem.* **1979**, *18*, 3475.
286. L. G. Vanquickenborne and A. Ceulenmans, *Inorg. Chem.* **1979**, *18*, 897.
287. A. D. Kirk, *Mol. Photochem.* **1973**, *5*, 127.
288. J. F. Endicott, *Comments Inorg. Chem.* **1985**, *3*, 349.
289. K. W. Hipps and G. A. Crosby, *Inorg. Chem.* **1974**, *13*, 1543.
290. L. Viaene, J. D'Olieslager, A. Ceulenmans, and L. G. Vanquickenborne, *J. Am. Chem. Soc.* **1979**, *101*, 1405.
291. V. M. Miskowski, H. B. Gray, R. B. Wilson, and E. I. Solomon, *Inorg. Chem.* **1979**, *18*, 1410.

292. G. Ferraudi, G. A. Argüello, and M. E. Frink, *J. Phys. Chem.* **1987**, *91*, 64.
293. D. J. Robbins and A. J. Thomson, *Mol. Phys.* **1973**, *25*, 1103.
294. M. Nishazawa and P. C. Ford, *Inorg. Chem.* **1981**, *20*, 294.
295. S. J. Milder, H. B. Gray, and V. M. Miskowski, *J. Am. Chem. Soc.* **1984**, *106*, 3764.
296. C. H. Langford, A. Y. S. Malkasian, and D. K. Sharma, *J. Am. Chem. Soc.* **1984**, *106*, 2727.
297. D. A. Sexton, P. C. Ford, and D. Magde, *J. Phys. Chem.* **1983**, *87*, 197.
298. T. J. Kemp, *Progr. React. Kinetics* **1980**, *10*, 301.
299. J. S. Svendsen and L. H. Skibsted, *Acta Chem. Scand. Ser. A* **1983**, *37*, 443.
300. L. Møsted and L. H. Skibsted, *Acta Chem. Scand. Ser. A* **1984**, *A38*, 535.
301. W. Weber, R. van Eldik, H. Kelm, J. DiBenedetto, Y. Docommun, H. Offen, and P. C. Ford, *Inorg. Chem.* **1983**, *22*, 623.
302. L. H. Skibsted, W. Weber, R. van Eldik, H. Kelm, and P. C. Ford, *Inorg. Chem.* **1983**, *22*, 541.
303. A. Ceulemans, D. Beyens, and L. G. Vanquickenborne, *Inorg. Chem.* **1983**, *22*, 1113.
304. J. L. Walsh and B. Durham, *Inorg. Chem.* **1982**, *21*, 329.
305. B. Durham, J. V. Caspar, J. K. Nagle, and T. J. Meyer, *J. Am. Chem. Soc.* **1982**, *104*, 4803.
306. N. S. Hush, in *Mechanistic Aspects of Inorganic Reactions* (*A.C.S. Symp. Ser.*) No. 198, Washington DC; American Chemical Society, **1982**, p. 301.
307. M. Talebinasab-Sarvari, A. Zanella, and P. C. Ford, *Inorg. Chem.* **1980**, *19*, 1835.
308. R. J. Watts, J. S. Harrington, and J. S. Van Houten, *J. Am. Chem. Soc.* **1977**, *99*, 2179.
309. W. A. Wickramsinghe, P. H. Bird and N. J. Serpone, *J. Chem. Soc. Chem. Commun.* **1981**, 1284.
310. P. J. Spellane, R. J. Watts, and C. J. Curtis, *Inorg. Chem.* **1983**, *22*, 4060.
311. P. S. Braterman, G. A. Heath, A. J. Mackenzie, B. C. Noble, R. D. Peacock, and L. J. Yellowless, *Inorg. Chem.* **1984**, *23*, 3425.
312. G. Nord, A. C. Hazell, R. G. Hazell, and O. Farver, *Inorg. Chem.* **1983**, *22*, 3429.
313. G. L. Geoffroy and M. S. Wrighton, *Organometallic Photochemistry*, New York: Academic, **1979**.
314. M. Wrighton, H. B. Gray, and G. S. Hammond, *Mol. Photochem.* **1973**, *5*, 165.
315. Y. S. Sohn, D. N. Hendrickson, and H. B. Gray, *J. Am. Chem. Soc.* **1970**, *93*, 3603.
316. M. S. Wrighton, D. I. Handeli, and D. L. Morse, *Inorg. Chem.* **1976**, *15*, 434.
317. H. Haas and R. K. Sheline, *J. Am. Chem. Soc.* **1966**, *88*, 3219.
318. R. A. Levenson and H. B. Gray, *J. Am. Chem. Soc.* **1975**, *97*, 6042.
319. Y. S. Sohn, D. N. Hendrickson, and H. B. Gray, *J. Am. Chem. Soc.* **1971**, *93*, 3603.

320. N. Rösch and K. H. Johnson, *Chem. Phys. Lett.* **1974**, *24*, 179.
321. M. S. Wrighton, D. L. Morse, and D. L. Pdungsap, *J. Am. Chem. Soc.* **1975**, *97*, 2073.
322. E. W. Abel and F. G. A. Stone, *Q. Rev. Chem. Soc.* **1970**, *24*, 498.
323. A. Vogler, in *Concepts in Inorganic Photochemistry* (A. W. Adamson and P. D. Fleischauer, Eds.), Chap. 6, Wiley: New York, **1975**.
324. M. S. Wrighton, H. B. Abrahamson, and D. L. Morse, *J. Am. Chem. Soc.* **1976**, *98*, 4105.
325. M. S. Wrighton and D. L. Morse, *J. Am. Chem. Soc.* **1974**, *96*, 998.
326. P. J. Giordano and M. S. Wrighton, *J. Am. Chem. Soc.* **1979**, *101*, 2888.
327. P. J. Giordano, S. M. Fredericks, M. S. Wrighton, and D. L. Morse, *J. Am. Chem. Soc.* **1978**, *100*, 2257.
328. M. A. Graham, R. N. Perutz, M. Poliakoff, and J. J. Turner, *J. Organomet. Chem.* **1972**, *34*, C34.
329. J. K. Burdett, M. A. Graham, R. N. Perutz, M. Poliakoff, A. J. Rest, J. J. Turner, and R. F. Turner, *J. Am. Chem. Soc.* **1975**, *97*, 4805.
330. R. N. Perutz and J. J. Turner, *Inorg. Chem.* **1975**, *14*, 262.
331. J. D. Black and P. S. Braterman, *J. Am. Chem. Soc.* **1975**, *97*, 2908.
332. R. N. Perutz and J. J. Turner, *J. Am. Chem. Soc.* **1975**, *97*, 4791.
333. J. K. Burdett, R. N. Perutz, M. Poliakoff, and J. J. Turner, *J. Chem. Soc. Chem. Commun.* **1975**, 157.
334. J. M. Kelly, D. V. Bent, H. Hermann, D. Schulte-Frohlinde, and E. Koerner von Gustorf, *J. Organomet. Chem.* **1974**, *69*, 259.
335. J. M. Kelly, D. V. Bent, H. Herman, D. Schulte-Frohlinde, and E. Koerner von Gustorf, *J. Organomet. Chem.* **1974**, *69*, 259.
336. M. Wrighton, *Inorg. Chem.* **1974**, *13*, 905.
337. M. Wrighton, G. S. Hammond, and H. B. Gray, *Mol. Photochem.* **1973**, *5*, 179.
338. W. Strohmeier and D. von Hobe, *Chem. Ber.* **1961**, *94*, 2031.
339. T. H. Whitesides and R. A. Budnick, *J. Chem. Soc. Chem. Commun.* **1971**, 1514.
340. W. Jetz and W. A. G. Graham, *Inorg. Chem.* **1971**, *10*, 4.
341. M. Wrighton and D. Bredesen, *J. Organomet. Chem.* **1973**, *50*, *C*35.
342. J. Nasielski, P. Kirsch, and L. Wilputte-Steinert, *J. Organomet. Chem.* **1971**, *27*, C13.
343. G. Platbrood and L. Wilputte-Seinert, *J. Organomet. Chem.* **1974**, *70*, 393.
344. I. Fischler, M. Budzwait, and E. A. Koerner von Gustorf, *J. Organomet. Chem.* **1976**, *105*, 325.
345. G. Platbrood and L. Wilputte-Steinert, *J. Organomet. Chem.* **1975**, *85*, 199.
346. D. Rietvelde and L. Wilputte-Steinert, *J. Organomet. Chem.* **1976**, *118*, 191.
347. D. J. Darensbourg and H. H. Nelson, *J. Am. Chem. Soc.* **1974**, *96*, 6511.
348. D. J. Darensbourg, H. H. Nelson, and M. A. Murphy, *J. Am. Chem. Soc.* **1977**, *99*, 896.
349. M. A. Schroeder and M. S. Wrighton, *J. Organomet. Chem.* **1974**, *74*, *C*29.

350. I. Fischler, R. Wagner, M. Budzwart, R. N. Perutz, and E. A. Koener von Gustorf, *Proc. Int. Conf. Organomet. Chem., 7th*, **1975**, p. 255.

351. M. Wrighton, G. S. Hammond, and H. B. Gray, *J. Organomet. Chem.* **1974**, *70*, 283.

352. D. K. Geiger and G. Ferraudi, *Inorg. Chim. Acta* **1985**, *101*, 197.

353. J. Schwartz, *J. Chem. Soc. Chem. Commun.* **1972**, 814.

354. G. N. Schrauzer and P. Glockner, *J. Am. Chem. Soc.* **1968**, *90*, 2800.

355. G. N. Schrauzer and S. Eichler, *Angew. Chem.* **1962**, *74*, 585.

356. D. J. Trecker, R. S. Foote, J. P. Henry, and J. E. McKeon, *J. Am. Chem. Soc.* **1966**, *88*, 3021.

357. R. G. Salomon and J. K. Kochi, *J. Am. Chem. Soc.* **1974**, *96*, 1137.

358. R. G. Salomon, K. Folting, W. E. Streib, and J. K. Kochi, *J. Am. Chem. Soc.* **1974**, *96*, 1145.

359. D. P. Schwendiman and C. Kutal, *Inorg. Chem.* **1977**, *16*, 719.

360. P. A. Grutsch and C. Kutal, *J. Am. Chem. Soc.* **1977**, *99*, 6460.

361. P. Borrell and E. Henderson, *J. Chem. Soc. Dalton Trans.* **1975**, 432.

362. O. Traverso, S. Sostero, and G. A. Mazzocchin, *Inorg. Chim. Acta* **1974**, *11*, 237.

363. J. Granifo and G. Ferraudi, *Inorg. Chem.* **1984**, *23*, 2210.

364. O. Traverso and F. Scandola, *Inorg. Chim. Acta* **1970**, *4*, 493.

365. T. Akiyama, P. Kitamura, T. Kato, H. Watanabe, T. Serizawa, and A. Sugimori, *Bull. Chem. Soc. Jpn.* **1977**, *50*, 1137.

366. O. Traverso, R. Rosi, S. Sostero, and V. Carassiti, *Mol. Photochem.* **1973**, *5*, 457.

367. L. H. Ali, A. Cox, and T. J. Kemp, *J. Chem. Soc. Dalton Trans.* **1973**, 1468.

368. G. Ferraudi, *Inorg. Chem.* **1978**, *17*, 2506.

369. M. Freiberg and D. Meyerstein, *J. Chem. Soc. Chem. Commun.* **1977**, 127.

370. T. S. Roche and J. F. Endicott, *Inorg. Chem.* **1974**, *13*, 1575.

371. S. Muralidahran and G. Ferraudi, *Inorg. Chem.* **1981**, *20*, 2306.

372. A. Vogler and R. Hirschmann, *Z. Naturforsch., B* **1976**, *B31*, 1082.

373. C. Y. Mock and J. F. Endicott, *J. Am. Chem. Soc.* **1977**, *99*, 1276.

374. C. Y. Mock and J. F. Endicott, *J. Am. Chem. Soc.* **1978**, *100*, 123.

375. G. Ferraudi and J. F. Endicott, *J. Am. Chem. Soc.* **1977**, *99*, 243.

376. E. A. Koerner von Gustorf, L. H. G. Lenders, I. Fischler, and R. N. Perutz, *Adv. Inorg. Chem. Radiochem.* **1976**, *19*, 65.

377. T. Inoue and J. F. Endicott, personal communication on $Rh(C_2H_5)(NH_3)_5^{2+}$ and $Rh(NH_3)_4(OH_2)C_2H_5^{2+}$.

378. N. W. Hoffman and T. L. Brown, *Inorg. Chem.* **1978**, *17*, 613.

379. B. H. Byers and T. L. Brown, *J. Am. Chem. Soc.* **1977**, *99*, 2527.

380. A. Camus, C. Cocevar, and G. Mestroni, *J. Organomet. Chem.* **1972**, *39*, 355.

381. G. L. Geoffroy and R. Pierantozzi, *J. Am. Chem. Soc.* **1976**, *99*, 8054.

382. M. J. Mays, R. N. F. Simpson and F. P. Stefanini, *J. Chem. Soc. A* **1970**, 3000.

383. S. Muralidharan, G. Ferraudi, M. A. Green, and K. Caulton, *J. Organomet. Chem.* **1983**, *244*, 47.
384. R. Pierantozzi, Ph.D. Thesis, Pennsylvania State University, University Park, **1978**.
385. C. Kutal, M. Weber, G. Ferraudi, and D. K. Geiger, *Organometallics* **1985**, *4*, 2161.
386. C. J. Ballhausen, *Introduction to Ligand Field Theory*, New York: McGraw-Hill, **1962**.
387. G. Racah, *Phys. Rev.* **1952**, *85*, 381.
388. J. S. Griffith, *The Theory of Transition Metal Ions*, Cambridge: Cambridge Univ. Press, **1961**, pp. 73–79.
389. C. E. Schäfler, *Structure and Bonding* **1968**, *5*, 68, and references therein.
390. R. S. Mulliken, *J. Chim. Phys.* **1949**, *46*, 497.
391. M. Wolfsberg and L. Helmholz, *J. Chem. Phys.* **1952**, *20*, 387.
392. C. J. Ballhausen and H. B. Gray, *Inorg. Chem.* **1962**, *1*, 111.
393. R. H. Hoffmann, *J. Chem. Phys.* **1963**, *39*, 1937.
394. T. L. Allen, *J. Chem. Phys.* **1957**, *27*, 810.
395. H. Basch, A. Viste, and H. B. Gray, *J. Chem. Phys.* **1966**, *44*, 10.
396. H. Basch, A. Viste, and H. B. Gray, *Theor. Chim. Acta* **1966**, *3*, 458.
397. R. F. Fenske, K. Caulton, D. D. Radtke, and C. C. Sweeney, *Inorg. Chem.* **1966**, *5*, 951.
398. R. F. Fenski and D. D. Radtke, *Inorg. Chem.* **1968**, *7*, 479.
399. W. A. Yeranos and D. A. Hasman, *Z. Naturforsch. A* **1967**, *22*, 170.
400. D. S. McClure, *Advances in the Chemistry of Coordination Compounds* (S. Kirschner, Ed.), New York: Macmillan, **1961**.
401. H. Yamatera, *Bull. Chem. Soc. Jpn.* **1958**, *31*, 95.
402. D. A. Johnson, *Some Thermodynamic Aspects of Inorganic Chemistry*, Cambridge: Cambridge Univ. Press, **1968**.
403. G. N. Lewis and M. Randall, *Thermodynamics*, 2nd ed., New York: McGraw-Hill, **1961**.

INDEX

INDEX